Indoor Environmental Quality

Indoor Environmental Quality

THAD GODISH

LEWIS PUBLISHERS
Boca Raton London New York Washington, D.C.

Library of Congress Cataloging-in-Publication Data

Godish, Thad.
    Indoor environmental quality/Thad Godish.
       p. cm.
    Includes bibliographical references and index.
    ISBN 1-56670-402-2 (alk. paper)
    1. Indoor air pollution. 2. Housing and health. 3. Industrial hygiene. I. Title.

TD883.17.G64   2000
628.5'3—dc21

           00-057400
           CIP

© 2001 by CRC Press LLC
Lewis Publishers is an imprint of CRC Press LLC

No claim to original U.S. Government works
International Standard Book Number 1-56670-402-2
Library of Congress Card Number 00-057400
Printed in the United States of America 1 2 3 4 5 6 7 8 9 0
Printed on acid-free paper

Preface

Indoor Environmental Quality is the third in a series of books written by the author over the past decade and focuses on environmental problems and issues associated with our homes, office buildings, schools, and other non-industrial indoor environments. This book differs in several ways from the author's previous works, *Indoor Air Pollution Control* (1989) and *Sick Buildings: Definition, Diagnosis, and Mitigation* (1995).

Most important, *Indoor Environmental Quality* reflects the success of research scientists and other investigators in defining the nature and causes of indoor environmental health and comfort problems, and the measures used to investigate and control them. It reflects an increasingly mature field of study. The published results of well-focused, careful research of colleagues around the world are the lifeblood of the author who labors to distill their findings and thoughts into a review article, reference book, or a text designed for use in the classroom.

The author has previously published review articles and reference books whose purpose was to describe major indoor air quality/indoor environment concepts and issues and associated research results. *Indoor Air Pollution Control* focused on the broad area of indoor air quality and the measures used to control indoor contaminants. *Sick Buildings: Definition, Diagnosis, and Mitigation* was more narrowly focused on problem/sick buildings, an area of intensive public health and scientific interest.

Indoor Environmental Quality is written in the style of a textbook, much like *Air Quality* (3rd edition), also by the author. It is anticipated that it will serve as the genesis for the establishment of indoor environment courses in environmental health and industrial hygiene programs in North America and other parts of the world.

Indoor Environmental Quality is intended as a primary resource for individuals who are entering, or are already in the field, whether their interest be research, governmental service, or private consulting. It accomplishes this purpose by defining the major issues and concepts and providing supporting facts in a highly readable manner. Its readability makes it suitable for use by educated laypersons who want to learn about specific indoor environmental problems and how to diagnose and mitigate them, or indoor environmental problems in general.

By its title, the book seeks to go beyond the historical focus on indoor air quality and inhalation exposures to indoor contaminants. Though most indoor environment health and comfort concerns are associated with the indoor air environment, in several major cases air appears not to be the primary route of exposure. This is particularly true in pediatric lead poisoning, which appears to be primarily due to exposures associated with hand-to-mouth transfer of lead-contaminated house dust and soil particles. Similar childhood exposures, including dermal exposures, may occur with pesticide-contaminated house dust. Exposures to office materials such as carbonless copy paper and other printed papers may cause indoor air quality-type symptoms that might be due to dermal and not inhalation exposures. As such, the book attempts to expand its focus beyond "indoor air quality" issues.

Readers of *Indoor Environmental Quality* will notice that many of the concepts and issues treated in previous reference works are included in this new work. That is due in good measure to the fact that concepts and principles continue to be important over time while the facts used to elucidate them may change.

About the author

Thad Godish is Professor of Natural Resources and Environmental Management at Ball State University, Muncie, Indiana. He received his doctorate from Pennsylvania State University, where he was affiliated with the Center for Air Environment Studies.

Dr. Godish is best known for his authorship of Lewis Publishers' *Air Quality*, a widely used textbook now in its third edition; two well-received reference books on indoor air quality: *Indoor Air Pollution Control* (Lewis, 1989) and *Sick Buildings: Definition, Diagnosis, and Mitigation* (Lewis, 1995); and his research, teaching, and public service activities in various areas of indoor air/indoor environmental quality. He maintains a weekly updated web site entitled Indoor Environment Notebook (www.bsu.edu/IEN), which provides expert answers and advice on a wide variety of indoor environmental quality concerns.

Dr. Godish continues to teach a variety of environmental science courses including air quality, indoor air quality management, occupational/industrial hygiene, asbestos and lead management in buildings, and hazardous waste operations and emergency response. He is a Fellow of the Air and Waste Management Association and the Indiana Academy of Science, as well as a member of the American Industrial Hygiene Association, American Conference of Governmental Industrial Hygienists, and International Society of Indoor Air Quality and Climate, and has served as chairman of the East Central section and Indiana chapter of the Air Pollution Control Association. He has been Visiting Scientist at Monash University, Gippsland, Australia, and at Harvard University, School of Public Health.

Contents

Dedication

To the scientists, engineers, architects, and other professionals whose efforts make our indoor environments healthier and more comfortable.

chapter one

Indoor environments

Humans in developed countries have, in the past few millennia, advanced from depending on rock shelters, caves, and rude huts to protect themselves from the elements to modern single- and multifamily dwellings and other buildings that provide amenities and conveniences far beyond the basic needs of shelter — conveniences that ensure comfort whatever the vagaries of weather and climate.

Our world is one of the structures that shelter our many activities: the small to grand shells that house a myriad of industrial processes and activities; institutional buildings such as schools, universities, hospitals, and government buildings; automobiles, trains, planes, and ships that provide transportation as well as shelter; shopping malls and office complexes where we trade goods and services; and cinemas, theaters, museums, and grand stadia that provide venues for entertainment.

Built environments comprise a diversity of functions, magnitudes, and, of course, forms. In addition to functional aspects, built environments reflect human aspirations and creativity. They also reflect more fundamental factors, such as the diversity and availability of construction materials, climate, cultural tastes, and human foibles.

We attempt to keep rain, snow, and wind out of our indoor environments; provide and maintain warm thermal conditions in seasonally cold climates; provide cooler and more acceptable conditions in hot climates; and mechanically ventilate our larger buildings to reduce odors and discomfort associated with human bioeffluents. Our ability to control thermal comfort and other aspects of indoor environments requires the application of a variety of climate-control technologies and a commitment to operate them properly.

The built environments of man are fragile artifacts. They are in constant peril from forces by which the earth renders all things unto itself. Just as water, ice, and wind level the mountains with time, so too do they act to level what man has built. Though the forms of ancient temples and buildings remain after millennia, they have long ceased to shelter humans and their activities. Wooden structures that housed humans for much of our history have been turned to mould. Indeed, the contagion of decay, fed by neglect and the forces of wind and water, constantly imperil even our newest structures. They may even affect our health and make our dwellings unclean. The book of Leviticus in the Old Testament of the *Bible* describes a "leprous" house and what is to be done about it.

> "If the priest, on examining it, finds that the infection on the walls of the house consists of greenish or reddish depressions which seem to go deeper than the surface of the wall, he shall close the door of the house for seven days. On the seventh day, the priest shall return to examine the house again. If he finds that the infection has spread on the walls, he shall order the infected stones to be pulled out and cast in an unclean place outside the city. The whole inside of the house shall be scraped, and the mortar that has been scraped off shall be dumped in an unclean place outside the city. Then new stones shall be brought and put in the place of old stones, and the new mortar shall be made and plastered on the house."

Though we design buildings and other structures to provide shelter from an often hostile outdoor environment, the shelter they provide is less than perfect. They are subject not only to the forces of nature, but also to the randomness inherent in the second law of thermodynamics or its derivative, the law of unintended consequences.

As we attempt to provide both shelter and those many amenities and conveniences that make life more comfortable, we, in many cases inadvertently and in other cases deliberately, introduce a variety of contaminants that have the potential to diminish the quality of our lives or pose moderate to significant health risks to occupants.

Indoor environments are often contaminated by a variety of toxic or hazardous substances, as well as pollutants of biological origin. When early humans discovered the utility of fire and brought it into rock shelters, caves, and huts, they subjected their sheltered environments to the enormous burden of wood smoke (not much different from modern cooking fires in developing countries) and attendant irritant and more serious health effects. Biological contaminants such as bacteria, mold, and the excretory products of commensal organisms (e.g., dust mites, cockroaches, mice, etc.) have caused human disease and suffering for most of human history. However, viewed within the context of infectious and contagious diseases such as tuberculosis and bubonic plague, illness caused by asthma and chronic allergic rhinitis can be seen as relatively minor.

In advanced countries, increasing concern has developed in the past several decades about contaminants in our building environments and potential exposure risks to occupants. These have grown out of previous and contemporary concern for the health consequences of ambient (outdoor) air pollution, water pollution, hazardous waste, and the general pollution of our environment and food with toxic substances such as pesticides, PCBs, dioxin, etc.

Other factors have also "conspired" to increase our awareness that contamination of built environments (particularly indoor air) poses potentially significant public health risks. These have included: (1) recognition of the health hazards of asbestos and its widespread presence in schools and many other buildings, and the regulatory requirements for inspection of public and private schools for asbestos as well as its removal prior to any building renovation/demolition; (2) recognition of the significant exposure to formaldehyde (HCHO) experienced by residents of mobile homes, urea–formaldehyde foam-insulated (UFFI) houses, and conventional homes in which a variety of formaldehyde-emitting urea–formaldehyde resin-containing products were used; (3) recognition that residential buildings and some schools have elevated radon levels (thought high enough to carry a significant risk of lung cancer); (4) the apparent consequences of implementing energy-reducing measures in response to increased energy prices in the mid 1970s, including reducing ventilation air in mechanically ventilated buildings, using alternative space heating appliances such as wood-burning stoves and furnaces and unvented kerosene heaters, and reduced air infiltration into buildings; (5) an eruption of air quality complaints in hundreds of buildings in the U.S. following changes in building operation practices; (6) progressive awareness of the problem of childhood lead poisoning and its association with house dust from lead-based paint; and (7) an increasing understanding that biological contaminants of the indoor environment, e.g., mold, dust mites, pet danders, cockroach excreta, etc., play a role in causing human asthma and chronic allergic rhinitis.

I. Indoor contamination problems

The contamination of indoor air and horizontal surfaces (by dusts) is common to all built environments. Such contamination is most pronounced in industrial environments where raw materials are processed and new products manufactured. These environments pose unique exposure concerns and are subject to regulatory control and occupational safety and health programs in most developed countries. Though industrial and other occupational exposures are significant, they are not included in discussions of indoor air quality and indoor environmental (IAQ/IE) contamination concerns in this book.

Indoor air quality as it relates to residential, commercial, office, and institutional buildings, as well as in vehicles of transport, is its own unique public health and policy issue, as is the contamination of building surfaces by lead, pesticides and other toxic, hazardous substances. As such, IAQ/IE concerns are, by definition, limited to nonindustrial indoor environments.

Indoor environment problems, as they are experienced in residential and nonresidential structures, tend to have their own unique aspects. In nonresidential buildings, occupants have little or no control over their environments, which are owned and managed by others. In theory, homeowners and lessees have some degree of freedom to modify (for better or worse) the environments in which they live. Because of the nature of activities conducted within, and how buildings are constructed and maintained, residential and nonresidential buildings often differ significantly in the nature of IAQ/IE problems and associated health risks. These building types also differ in how problem investigations are conducted and, in many cases, who conducts such investigations. Because of the differences described above, IAQ/IE problems treated here are described in the context of both residential and nonresidential built environments.

II. Characteristics of residential buildings

Residential buildings can be characterized in the context of (1) the population they serve, (2) ownership status, (3) building types, (4) construction characteristics, (5) heating and cooling systems, (6) site characteristics, (7) occupants and occupant behavior, and (8) exposure concerns.

A. Population served

Residential dwellings are different from other built environments because they must provide shelter for everyone, i.e., an enormous population. This includes individuals ranging in age from infants to the elderly, individuals whose health status varies from healthy to a variety of ailments, illnesses, and infirmities, and who spend anywhere from a few to 24 hours per day indoors. In the U.S., on average, individuals spend 22 hours/day indoors, with approximately 14 to 16 hours at home.

Those who spend the most time at home are the very young, very old, ill or infirm, or those not employed outside the home.

B. Ownership status

Approximately 70% of the U.S. population resides in occupant-owned dwellings, while 30% lease their residence from private individuals or government agencies. This significant private ownership of individual dwellings is unique among nations.

Ownership status is an important factor as it relates to IAQ/IE concerns. It is widely accepted that home ownership carries with it both individual responsibility and pride. Such responsibility and pride can be expected to result in better building maintenance, reducing the potential for problems such as extensive water damage and mold infestation. On the other hand, home ownership can, in many cases (because of human attitudes and foibles), increase the probability that home contamination problems will occur (e.g., indiscriminate pesticide application or storage of toxic/hazardous materials; or engaging in commercial activities or hobbies, e.g., hair dressing salons, silk screening, or wood refinishing, that could cause significant air or building surface contamination).

Home ownership is a significant decision-determining factor when IAQ/IE issues arise. If a dwelling is discovered to have excessively high radon levels, significant mold infestation, or a high potential for lead dust exposure to young children, homeowners have the opportunity to mitigate such problems at their own expense. If the dwelling is owned by a second party, occupants must convince an often reluctant lessor to mitigate the problem, seek alternative housing, or "live with it."

C. Building types

There are two basic types of residential structures: single-family and multiple-family dwellings. Typically, single-family dwellings (Figure 1.1) are detached from other residential structures (although some row houses and condominiums blur the line); multifamily dwellings are constructed as single large structures that provide 2 to >1000 leased individual apartments. Single-family dwellings are characteristic of American rural and suburban areas and older parts of cities. Multiple-family dwellings are characteristic of urban areas and are becoming increasingly common in other areas as well. Because of the limited availability of building sites, multifamily dwellings are the primary form of housing used by families in cities and densely populated countries.

Single-family residences may be site-built or manufactured and placed on site. In the U.S., manufactured houses (Figure 1.2) comprise approxi-

Figure 1.1 Single-family owner-occupied home.

Figure 1.2 Mobile or manufactured home.

mately 10 million housing units. These are often described as trailers, mobile homes, double-wides, modulars, and, increasingly, prebuilts. Most are described as mobile homes because they are transported on a frame and wheels which are part of the structure. The construction of manufactured houses differs significantly from that of site-built houses because the former are designed to provide lower cost, more affordable housing. They often employ lower cost materials and have, in the past, been less well-constructed than site-built houses. They are more vulnerable to wind and weather-related damage and are usually less well-insulated. Prebuilts are erected on substructures and differ from site-built homes primarily in their simplicity of design.

Multifamily dwellings (Figure 1.3) vary from single-story to multistory structures. In most instances, ownership is second-party. Multifamily dwellings are always site-built, with building materials that reflect cost and engineering considerations.

Figure 1.3 Multiple-family dwelling.

D. Construction characteristics

Residential buildings vary enormously in their construction characteristics, including size, design, building materials used, substructure, cladding, use of insulation, quality of construction, and site conditions. They vary in size from simple shanties, to nice single- and multifamily homes, to palatial mansions. They vary in design from the simple rectangular boxes of manufactured houses, to the diversity of home designs of middle-income individuals, to the more complex and architecturally inspiring homes of the Victorian era and the present.

All residential buildings have similar construction requirements. They include a substructure, sidewalls, flooring, windows, roofing, attic and crawlspace ventilation, plumbing, electrical wiring, attic and wall insulation (depending on climate), and roof and site drainage. They also include interior furnishings such as storage cabinets, closets, and finished wall and floor surfaces. These reflect construction practices that depend on regional climate, site characteristics, design preferences, and availability of construction materials. They also reflect evolving builder and homeowner preferences and new amenities in the marketplace. Construction characteristics are much influenced by cost, the most important factor in residential building construction.

1. Substructures

Most residential buildings rest on a substructure that supports their weight and anchors them to the ground. There are three common types of substructure: slab-on-grade, crawlspace, and basement. Some residences have combinations of these.

House substructures reflect regional preferences (in general, basements are preferred in the northeastern U.S.); contractor preferences (assuming equal costs, some contractors prefer to build houses on crawlspaces, while others prefer slab-on-grade); soil characteristics (poorly drained clay soils are unsuitable for basements); and cost and construction time (this is a major contributor to the increasing construction of slab-on-grade, single-family dwellings).

Substructure type often has significant effects on building IAQ/IE problems. Houses with basements or slab-on-grade tend to have higher radon levels (given the same soil radon-emitting potential). Basements tend to have problems with water penetration and excess humidity, factors that contribute to mold infestation and attendant exposure and health risks. Such health risks may also exist in dwellings with crawlspace or slab-on-grade substructures when constructed on poorly drained sites (as is often the case).

2. Roofing

Roofs are constructed to protect building interiors from rain, snow, and wind. They are designed to intercept rain and snow and carry their waters from the roof edges to the ground, either directly or through guttering. Climatic factors determine the nature of roof construction and the use of guttering.

Roof construction also reflects resource availability and cultural preferences. In new U.S. residential construction, roofs are typically constructed with oriented-strand board decking on wood trusses. Decking is then covered with asphalt felt and shingles. In the southwestern U.S., as in some other parts of the world (Europe, Southeast Asia, Japan, Australia), terra cotta roofing is preferred. In some parts of the U.S. (South) and Australia, painted galvanized steel is the most common roofing material.

Roof construction and materials used are important. The roof must carry away water without leakage, lest significant internal structural damage and mold infestation occur. In cold, snowy climates, the roof must be strong enough to support the weight of heavy snow. In regions with severe storms, roofs must be securely anchored lest they experience serious damage. The cavity between the roof and ceiling timbers must be adequately ventilated to prevent the build-up of excessive moisture, which in cool/cold climates may result in condensation and even freezing. Poorly ventilated attics may result in structural damage and mold infestation.

3.   Sidewalls and walls

The exterior sidewalls of dwellings are typically constructed using structural timbers, fiberglass insulation, Styrofoam or polyurethane sheeting to provide additional low-cost insulation, oriented-strand board sheeting in corners and around windows to provide extra strength, an external semipermeable membrane (e.g., Tyvek), and one or more types of external cladding. Typical cladding includes, or has included, aluminum or vinyl siding, wood or fibrocement weatherboard, stucco over concrete block, and brick or stone veneer. Cladding is an important factor in protecting the building from the vagaries of weather and climate. All cladding types indicated above provide reasonable protection from wind, water, and snow. From a structural standpoint, houses constructed with brick/stone veneer are less prone to damage from wind gusts which can tear off small to large pieces of vinyl and aluminum siding. Wood weatherboard must be painted repeatedly and, with time, can deteriorate as a result of weathering and inadequate maintenance. In many older houses, wood weatherboard was painted with lead-based paint and represents a potentially significant source of lead contamination of the soil surrounding the building, as well as interior dust. Old weatherboard-clad houses are often a major public health concern because of their potential to cause lead poisoning in young children.

As in the story of the three little pigs, an all-brick or stone house would seem to provide the best shelter. However, such houses are not without problems. Brick/stone veneer houses constructed on unstable soils develop small to large settlement cracks which provide an avenue for rain to enter building cavities. Here both liquid water and water vapor can cause structural damage and mold infestation. In the absence of settlement cracks, many brick/stone veneer facades pass water through porous mortar and brick, and through small holes. If constructed properly, rain water will drain down the interior surface of the mineral facade and seep out through properly functioning weep holes at the bottom. If brick/stone veneer facades are poorly constructed (without weep holes and the removal of excess mortar), rain water will be carried into walls, again causing mold infestation and structural damage.

Timbers on wall interiors are covered by polyethylene plastic, which serves as a vapor barrier. It is designed to prevent warm, moist air from passing into building cavities where it may condense and cause structural damage and mold infestation.

4. Windows

Windows in dwellings differ in style, size, placement, and materials. They are designed to keep wind, rain, and snow out, allow light in, and provide a means of natural ventilation during moderate to hot weather. Windows are a major source of energy loss because of their thermal energy transmitting properties. On single-pane windows (found in older houses), moisture on interior surfaces cools and condenses, causing damage to interior window surfaces (and in many cases significant mold infestation). Windows also break the continuity of building cladding. These breaks must be provided with flashing or be caulked to prevent water from penetrating wall cavities during heavy rains. Water penetration into wall cavities around windows is common as houses age and maintenance is neglected.

5. Flooring

Materials used in both exterior and interior house construction change with time. In older houses (>40 years), softwood boards were commonly used to construct floors. In many cases these were overlain with hardwood oak flooring. Because of the high cost of such flooring, it became common to construct floors using CDX plywood sheeting. Later, contractors used a combination of softwood plywood sheeting as a base, with $5/8''$ (1.6 cm) particle board underlayment above it. This was inexpensive and provided a smooth surface for attachment of wall-to-wall carpeting. Between 1960 and 1990, over 10 million homes were constructed in the U.S. using particle board underlayment, a very potent source of formaldehyde (HCHO). Emissions of HCHO from underlayment have significantly declined in the last decade or so (1988 to 2000). It is little used in modern site-built construction and has declined to approximately 50% of new manufactured house construction. Particle board flooring has been displaced by oriented-strand board (OSB), a composite wood material that has better structural properties and very low HCHO emissions.

The main floor surface of slab-on-grade houses is, of course, concrete, with wall-to-wall carpeting and other floor coverings overlaying it. This concrete–ground contact provides a cool surface, which may result in optimal humidity levels for the development of high dust mite populations. Slab-on-grade substructures also provide (through cracks) a mechanism for the conveyance of radon and other soil gases (most notably water vapor) into building interiors.

6. Decorative wall and ceiling materials

A variety of materials are used to finish interior walls and ceilings. Base materials have historically included plaster over wood or metal lath, or gypsum board panels. Because of cost factors and ease of installation, gypsum board has dominated the construction market for interior wall and ceiling covering for the past four decades, with plaster found primarily in older homes. Gypsum board in itself appears to pose no direct IAQ/IE concerns. However, during installation, spackling materials are used to cover gaps between individual panels. Prior to 1980, most spackling compounds contained asbestos; therefore, many older homes with gypsum board wall covering contain a limited amount of asbestos fibers. In other older homes, acoustical plaster containing 5 to 10% asbestos was sprayed on ceiling surfaces to provide a decorative finish with sound-absorbing properties.

Gypsum board has become increasingly associated with *Stachybotrys chartarum* infestations in residences and other buildings. *S. chartarum* is a fungus that produces a potent mycotoxin (see Chapter 6 for an expanded discussion of *S. chartarum*). It grows readily on the cellulose face of gypsum board when it has been subjected to a significant or repeated episodes of wetting.

Residential buildings have a variety of exterior and interior surfaces that have been coated with paints, stains, varnishes, lacquers, etc. These coatings may have significant emissions of volatile organic compounds (VOCs) and semivolatile organic compounds (SVOCs), particularly when newly applied. Old leaded paints (pre-1978) may pose unique indoor contamination problems (see Chapter 2).

Base gypsum board materials are usually finished with the application of latex, or in some cases oil-based, paints. In the early history of a dwelling, latex paints, though water-based, emit a variety of VOCs and SVOCs, with significant emissions of VOCs and polyvalent alcohols from oil-based and latex paints, respectively. Though these emissions diminish with time, high initial emissions from walls and other painted surfaces represent a significant source of odor, and in some cases irritant effects, in the early days of home occupancy. Notably, some manufacturers have recently included biocides in their latex paint formulations that emit significant quantities of formaldehyde in the first weeks after application.

In some dwellings the base gypsum board may be covered, in whole or in part, with decorative materials other than paint. These may include wallpaper, hardwood plywood paneling, hardboard, vinyl, fabric, etc. Hardwood plywood paneling may be used to cover walls in single rooms or, as was the case in mobile homes, most rooms. Though gypsum board panels with a paper or wallpaper overlay are now used most often, decorative hardwood plywood paneling covered most interior walls of mobile homes constructed in the U.S. prior to 1985. Hardwood plywood paneling was a potent source of HCHO and a major contributor to elevated HCHO levels reported in mobile homes constructed in the U.S. in the 1970s and early 1980s.

7. Energy conservation

Modern dwellings are being constructed to be more energy efficient. Energy efficiency is achieved, in part, by using insulating materials such as fiberglass batting in sidewalls and attics and blown-in cellulose in attics and wall cavities. In the latter case, wet cellulose is often used to insulate wall cavities in new construction. Intuitively, the practice of applying wet cellulose to building sidewalls could cause significant mold infestation problems and even structural decay. In some cases, wall cavities are being insulated with a combination of foamed-in-place polyurethane and fiberglass batts. Contamination of building interiors with diisocyanate and other compounds from polyurethane foam has been reported.

In addition to the use of insulation, dwellings are being constructed more tightly; i.e., modern construction practices are designed to reduce infiltration of cold air in the cold season and exfiltration of cool air during the cooling season. Concerns have been raised that such construction practices result in reduced air exchange and, as a consequence, increased contaminant concentrations and attendant health risks.

8. Furnishings

Modern dwellings are provided with a variety of furnishings and amenities. These include wall-to-wall carpeting, floor tile, furniture, decorative wall and ceiling materials, fireplaces, etc. The use of wall-to-wall carpeting in modern dwellings in the U.S. is now nearly universal.

Wall-to-wall carpeting is a highly attractive home furnishing. It absorbs sound, diminishes the perception of cold floor surfaces, is aesthetically attractive, and provides a comfortable playing surface for children. Wall-to-wall carpeting has negative attributes that are not as apparent as its attractions. Until recently, new carpeting was characterized by emission of a variety of volatile and semivolatile compounds that have caused odor problems (e.g., the rubbery smell of 4-PC associated with latex binders) and, in some cases, health complaints.

Wall-to-wall carpeting is an excellent reservoir for a variety of inorganic and organic particles, particles that are often difficult to remove even with regular cleaning. These include human skin scales, which serve as a food source for a variety of mold species and dust mites, and mite excretory antigens, which are the most common cause of chronic allergic rhinitis and asthma. They also include cat and dog dander, cockroach antigens, etc. These antigens are significant causes of inhalant allergies and many cases of asthma.

In addition to being a reservoir for a variety of dirt particles which are allergenic, carpeting produces a microenvironment favorable to dust mites and a variety of mold species. The high relative humidity needed to sustain development of large dust mite populations (see Chapter 5) is present in homes as a result of favorable conditions produced by the combination of carpeting and cool floor surfaces. Environmental concerns associated with carpeting are described in detail in Chapter 7, Section F.

Most houses are furnished with wood furniture. Most modern wood furniture is constructed with HCHO-emitting pressed-wood materials, and even solid wood furniture is coated with HCHO- and/or VOC-emitting finish coatings.

9. Storage

Residential building interiors are designed to provide a variety of storage capabilities. These include bedroom and hallway closets as well as kitchen and bathroom cabinetry. Most modern cabinetry is constructed with various pressed-wood products. These include hardwood plywood, particle board, medium-density fiber board (MDF) and OSB. With the exception of OSB and softwood plywood used for shelving and counter tops, respectively, most wood components are constructed from urea–formaldehyde resin-bonded wood materials that have the potential to emit significant quantities of HCHO. Formaldehyde emissions also occur from acid-cured finishes used on exterior surfaces of hardwood cabinets and good quality furniture.

10. Attached garages

Many single-family residences have attached garages. These provide an enclosure for motor vehicles and utilities such as furnaces, and a storage area for the varied needs of the building's occupants.

Because of diverse uses and their physical attachment to occupied spaces, garages may be a source of a variety of contaminants. Because occupied spaces are negatively pressurized relative to garages, motor vehicle emissions, gasoline and solvent vapors, etc., are readily drawn into living spaces.

11. Heating/cooling systems

In many parts of the world as well as the U.S., seasonal changes in outdoor temperature require that some form of heating and cooling appliance or system be used to provide more acceptable thermal conditions than occur outdoors. In addition, appliances provide hot water for bathing and other washing activities.

Energy sources and appliances used to heat residences or provide other heating needs (such as for cooking and supplying hot water) vary widely. In developing nations where population densities are high and resources limited, building occupants rely on biomass fuels to cook food over poorly vented fires, which during cold seasons also provide some degree of warmth.

In developed countries, a variety of manufactured appliances are used for cooking and others for space heating. Cooking appliances include natural gas or propane-fueled stoves and ovens, electric stoves and ovens, and microwave devices. Gas stoves and ovens are not vented to the outdoors and, as such, are a potentially significant source of indoor air contamination.

Single-family dwellings in the U.S. are typically heated by some form of appliance. These include vented furnaces fueled in most cases by natural gas, propane, or oil, or, less commonly, wood or coal. Other vented fuel-fired appliances include wood or coal stoves and, in parts of northern Europe, fireplaces. In the U.S., fireplaces serve primarily an aesthetic and decorative function.

Home space heating is accomplished primarily by the use of modular, freestanding, unvented natural gas or kerosene space heaters in the warmer regions of the U.S. and in countries such as Japan and Australia. These devices are designed to emit only limited quantities of carbon monoxide (CO) and do not pose an asphyxiation hazard. They may cause significant indoor air contamination with a variety of combustion by-products (see Chapter 3). Such space heaters are commonly used to "spot heat" individual rooms to reduce energy costs.

Electrical devices or systems are often used for home space heating in the U.S. These include cable heat, with elements in the ceiling, or electric heat pumps. In the latter case, energy is extracted from outside air or ground-water, with a heating coil supplement during very cold weather. Electric heating devices do not produce any combustion by-products and do not, in theory, pose any indoor air contamination risks.

Indoor space heating can be provided by central systems which forcibly or passively distribute heat from a combustion or electrical appliance to attain and maintain desired thermal conditions. Such systems heat all spaces, including those that are unoccupied. In forced air systems, heat is distributed through duct systems. In radiant heat systems, hot water is pumped to radiators distributed in various parts of the house.

In forced air systems, a fan draws air through a filter into the appliance, where it is heated and then delivered through ducts to building spaces through supply air registers. Air is returned to the furnace to be reheated through a second duct system described as a cold air return.

Duct systems in residences can cause or contribute to indoor air contamination problems. Historically, ducts (both supply and return) were constructed of galvanized steel. Increasingly, ducts are being fabricated from fiberglass materials. These include duct board, which is fashioned into supply and return air trunklines on-site, with polyethylene-lined, fiberglass-insulated flex duct serving to deliver conditioned air to supply air registers. Duct board may release contaminants such as methylamine, which is both an odorant and an irritant. Porous surfaces of duct board are deposition sites for organic dust, which may serve as a medium for mold growth and subsequent indoor air contamination.

Return air ductwork, which is under a high negative pressure, is typically located in attic, crawlspace, or basement areas, or attached garages. Wet crawlspaces and basements are often heavily contaminated with mold, and the usually leaky ductwork located in these spaces serves as a conduit for mold spores, moisture, and even radon into living spaces.

In many dwellings, cooling is provided either by window or whole-house systems that are integrated into heating systems.

12. Plumbing

Most houses have plumbing systems that carry water into them, then heat (and in many cases soften) it and distribute it to kitchen and bathroom sinks, toilets, showers/tubs, laundries, and external faucets for lawn and garden use. They also carry away cooking and bathing waters, toilet wastes, condensate from air conditioners and high-efficiency furnaces, and, in some cases, food wastes. In most cases, plumbing systems perform their functions well. Good maintenance, however, is required to prevent damage from leaks (which are relatively common) and entry of sewer gases through drains that develop dry traps. Plumbing-related problems that result from improper installation or maintenance are common in many residences.

13. Other utilities

Other utilities that are integral parts of housing structures include electrical wiring and, in many homes, pipe systems for natural gas or propane. Except for the potential to cause structural fires, electrical wiring poses no environmental concerns. Gas utility systems are subject to leakage and may cause an odor problem; the odor is designed to warn homeowners that leaking fuel gas may pose an explosion hazard.

E. Age and condition

Buildings vary in age and condition, so they vary in the types and magnitude of IAQ/IE problems associated with them. Because lead-based paint was used to cover exterior surfaces prior to 1978, and both interior and exterior surfaces prior to the 1950s, older houses are more likely to be contaminated with lead-containing dusts. This problem is exacerbated by the fact that many older houses are not well maintained; many are dilapidated.

Older houses are more likely to have had problems such as water intrusion, flooding, and condensation on windows and walls during their history than new houses. As a consequence, they are at much higher risk of being infested with mold.

Older houses are more likely to be less insulated, and therefore better ventilated, than newer houses. As such, they have higher air exchange rates. They are also more likely to have hardwood floors and less likely to have wall-to-wall carpeting.

F. Site characteristics

Inadequate site drainage bedevils many homeowners. Many building sites were historically poorly drained and remain so after building construction. Dwellings constructed on such sites are subject to a variety of water-related problems, including: basement seepage or flooding; episodically wet crawlspaces; water in heating/cooling ducts in slab-on-grade houses; and infestation of slab-on-grade ducts with moisture-loving crustaceans (sow bugs), spiders, insects, and mold. The periodic incursion of moisture into basements, crawlspaces, and slabs poses major IE problems. It may cause wetting of materials and subsequent mold infestation; high indoor humidity which increases the risk of condensation on cold window surfaces and a variety of mold infestation problems; and a favorable indoor climate for development of large dust mite populations, with their associated antigen production and exposure risks.

Other site characteristics also contribute to moisture and IE concerns. Heavily shaded sites tend to retard drying of exterior building surfaces as well as the site itself. Well-drained sites with sandy or gravelly soils reduce risks associated with moisture-requiring biological contaminants. On the other hand, high soil permeability may be associated with elevated radon levels.

G. Occupants and occupant behavior

Once a dwelling is occupied, it is subject to a number of contaminant-generating activities. These include: production of bioeffluents by occupants as well as odors associated with food preparation and use; emissions from personal care and clothing/home cleaning products; smoking of tobacco products (and possibly other weeds); emissions/by-products of hobbies, crafts and in-home enterprises; fragrant emissions from candles, potpourri and decorative items, as well as combustion by-products from the frequent use of candles; pet odors and danders; production of organic debris that serves as food for antigen-producing dust mites, cockroaches, mice, etc.; both proper and improper use of pesticides to control common household pests; building interior renovation activities; improper/ inadequate care and maintenance of building combustion, plumbing and other systems; introduction of furniture and other materials which may be a significant source of contaminants; and introduction of particulate-phase contaminants on shoes and clothing from the building site (e.g., lead-based paint dusts and pesticides) as well as work environment (e.g., industrial dusts, starch and talc from hair care establishments, etc.).

H. Exposure concerns

Occupants of dwellings are exposed to indoor air and other environmental contaminants in ways that are different from other nonindustrial buildings. As indicated previously, occupants of residences can be exposed to potentially toxic indoor contaminants 12 to 24 hours/day. Exposed populations include infants/children, healthy adults, the aged, and the infirm.

Contaminant exposure concerns that are unique in residences include: radon; HCHO; environmental tobacco smoke (ETS); pesticides; unvented combustion appliances; biocontaminants such as dust mite and cockroach antigens, mold, and animal danders; lead-based paint-contaminated dust; emissions from personal and home-care products, and arts and crafts activities; and contagious disease.

Exposure concerns in residences are increased due to limited dilution/ventilation potential when windows are closed during heating and cooling seasons. Residences, unlike other nonindustrial buildings, are not, in most cases, mechanically ventilated.

Figure 1.4   Retail building.

III. Characteristics of nonresidential buildings

Nonresidential buildings can be characterized by (1) the functions and populations they serve, (2) access and ownership status, (3) building types and construction characteristics, (4) building operation and maintenance, (5) occupant density/activities, and (6) health and other exposure concerns.

A. Building functions and populations served

Nonresidential buildings are designed and constructed to serve a variety of human needs. These include retail and other commercial activities (Figure 1.4), private and public office space, education, health care, imprisonment/detention, worship, entertainment, etc. These buildings vary in the populations they serve. In office buildings (Figure 1.5), prisons, colleges and universities, and many commercial establishments, most of the building

Figure 1.5   Office building.

Figure 1.6 School building.

population is comprised of adults; in schools (Figure 1.6), children age 5 to 18 dominate; in health-care facilities (Figure 1.7), the building population consists of the infirm, as well as care givers and a variety of service personnel; entertainment and sport facilities serve populations that reflect the popularity of the entertainment provided to various age groups. Unlike residences, nonresidential buildings tend to serve diverse populations. As such, IAQ/IE exposure concerns differ in many cases from those that occur in residential buildings.

B. Access and ownership status

Although many nonresidential buildings are owned and operated by government and a variety of not-for-profit entities, most are privately owned by individuals or corporations. Because these buildings are open to the public for at least a portion of the day, they can be described as public-access

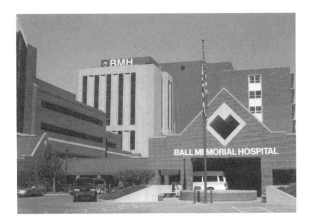

Figure 1.7 Health-care facility.

buildings. Occupants of public-access buildings as well as "visitors" depend on building management to provide a comfortable and low-health-risk environment. Occupants and visitors usually do not have any control over building environment conditions.

C. Building types and construction characteristics

Building type, design, and construction characteristics reflect the needs served by individual buildings, resources and desires of the owner, preferences of architects and contractors, resource and material availability, climate, etc. Office buildings, for example, vary in size from small structures no larger than a residence to giant towers providing hundreds of thousands to millions of square feet of floor space. They vary from wood-framed with a variety of cladding types to structures with ribs of steel and facades of glass and stone.

Like residences, nonresidential buildings have similar basic structural and furnishing requirements. These include a substructure or building base; a skeletal frame; external cladding; windows; a roof; insulating materials; interior wall coverings; flooring; finish coatings; and interior furnishings such as floor coverings, furniture, storage cabinets, room dividers, etc.

1. Substructure/structure

Because nonresidential structures tend to be large, they are constructed to reflect structural demands. The building base must support the weight of the building whose structural components are steel and concrete. Substructures therefore are slab-on-grade or have one or two subgrade levels, with structural members often anchored to bedrock or stabilized ground. The building frame may be constructed of steel or reinforced concrete columns. In multistory buildings, structural steel is sprayed with fireproofing insulation, which reduces the risk of warping and building collapse in a fire. Before 1973, such fireproofing contained asbestos.

2. Walls

Exterior walls may include extensive glass or cladding of limestone building stone, brick, etc. As in residences, they are often insulated (depending on climate) with materials manufactured for the walls and roofs of large buildings. Such buildings may or may not have vapor barriers. In warm climates, such as Florida, severe mold infestations on interior wall materials have occurred as a result of thermal-enhanced movement of water through the building envelope (without vapor barrier) and subsequent condensation on cooled wall surfaces. Interior walls are typically covered with gypsum board, with other materials used for decorative purposes.

3. Flooring/floor covering

Floors of nonresidential buildings are usually poured concrete, often covered with vinyl tile, terrazzo, or carpet. Vinyl asbestos tile was widely used in schools and other nonresidential buildings prior to 1980. Increasingly, in new buildings, floor coverings are glued-down industrial-grade carpeting. Emissions from carpeting and associated adhesives have been the subject of IAQ complaints in a number of buildings. As in residences, carpeting in nonresidential, nonindustrial buildings becomes a sink for a wide variety of organic particles.

4. Ceilings
Ceilings of many nonresidential buildings serve several functions. They need to provide an aesthetically acceptable appearance and, in many cases, a cavity through which utilities such as wiring, plumbing, and mechanical systems are extended. These cavities often serve as plenums through which return air is conveyed to air-handling units (AHUs) to be reconditioned. Suspended ceiling tile commonly serves as the plenum base. In other cases, decorative acoustical plaster may be sprayed on ceiling surfaces. Prior to 1978, acoustical plaster containing upwards of 10% chrysotile asbestos was commonly used in foyers and hallways in schools, auditoria, and other buildings.

5. Roofs
Roofs of a large percentage of nonresidential buildings are flat. A flat roof is considered more aesthetically pleasing by architects and serves as a platform for heating, ventilation and air conditioning (HVAC) system AHUs, exhaust vents, etc. Flat roofs require design and construction care to assure proper drainage of rain and snow melt, and maintenance to prevent water intrusion into building interiors. Flat roofs are often plagued by water leaks that damage ceiling tiles and other interior materials. Such water intrusion is a common problem in school buildings. Because HVAC system AHUs as well as exhaust vents are often located on flat roofs, re-entry of flue and other exhaust gases is also a common problem.

6. Furnishings/equipment
Nonresidential buildings are provided with a variety of furnishings, e.g., chairs, desks, storage cabinets, office dividers, etc. These furnishings can emit a variety of VOCs and SVOCs which contaminate indoor spaces. Steel desks and storage cabinets have low emissions. Wooden desks, storage cabinets, counter tops, and office dividers may be constructed from HCHO-emitting pressed-wood products, and thus serve as a source of HCHO and potential irritant effects.

Nonresidential buildings also contain a variety of equipment types including computers, printers, photocopying machines, etc. Such equipment can be a source of indoor contaminants (see Chapter 7).

7. Heating, cooling, and ventilation systems
Nonresidential, nonindustrial buildings vary considerably in how they are climate-controlled and ventilated. Climate control in seasonally colder cli-

mates may be limited to providing heat by radiant heating elements, forced air furnaces, boilers plus forced air heating coils, heat pumps, etc. The systems used reflect the building's needs and often its age. Ventilation may be provided by mechanical systems or, as in many older buildings, by opening windows (natural ventilation).

In seasonally hot climates, the primary focus of climate control is cooling. Again, ventilation may be provided by mechanical means or by opening windows.

The trend in most developed countries is to design and construct buildings with year-round climate control. In these buildings, windows are sealed and cannot be opened to provide natural ventilation. Ventilation must be provided by mechanical systems. Because ventilation is integrated into the heating and cooling systems, they are described by the acronym, HVAC (heating, ventilating, air conditioning). HVAC systems vary in design and operation (see Chapter 11).

HVAC systems control thermal conditions and air exchange with the ambient environment, so their operation is a major determinant of occupant comfort and satisfaction with the indoor environment. Well-designed and operated HVAC systems are essential to provide occupants with ventilation air sufficient to dilute human bioeffluents to acceptable levels and, to a limited degree, control levels of other contaminants as well. Inadequacies in design and operation of HVAC systems are the primary cause of air quality complaints in mechanically ventilated buildings.

8. Plumbing

Plumbing systems are designed to provide a potable water supply, heated (hot water or steam) or chilled water lines serving AHUs, sprinkler water supply for fire suppression, static water supply for emergency fire use, and waste water lines. Heated or chilled water lines are typically insulated. A variety of insulating materials have been used, including molded gypsum-containing asbestos. In many older buildings, the plumbing system is the major site of asbestos-containing materials. Modern plumbing is insulated with a variety of materials including fiberglass, foamed rubber, molded gypsum, etc.

Plumbing in nonresidential buildings is subject to a variety of leakage problems that cause minor stains (and mold infestation) to major flooding. The former are common; the latter rare. Both are environmental quality concerns.

9. Other utilities

In addition to heating, cooling, ventilating, and plumbing, nonresidential buildings are provided with a variety of utilities including lighting fixtures and electrical, telephone, and computer wiring. Wiring is typically arranged to minimize space and resource requirements and is co-located in wiring runs, pipe chases, hallway plenums, etc.

D. Building operation and maintenance

Nonresidential buildings such as office, commercial, and institutional build-ings are typically large, with relatively complex systems (plumbing, lighting, HVAC) that need to be properly operated and maintained. Many buildings and their systems are often poorly operated and maintained and are therefore subject to a variety of problems, including poor thermal control, inadequate ventilation, inadequate cleaning, recurring roof and other structural leaks, and mold infestation. Poor building and building systems operation and maintenance may be due in part to the complexities involved (particularly HVAC systems), inadequately trained or motivated facilities service staff, lack of commitment by building management, and inadequate building operation/maintenance resources. School buildings, in poor (and sometimes not-so-poor) school districts, are particularly subject to resource limitations. In such instances maintenance is often deferred. As a result, poorly operating mechanical systems, water damaged/mold infested materials, and inade-quately cleaned surfaces are common.

E. Occupant densities and activities

Nonresidential buildings are distinguished by varied occupant densities and activities. Projected occupant densities are a major building design factor. They determine space requirements and ventilation needs. Highest occupant densities occur in school buildings and sports arenas.

Occupant activities vary from building to building, as well as within a single building. In office buildings, these may include general clerical work, using office equipment, preparing/serving/eating food, printing, etc. They may also include maintenance activities such as cleaning floors and other horizontal surfaces, repainting, repairing problem systems, and pest control, among others. In schools, they include teaching/learning, clerical/adminis-trative work, food preparation/eating, athletic activities, art and shop projects, and maintenance activities such as floor waxing and pest control.

Occupant activities may, in many cases, be a source of contaminants that affect IAQ and the cleanliness of building surfaces. They may also affect the health and well-being of occupants engaged in such activities and/or the general building population.

F. Exposure concerns

Nonresidential buildings are subject to a number of contamination, exposure, and health concerns. These include: elevated bioeffluent levels associated with high occupant densities and inadequate ventilation; emissions from office equipment and materials; cross-contamination from contaminant-gen-erating areas; re-entry of building exhaust gases; entrainment of contami-nants generated outdoors; contamination of AHUs by organisms/biological products that can cause illness, e.g., hypersensitivity pneumonitis, humidifier fever, and Legionnaires' disease; transmission of contagious diseases such as flu, colds, and tuberculosis; exposure to resuspended surface dusts; exposure to ETS where smoking is not restricted; etc. With some exceptions, radon, unvented combustion appliance emissions, pesticides, and lead-based-paint-contaminated dusts are not major exposure concerns; radon, pesticides, and lead-contaminated dusts are, however, concerns in school buildings.

Health concerns, as indicated above, include diseases such as Legionnaires' disease, hypersensitivity pneumonitis, and illness symptoms often described as "sick building syndrome."

IV. Other indoor environments

Because humans spend so much time indoors, most IE concerns have focused on buildings. Nevertheless, exposures to airborne or resuspended surface contaminants occur in other environments as well. These include interiors of motor vehicles, airplanes, trains, ships, submarines, and space capsules.

A. Motor vehicles

Motor vehicles represent unique IAQ/IE concerns. Contaminants such as VOCs and SVOCs may be emitted from materials used in vehicle interiors, e.g., vinyl plastics. They may also become entrained in the interior compartment from the vehicle's own exhaust or the exhaust of other vehicles. Air-conditioning systems may also be a source of contamination. Exposures may be brief, varying from minutes to hours per day, and possibly repeated daily.

B. Commercial airplanes

Travel in commercial aircraft (Figure 1.8) is a relatively infrequent occurrence for most individuals (except flight crews). Airplanes are in some ways similar to nonresidential buildings. They are characterized by high occupant densities (2 m^3/person) and mechanical systems that use both recirculated and

Figure 1.8   Commercial aircraft.

outside air; uniquely they have interior pressures comparable to an altitude of 8000 feet. Despite the fact that ventilation standards for aircraft have been revised to increase outside air, aircraft designs limit outside air delivery capacities to environmental control systems. In a filled-to-capacity airplane, the amount of outside air provided to passengers and crew is half that recommended for office buildings.

Commercial aircraft personnel and passengers are subject to a variety of contaminant exposures. These include human bioeffluents, VOC emissions from seats and other interior materials, entrained fuel combustion by-products, ETS on smoking-permitted flights [respirable suspended particulate (RSP) levels are approximately 20 times those on nonsmoking flights], and elevated ozone when flying at high altitudes (circa 30,000 to 40,000 feet). Passengers and crew are often subject to low relative humidity (5 to 25%) as well.

C. Trains

Trains are used for both surface and underground transportation. Tens of millions of individuals, in North America and other parts of the world, use train transportation daily. Such use is particularly heavy in large cities. Underground transportation includes two indoor environments, the train and underground tunnels and platforms. Underground systems require ventilation that occurs passively or by mechanical means.

Contaminant exposures in train compartments may include human bioeffluents, emissions from interior materials, entrainment from combustion-driven systems, entrainment of contaminants from underground sources, etc.

D. Ships

A ship can be likened to a hotel. It contains sleeping/living quarters, dining areas, food-handling areas, lounges, theaters, etc. It also includes an on-board, combustion-driven propulsion system, and waste handling and storage systems.

Many modern passenger ships are mechanically ventilated and air conditioned. Like land-based nonresidential buildings, such systems vary in the degree of thermal comfort provided, as well as ventilation adequacy.

Exposure concerns on ships may include human bioeffluents, emissions from ship materials, ETS in smoking-permitted areas, entrainment from combustion systems, and cross-contamination from high source areas. Of special concern have been the transmission of contagious diseases, such as influenza, and outbreaks of Legionnaires' disease.

E. Submarines and space capsules

Submarines and space capsules represent truly unique indoor environments and exposure concerns. In both cases, ventilation is not possible. Air/oxygen, which is quite limited, has to be continuously recycled. In submarines, special air cleaning systems must be used to maintain acceptable CO_2 (<1%) and VOC levels.

Exposures to crew members in both cases include human bioeffluents, emissions from interior materials, and combustion by-products. Crew members in early space flights reportedly experienced symptoms similar to those associated with poor IAQ in buildings.

Readings

Godish, T., *Sick Buildings. Definition, Diagnosis, and Mitigation*, CRC Press/Lewis Publishers, Boca Raton, 1995.

Lstiburek, J.W., *Contractor's Field Guide*. Building Science Corporation, Chestnut Hill, MA, 1991.

Lstiburek, J.W. and Carmody, J., *Moisture Control Handbook. Principles and Practices for Residential and Small Commercial Buildings*, Van Nostrand Reinhold, New York, 1994.

Maroni, M., Seifert, B., and Lindvall, T., *Indoor Air Quality — A Comprehensive Reference Book*, Elsevier, Amsterdam, 1995.

National Research Council, *The Airliner Cabin Environment. Air Quality and Safety*, National Academy Press, Washington, D.C., 1986.

Samet, J.M. and Spengler, J.D., Eds., *Indoor Air Pollution. A Health Perspective*, Johns Hopkins University Press, Baltimore, 1991.

Spengler, J.D., Samet, J.M., and McCarthy, J.C., Eds., *Indoor Air Quality Handbook*, McGraw-Hill, New York, 2000, chaps. 6, 64, 66–68.

Questions

1. What factors have contributed to our recent concerns about IAQ/indoor environment problems?
2. How are IAQ/indoor environment problems in residences different from those experienced in nonresidential buildings?
3. What is the significance of ownership status relative to preventing and mitigating indoor environment problems?
4. Describe differences in substructure types and their potential roles in indoor environment concerns.
5. How do site-built homes differ from manufactured homes?
6. Describe sidewall components and their function.
7. Describe problems experienced with cladding that might cause structural and indoor environment problems.
8. Describe environmental problems in residential buildings associated with windows.
9. Describe residential flooring materials and potential indoor environment problems associated with them.
10. Describe energy conservation practices and potential associated indoor environment problems.
11. What kind of environmental problems have been associated with the use of gypsum board?

12. Describe the nature of storage in residences and how it may be related to indoor environment problems.
13. How are site characteristics related to the indoor environment of residences?
14. A variety of mechanisms/appliances are used to heat our food and make our dwellings comfortable. How may these affect the indoor environment and human health?
15. How do occupants in residential buildings affect their indoor environment?
16. Describe contaminant exposure concerns in residential buildings.
17. What are public-access buildings?
18. How do roofs on nonresidential buildings differ from those on residential buildings? What unique problems are associated with roofs?
19. Describe the nature of ceilings in many nonresidential buildings.
20. Describe the use and operation of HVAC systems in public-access buildings and how they differ from residential buildings.
21. Describe the nature of building operation and maintenance concerns in non-residential, nonindustrial buildings.
22. Describe health and exposure concerns in nonresidential, nonindustrial buildings.
23. How do occupant activities affect the indoor environment in nonresidential, nonindustrial buildings?
24. What human exposure concerns are associated with the interiors of motor vehicles?
25. What human exposure concerns are associated with commercial air travel?

chapter two

Inorganic contaminants: asbestos/radon/lead

Inorganic substances such as asbestos, radon, and lead are major indoor contaminants. Though very different, they have in common a mineral or inorganic nature. Exposures may pose significant health risks.

Lead is of concern because it is a common surface contaminant of indoor spaces, and contact with lead-contaminated building dust is the primary cause of elevated blood levels in children under the age of six.

I. Asbestos

Potential airborne asbestos fiber exposures in building environments and associated public health risks were brought to the nation's (United States) attention in the late 1970s by both public interest groups and governmental authorities. This attention was a logical extension of exposure concerns associated with the promulgation of a national emission standard for asbestos as a hazardous pollutant (NESHAP) by the United States Environmental Protection Agency (USEPA) in 1973. The asbestos NESHAP banned application of spray-applied asbestos-containing fireproofing in building construction; there was a subsequent ban of other friable asbestos-containing building products in 1978. Under NESHAP provisions, friable (crushed by hand) asbestos-containing building materials (ACBM) must be removed prior to building demolition or renovation. Such removal must be conducted in accordance with Occupational Safety and Health Administration (OSHA) requirements to protect construction workers removing asbestos, as well as building occupants. As a consequence of these regulatory actions, asbestos in buildings, particularly in schools, became a major indoor air quality (IAQ) and public health concern.

The ban on friable asbestos-containing materials used in building construction and requirements for removal prior to demolition or renovation were intended to minimize exposure of individuals in the general community to contaminated ambient (outdoor) air. Potential exposures to building occupants from fibers released from building products in the course of normal activities had not been addressed. In 1978, public attention was drawn to the large quantities of friable or potentially friable ACBM that was used in school construction as well as other buildings.

A. Mineral characteristics

Asbestos is a collective term for fibrous silicate minerals that have unique physical and chemical properties that distinguish them from other silicate minerals and contribute to their use in a wide variety of industrial and commercial applications. These include thermal, electrical, and acoustic insulation properties; chemical resistance in acid and alkaline environments; and high tensile strength, which makes them useful in reinforcing a variety of building products.

Asbestos comprises two mineral groups which are distinguished by their crystalline structure: serpentine and amphiboles. Serpentine chrysotile (Figure 2.1), the most widely used asbestos mineral, has a layered crystalline structure with the layers rolling up on each other like a scroll or "tubular fibrils." The amphiboles, which include amosite, crocidolite, anthophyllite, actinolite, and tremolite, have a crystalline structure characterized by double-chain silicate "ribbons" of opposing silica tetrahedra linked by cations.

Individual asbestos fibers have very small diameters, high aspect (length:width) ratios, and smooth parallel longitudinal faces. Asbestos fibers are defined for exposure monitoring as any of the minerals in Table 2.1 that have an aspect ratio ≥3:1, lengths >5 μm and widths <3 μm. In actual practice,

Figure 2.1 Chrysotile asbestos fibers under microscopic magnification. (Courtesy of Hibbs, L., McTurk, G., and Patrick, G., MRC Toxicology Unit, Leicester, U.K.)

Table 2.1 Asbestos Minerals Used Commercially or Found in
Asbestos Products Used in Buildings

Mineral	Commercial name	Chemical formula	Building occurrence
Chrysotile	Chrysotile	$(Mg)_6(OH)_8S_{14}O_{10}(\pm Fe)$	*
Grunerite	Amosite	$Fe_7(OH)_2S_{18}O_{22}(\pm Mg, Mn)$	**
Rubeckite	Crocidolite	$Na_2(Fe^{3+})_2(Fe^{2+})_3(OH)_2S_{18}O_{22}(\pm Mg)$	X
Anthophyllite	Anthophyllite	$(Mg, Fe)_7(OH)_2OS_{18}O_{22}$	***
Actinolite	Actinolite	$Ca_2Fe_5(OH)_2S_{18}O_{22}(\pm Mg)$	***
Tremolite	Tremolite	$Ca_2Mg_5(OH)_2S_{18}O_{22}(\pm Fe)$	***

* Very commonly found in ACM products.

** Commonly found.

*** Uncommonly found.

X Typically not used in ACM in North America.

Source: From Health Effects Institute–Asbestos Research, *Asbestos in Public and Commercial Buildings: A Literature Review and Synthesis of Current Knowledge,* Cambridge, MA, 1991. With permission.

asbestos fibers have the following characteristics when viewed by light microscopy: (1) particles typically having aspect ratios from 20 to 100:1 or higher, and (2) very thin fibers (typically <0.5 μm in width). The parallel fibers often occur in bundles. The very fine individual fibers are best seen using transmission electron microscopy. Chrysotile asbestos fiber diameters have been reported to range from 0.02 to 0.08 μm, amosite between 0.06 and 0.35 μm, and crocidolite between 0.04 and 0.15 μm. The smaller the diameter, the higher the tensile strength.

B. Asbestos-containing building materials

Commercial and industrial use of asbestos has a relatively long history. Asbestos fibers have been used extensively, with well over 3000 applications. Generic uses have included fireproofing, thermal and acoustical insulation, friction products such as brake shoes, and reinforcing material.

Materials made of asbestos, or having asbestos within them, are described as asbestos-containing materials (ACM). When used in building construction, they are identified as asbestos-containing building materials (ACBM). Types of ACBM, their characteristics, asbestos content, and time period of use are given in Table 2.2.

1. ACM in nonresidential buildings

For regulatory purposes, asbestos-containing building materials are classified as surfacing materials (SM), thermal system insulation (TSI), and miscellaneous materials (MM). Surfacing materials include spray-applied fireproofing (Figure 2.2) and spray-applied or troweled-on acoustical plaster. Asbestos-containing fireproofing was sprayed on steel I beams in multistory buildings to keep buildings from collapsing due to structural fires. Acoustical

Table 2.2 Some Asbestos-Containing Materials Used in Buildings

Category	Characteristics	Asbestos (%)	Dates of use
Surfacing material	Sprayed on	1–95	1935–1970s
	Troweled on		
Thermal system insulation (preformed)	Batts, blocks, pipe covering		
	85% magnesia	15	1926–1949
	Calcium silicate	6–8	1949–1970s
Textiles	Curtains (theater)	60–65	1945–present
Cementitious concrete-like products	Flat panels	40–50	1930–present
	Corrugated panels	20–45	1930–present
	Pipe	~20	1930–present
Paper products	Corrugated		
	High temperature	90	1935–present
	Moderate temperature	35–70	1910–present
	Indented	98	1935–present
	Millboard	80–85	1925–present
Asbestos-containing compounds	Caulking putties	30	1930–present
	Adhesive	5–25	1945–present
	Joint compound		1945–1975
	Spackling compound	3–5	1930–1975
	Insulating cement	20–100	1900–1973
	Finishing cement	55	1920–1973
Flooring tile/sheet goods	Vinyl asbestos tile	21	1960–present
	Asphalt/asbestos tile	26–33	1920–present
	Resilient sheeting	30	1950–present
Paints/coatings	Roof coating	4–7	1900–present
	Airtight asphalt coating	15	1940–present

Note: Information in this table was based on a 1985 study by the USEPA. Many ACM products have been phased out or discontinued. Use period "present" indicates 1985.

Source: From Health Effects Institute–Asbestos Research, *Asbestos in Public and Commercial Buildings: A Literature Review and Synthesis of Current Knowledge*, Cambridge, MA, 1991. With permission.

plaster was widely used in foyers, hallways, school gymnasia, classrooms, etc., as a decorative surface and sound absorption medium. Surfacing materials are very friable and have a significant potential for releasing asbestos fibers into the general building environment when disturbed.

Thermal system insulation was widely used to insulate mechanical room boilers, associated equipment, and steam/hot water lines (Figure 2.3). Occasionally it was used for cold water lines to prevent condensation. In most cases, TSI was wrapped with a protective cloth and poses an exposure risk to service/maintenance workers only when the protective cloth (lagging) is damaged or disturbed. Thermal system insulation was applied to boilers as blocks or batts and to steam/hot water lines as preformed pieces.

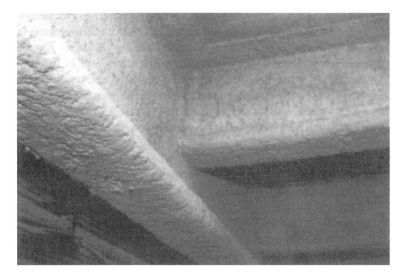

Figure 2.2   Asbestos fireproofing sprayed on building I beams.

Miscellaneous materials include all other asbestos applications in buildings, such as ceiling tile, vinyl asbestos floor tile, adhesives/mastics, spackling compounds, asbestos–cement products, etc.

Chrysotile, as seen in Table 2.2, has been the most widely used asbestiform mineral in products used in buildings. It has been reported that chrysotile accounts for 95% of asbestos used in the U.S. and is the predominant fiber in ACBM. However, in building inspections, the pattern of asbestos minerals in TSI and SM reflects different proportions of serpentine and

Figure 2.3   Partially damaged thermal system insulation containing asbestos.

amphibole fibers. In a study of U.S. municipal buildings, TSI contained asbestos fibers in the following proportions: chrysotile only — 60%, mixed chrysotile and amphibole — 35%, and amphibole only — 7%. For SM, proportions were: chrysotile only — 73%, mixed fibers — 10%, and amphibole only — 17%. In the latter case, ceiling tile was classified as SM rather than MM. This is significant in that most of the amphibole-only surfacing material was found in asbestos-containing ceiling tiles.

Various surveys have been conducted to assess the prevalence of buildings with friable ACBM (SM, TSI, and ceiling tile). In 1988, USEPA estimated that approximately 700,000 U.S. public and commercial buildings (about 20%), out of a population of 3.5 million, contained some type of friable ACBM. A study conducted by the Philadelphia Department of Health found that 47% of 839 municipally owned or occupied buildings contained friable ACBM. In a California study, 78% of its public buildings constructed before 1976, and 56% of all public buildings, were estimated to contain ACBM. A similar study estimated that 67% of the 800,000 buildings in New York City contained ACBM. Most of this material (84%) was TSI, 50+% of which was found in mechanical rooms. Eighty +% of this material was assessed as being moderately to severely damaged.

The percentage of buildings containing ACBM increases considerably when other nonfriable or mechanically friable materials, such as vinyl asbestos tile, asbestos cement board, mastics, and drywall taping products, are included. Asbestos fibers in floor tile, cement board, and mastics are bound in a hard material that prevents them from being easily released. As such, they are not hand-friable. They are, however, mechanically friable (broken, cut, drilled, sanded, or abraded in some way). Mechanically friable ACBM can pose an exposure hazard under certain conditions and activities. Consequently, such activities are regulated under federal and state demolition and renovation requirements.

2. ACM in residences and other structures

Asbestos in residences has received relatively limited regulatory attention. This has been due, in part, to the fact that ACBM was not as widely used in residences (except large apartment houses) as it was in large institutional and commercial buildings. ACBM in residences includes a variety of products, e.g., TSI around hot or cold water lines, asbestos paper wrap around heating ducts, cement board around furnaces/wood-burning appliances, cement board (Transite) siding, cement board roofing materials, asbestos-containing asphalt roofing, wallboard patching compounds, asbestos-containing ceiling materials that were spray-applied or troweled on, and vinyl asbestos tiles.

With the exception of TSI and SM used on ceilings, most ACBM in residences contains asbestos in a bound matrix. It is therefore mechanically friable and should only produce an exposure risk if significantly disturbed. Materials used on building exteriors should also pose little risk of human exposure.

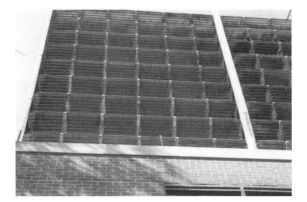

Figure 2.4 Cooling tower constructed using asbestos fibrocement panels.

Asbestos-containing fibrocement materials were once widely used in the construction of farm and other utility buildings and mechanical-draft cooling towers. In the latter case, asbestos-containing cement board was extensively used in external and internal cooling tower components (Figure 2.4).

C. Asbestos exposures

Because of the many desirable properties of asbestos and its widespread use in ACM, it is a ubiquitous contaminant of both indoor and outdoor air. A number of studies have been conducted to assess levels of asbestos fibers/structures in indoor and ambient air. Unfortunately, these studies used a variety of optical methodologies to determine fiber concentrations in collected samples. Early studies are based on phase-contrast microscopy used in occupational exposure monitoring. Phase-contrast microscopy cannot distinguish between asbestos and non-asbestos fibers, and fibers with small diameters (<0.5 μm) cannot be easily seen. Asbestos fibers are best analyzed using transmission electron microscopy (TEM) and direct sample preparation techniques.

1. Units of measurement

Concentrations of airborne asbestos have been expressed in a variety of ways. These include: (1) fibers with aspect ratios of ≥3:1 or ≥5:1 reported as fibers per cubic centimeter (f/cc or f/ml), (2) structures (≥5 μm) per liter (s/l), and (3) fiber mass per unit volume (ng/m^3). The term "structure" refers to fibers, clusters, bundles, and matrices.

Only concentrations of asbestos fibers with lengths ≥5 μm are used for risk assessment calculations since epidemiological studies have shown that asbestos-related disease increases significantly with exposure to asbestos fibers ≥5 μm. Though there is no sharp demarcation of asbestos toxicity associated with decreasing fiber length, occupational exposure standards are based on fibers ≥5 μm. Such concentrations are best characterized as an index of exposure. In most cases, concentrations of fibers <5 µm are greater than those ≥5 µm.

2. Persons exposed to asbestos in buildings

A variety of individuals may be exposed to airborne asbestos fibers. These include general building occupants such as teachers, students, office workers, and visitors; housekeeping/custodial employees who may come in contact with or disturb ACBM or contaminated settled dust during their work activities, and maintenance/construction workers who may disturb ACBM during repair or installation activities. Asbestos abatement/remediation workers and emergency personnel such as firefighters may also become exposed.

3. Ambient (outdoor) concentrations

Samples collected from Antarctic ice indicate that chrysotile asbestos has been a ubiquitous contaminant of the environment for at least 10,000 years. Snow samples in Japan have shown that ambient background levels are one to two orders of magnitude higher in urban than in rural areas. Higher concentrations of airborne asbestos fibers are reported in urban areas where there is more ACM and mechanisms of release (vehicles braking and weathering of asbestos cement materials); concentrations in the range of 1 to 20 ng/m^3 have been reported. Fibers longer than 5 µm are rarely found in rural areas. Ambient concentrations using TEM analysis have been based on mass measurements.

4. Asbestos concentrations in building air

Asbestos concentrations in buildings have been measured using a variety of techniques. Representative samplings of asbestos fiber concentrations (f/cc) determined by TEM with direct sample analysis are summarized in Table 2.3. These studies indicate that asbestos concentrations vary from below the limit of detection to maximum concentrations approximately 1.5 to 2+ orders of magnitude greater than the current 8-hour OSHA TWA occupational standard of 0.1 f/cc. Average concentrations are 2 to 3 orders of magnitude lower than the occupational permissible exposure limit (PEL).

Average building asbestos concentrations ranging from 0.00004 to 0.00243 f/cc have been reported in a study of 198 randomly selected ACBM-containing buildings. Mean concentrations for schools, residences, and public/commercial buildings were 0.00051, 0.00019, and 0.0002 f/cc, respectively, with 90 percentile concentrations of 0.0016, 0.0005, and 0.0004 f/cc. The higher asbestos fiber concentrations observed in school buildings may be due to the greater activity there that disturbs ACBM and resuspends asbestos fibers. The concentration of airborne asbestos fibers in buildings of all types appears to be associated with the presence of occupants and their level of activity. These data also indicate that asbestos fibers longer than 5 µm represent only a small fraction of the total number of airborne asbestos fibers.

Table 2.3 Airborne Asbestos Concentrations in Buildings Determined by Transmission Electron Microscopy

Description	Sample #	Asbestos fibers >5 µm (f/cc) Range	Mean
Nonlitigation			
19 Canadian buildings with spray-applied ACBM	63	ND–0.003	0.00042
12 United Kingdom nonresidential buildings with ACBM	96	ND–0.0017	0.00032
37 U.S. public buildings with damaged ACBM	256	ND–0.00056	0.00005
19 U.S. schools with ACBM	269	ND–0.0016	0.0002
Litigation			
121 schools and universities	1008	ND–0.0017	0.00046

Source: From Health Effects Institute–Asbestos Research, *Asbestos in Public and Commercial Buildings: A Literature Review and Synthesis of Current Knowledge,* Cambridge, MA, 1991. With permission.

Data based on arithmetic mean averages are likely to overestimate asbestos exposure in buildings. Asbestos fiber concentrations are not normally distributed and, as a result, geometric mean or median values are more appropriate than arithmetic means. Arithmetic means have often been reported in studies because 50% of airborne building asbestos values are below the limit of detection; as a consequence, the median value would be zero. Arithmetic means are very sensitive to a few very high values and thus are likely to overestimate occupant exposure.

Higher exposures can be expected for custodial workers whose activities may resuspend settled asbestos fibers and structures on a regular basis, and disturb ACBM on occasion. Comprehensive exposure studies associated with custodial activities have not been reported. Higher exposures can also be expected for maintenance workers who damage ACBM during their work. Elevated episodic exposure concentrations of >1 f/cc (determined by phase contrast microscopy) have been reported for a variety of maintenance activities.

5. Factors contributing to asbestos fiber release and potential airborne exposure

When fibers or asbestos structures from ACM become airborne, the process is called primary release. Primary release mechanisms include abrasion, impaction, fallout, air erosion, vibration, and fire damage. Secondary release occurs when settled asbestos fibers and structures are resuspended as a result of human activities. In unoccupied buildings or during unoccupied periods, fiber release typically occurs by fallout or is induced by vibration or air erosion.

Impaction and abrasion are likely to be the major causes of increased airborne fiber levels. Fallout occurs when cohesive forces that hold ACM together are overcome. For small particles, both cohesive and adhesive forces are very strong, but mechanical vibration may produce sufficient energy to overcome these forces. Release of fibers by air erosion, even in return air plenums with spray-applied ACBM, has been shown to be minimal.

Several studies have indicated that resuspension of surface dust is the main source of airborne asbestos fibers indoors. Other studies have suggested that the resuspension of asbestos-containing surface dust is of minor, if not negligible, importance. Resuspension requires sufficient disturbance to overcome the strong adhesive forces that exist between particles and surfaces.

6. *Indirect indicators of potential exposure*

Measurements of indoor asbestos fiber concentrations are often made by building managers in response to occupant asbestos exposure concerns. Such one-time measurements are, at best, a snapshot of potential exposure in sampled spaces. Concentrations vary significantly over time, depending on the amount of ACBM/asbestos-containing dust disturbance. Consequently, USEPA does not recommend, and even discourages, use of airborne asbestos sampling to determine potential asbestos exposures in buildings.

Under USEPA regulatory requirements, asbestos hazard determinations for school buildings are based on detailed inspections which include identifying potentially hand friable ACBM, collecting bulk samples, assessing the extent of damage, and determining the potential for future damage.

Building asbestos hazard assessments are used to select abatement priorities. Assessment methods in current use consider the accessibility and condition of the ACBM. Assessment is based on the following premises: (1) the likelihood of disturbance increases with accessibility, (2) damaged ACBM is evidence of past disturbance and the potential for future disturbance, and (3) damaged ACBM is more likely to release fibers when disturbed. In the USEPA decision tree, Figure 2.5, ACBM is given exposure hazard rankings

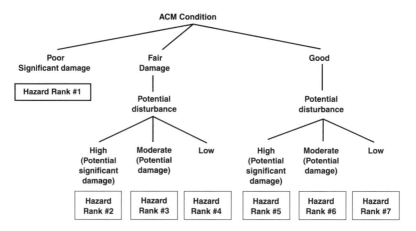

Figure 2.5 USEPA building asbestos hazard assessment decision tree.

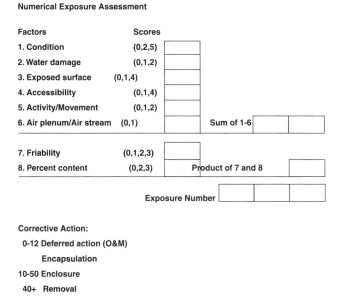

Numerical Exposure Assessment

Factors	Scores
1. Condition	(0,2,5)
2. Water damage	(0,1,2)
3. Exposed surface	(0,1,4)
4. Accessibility	(0,1,4)
5. Activity/Movement	(0,1,2)
6. Air plenum/Air stream	(0,1)

Sum of 1-6

| 7. Friability | (0,1,2,3) |
| 8. Percent content | (0,2,3) |

Product of 7 and 8

Exposure Number

Corrective Action:
0-12 Deferred action (O&M)
    Encapsulation
10-50 Enclosure
40+ Removal

Figure 2.6   Asbestos exposure hazard assessment algorithm.

from 1 to 7. Lowest numbers indicate a potentially high risk of exposure and thus a high abatement priority; highest values represent relatively low asbestos exposure hazards. The USEPA decision tree takes both the condition of ACBM and its potential for disturbance (accessibility, vibration, and air erosion) into account.

In algorithm methods, a numerical score is assigned to a number of factors (Figure 2.6) which may affect exposure. These include ACBM condition, water damage, exposed surface area, accessibility, activity/movement, friability, asbestos content, and the presence of an open air plenum or direct airstream. The significance of the algorithm in Figure 2.6 is that it considers important factors that the USEPA's relatively simple decision tree ignores. Intuitively, one would expect that the surface area exposed and percent asbestos content would increase the potential for exposure. Water damage would also be a significant potential exposure factor since it is one of the major causes of damage to SM such as acoustical plaster, often causing delamination from the substrate and fiber release episodes. Though visual inspections and the use of USEPA's decision tree are standards for exposure hazard assessments in school buildings, and occasionally in other buildings, such assessments have not been successful in predicting airborne asbestos fiber concentrations.

D. Health effects

Exposure to elevated airborne levels of asbestos fibers have definitively been shown to cause disease. These include fibrotic disease of the lungs and pleura (lining and coverings of the chest cavity) and cancers of the lung, pleura, peritoneum (lining and coverings of the abdominal cavity), and possibly other sites.

1. Asbestosis

Asbestosis is a progressive, debilitating disease of the lungs characterized by multisite fibrosis (scarring). Asbestosis has only been reported in asbestos workers, and it appears that high-concentration, long-term exposures are necessary for the development of clinical disease. Asbestosis, as a consequence, is not a major health risk for exposures that occur in building environments. As a result, occupants and those service personnel who occasionally disturb ACBM are highly unlikely to develop asbestosis.

2. Pleural disease

Exposure to asbestos fibers can result in physical changes in the lining and coverings of the chest cavity (pleura). Such changes are described as pleural plaques and diffuse thickening of pleural tissue. Pleural plaques are distinct areas where pleura have developed a fibrotic thickening. Pleural plaques impair lung function and produce respiratory symptoms. They are causally related to asbestos exposure but do not appear to contribute to cancer development.

Fibers from all asbestos minerals appear to cause pleural disease. The development of plaques requires a latency period of more than 15 years. Their prevalence increases with dose (number of fibers inhaled) and number of years since initial exposure.

Radiological studies of building service workers, including carpenters, sheet metal workers, and school custodians, have shown that these workers are at risk of developing pleural plaques. This risk appears to be associated with disturbing ACBM and/or resuspending asbestos fibers.

3. Cancers

Exposure to asbestos fibers increases the incidence of bronchogenic carcinoma, i.e., lung cancer. Tumors, however, are indistinguishable from those caused by exposure to tobacco smoke or radon decay products. Asbestos-associated lung cancer has been reported in both smokers and nonsmokers. Combined exposures to tobacco smoke and asbestos fibers result in a synergistic response. On average, smokers have a lung cancer risk that is 10 times greater than that of nonsmokers; the lung cancer risk in nonsmokers heavily exposed to asbestos fibers is approximately 5 times greater than in nonsmoking, nonexposed workers. A smoker exposed to high levels of asbestos has an increased risk of lung cancer that is approximately 50 to 55 times greater than that of a non-asbestos-exposed, nonsmoking worker.

The risk of developing lung cancer from asbestos fiber exposure is dose dependent, with risks greater for those with greater exposure. The latency period is typically 20 years or longer.

Lung cancer has been a major public health concern in the context of potential exposures that may be experienced by building occupants, such as school children, teachers, and service personnel (custodians, maintenance workers). This concern reflects the fact that there is apparently no threshold for asbestos exposure and the induction of lung cancer.

Asbestos exposure can cause mesothelioma, a rare cancer of the mesothelium (i.e., the tissue which comprises pleura and peritoneum) of the chest and abdominal cavities. Pleural mesothelioma is 5 times more common than abdominal or peritoneal mesothelioma. The annual incidence of mesothelioma in the U.S. has been estimated to be in the range of 1500 to 2500 cases. Most mesotheliomas appear to be associated with industrial/occupational exposures or household contact with an asbestos worker. However, in 10 to 30% of reported cases, there is no evidence of exposures to asbestos or talc (which often contains asbestos fibers). Mesothelioma is incurable and, as a consequence, has a 100% fatality rate.

The scientific literature indicates that mesothelioma risk is greater among workers exposed to amphiboles, such as crocidolite, than to chrysotile, the most widely used asbestos fiber in the U.S. There is considerable uncertainty as to whether chrysotile asbestos can cause mesothelioma. A potential causative role for the amphibole, tremolite, has been proposed for mesothelioma among some chrysotile-exposed workers. Tremolite is commonly reported as a minor constituent of chrysotile mineral extracted and used for ACBM and other products.

The risk of developing mesothelioma increases with increasing cumulative exposure and is independent of age or smoking history. The earlier one is exposed, the higher the probability of developing mesothelioma in one's lifetime. This phenomenon is due to the fact that early-in-life exposures mean that an individual has a longer period of time in which to develop the disease. Such potential early-in-life exposures have, in part, underlain asbestos exposure concerns for children in ACBM-containing school buildings in the U.S. Additional concern lies in the fact that mesothelioma may also be caused by brief high exposures.

Epidemiological studies of asbestos-exposed workers have reported increased prevalence rates of cancers of the larynx (voice box), oropharynx (mouth/throat), and upper and lower digestive tract. The risk of such cancers in asbestos workers appears to be small; the risk to building occupants is several orders of magnitude smaller.

4. Cancer risks associated with building asbestos exposures

During the mid- to late 1980s, public health concern focused on potential asbestos fiber exposures of building occupants and workers in buildings containing ACBM and their risks of developing lung cancer or mesothelioma. As a consequence, the Health Effects Institute (Cambridge, MA) convened a panel to evaluate the lifetime cancer risk of general building occupants as well as service workers. Unlike general building occupants,

Table 2.4   Estimated Lifetime Cancer Death Risk for Environmental
Asbestos Fiber Exposure

Exposure conditions	Premature cancer deaths per million exposed individuals
Continuous Outdoor, Lifetime Exposure	
0.00001 f/cc from birth (rural)	4
0.0001 f/cc from birth (high urban)	40
School with ACBM Exposure, from Age 5–18 Years	
0.0005 f/cc (average)	6
0.005 f/cc (high)	60
Public Building with ACBM Exposure, from Age 25–45 Years	
0.0002 f/cc	4
0.002 f/cc	40
Occupational Exposure from Age 25–45	
0.1 f/cc (OSHA PEL)	2000

Source: From Health Effects Institute–Asbestos Research, *Asbestos in Public and Commercial Buildings: A Literature Review and Synthesis of Current Knowledge*, Cambridge, MA, 1991. With permission.

custodial, maintenance, and renovation workers may experience peak exposure episodes resulting from disturbance of or damage to ACBM or disturbance of asbestos-containing dust. Service worker exposures are in fact significantly higher. Based on an evaluation of asbestos fiber measurements conducted in buildings (described previously) and estimated risk based on linear extrapolation from effects in workers with heavy occupational exposure, the Institute asbestos panel concluded that the lifetime cancer risk (both lung cancer and mesothelioma) among general building occupants was relatively low: for 20 years of exposure, 4 per million. For workers exposed to 20 f/cc for 20 years, the lifetime risk was estimated to be 1 in 5. Estimated lifetime cancer risks for different asbestos fiber exposures are summarized in Table 2.4.

II. Radon

Radon is a naturally occurring, gas-phase element found in the earth's crust, water, and air. Like helium, argon, neon, xenon, and krypton, radon is a noble gas and does not react with other substances. Radon-222 is an isotope produced as a result of the decay of radium-226. Radon-222 has a half-life (the time period in which one-half of a given quantity of any radioactive element will decay to the next element in a decay sequence) of 3.8 days. On radioactive decay, radon-222 produces a series of short-lived decay products until lead-210, a stable (long-lived) lead isotope, is produced. This decay series, with characteristic half-lives and emissions of alpha (α) and beta (β) particles and gamma (γ) rays, is summarized in Figure 2.7.

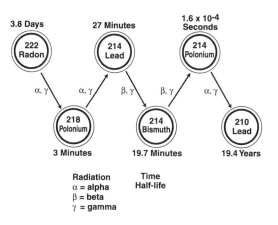

Figure 2.7   Radon radioactive decay series.

In the first two radioactive decays, α-particles and γ-rays are emitted to produce polonium-218, and then lead-214. An α-particle, which is equivalent to a helium nucleus (2 protons, 2 neutrons), carries a significant positive charge. Because of their large mass and limited potential to travel in air, α-particles pose almost no external exposure hazard. Inside the human body or other living organisms, they have the potential to cause significant ionization and therefore can damage exposed tissues. The relative biological effectiveness of α-particles inside the body is approximately 5 times greater than that of X- or γ-rays. Lead-214 decays to bismuth-214 and then to polonium-214, with the release in each case of a β-particle (nuclear electron) and γ-rays. Polonium-214 decays to lead-210 by releasing a third α-particle. Lead-210 has a half-life of 20 years, and ultimately decays to lead-206.

The radioactive decay of radon-222 to lead-210 is notable in several respects. This includes the relatively short half-lives of radon-222 and its progeny, and the emission of three α-particles, two β-particles, and associated γ-ray energy. Because of the emission of charged particles, radon decay products (RDPs) are electrically charged. As a consequence, they readily adhere to suspended dust particles or other surfaces (termed "plateout"). Attached (to dust particles) or unattached RDPs may be inhaled and deposited in respiratory airways or lung tissue, where subsequent radioactive decay and tissue irradiation (particularly by α-particles) occurs. Tissue exposure to α-particles is of considerable biological significance.

A. Soil sources/transport

Radon is produced by the radioactive decay of radium-226, which is found in uranium ores; phosphate rock; shales; metamorphic minerals such as granite, gneiss, and schist; and, to a lesser degree, in common minerals such as limestone. As a consequence, the primary sources of radon and RDPs in buildings are the soil beneath and adjacent to buildings, domestic water

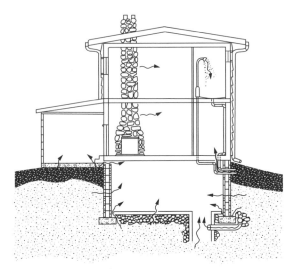

Figure 2.8 Radon entry pathways into a residential building.

supplies, and building materials. Radon can enter building environments, particularly residences, by pathways illustrated in Figure 2.8.

The dominant source of elevated radon levels in buildings is the soil beneath and adjacent to the building. The potential for soils to emit radon-222 depends on concentrations of uranium-238, thorium-232, and radium-226, which are usually proportional (in concentration) to each other based on the uranium decay series. The world average concentration for uranium-238 and thorium-232 in soil is approximately 0.65 picocuries (pCi) per gram. Local concentrations, however, vary widely from this average. The radon source potential under an individual dwelling reflects concentrations of radioactive parent isotopes in site soils.

As radon is produced, it moves through air spaces between soil particles. The emanation power (radon that enters soil pores) of radon formed on or in soil fragments depends on the soil type, pore volume, and water content. Reported emanation powers vary from 1 to 80%. Soil gas measurements of a few hundred to several thousand pCi/L have been reported.

Movement of radon through soil pores occurs by diffusion, convection, or both. The movement of radon into building structures appears to be primarily due to convection-induced pressure flows associated with indoor–outdoor temperature differences and pressures associated with wind speed.

Radon entry into buildings through building substructures is affected by (1) soil radon production rates, (2) soil permeability, (3) cracks/fissures in underlying geology, (4) the nature of the building substructure, and (5) meteorological variables that cause indoor/outdoor pressure differences.

In general, higher radon levels can be expected in buildings constructed on soils with high radon production rates. Though production rates are important, indoor concentrations are more directly affected by soil/ground emanation power. The emanation rate is influenced by both radon production and soil porosity. Typical radon emanation rates for U.S. soils have been reported in the range of 1 to 4 × 10^{-5} becquerels (37 becquerels = 1 pCi)/kg/sec, with radon in soils in the concentration range of 20 to 30 to >100,000 pCi/L. Most soils in the U.S. have radon concentrations between 200 and 2000 pCi/L. Buildings on sandy/gravelly soils typically have significantly higher radon levels than those on clay soils.

Radon entry into buildings occurs through substructures. In basements, radon-laden soil gas flows through cracks in the floor slab and walls, block wall cavities, plumbing connections, and sump wells. In slab-on-grade buildings, cracks in the slab and penetrations associated with plumbing are the primary avenues of soil gas flow. In houses constructed on crawlspaces, soil gas must move through the airspace between the ground and building floor.

Crawlspaces are of two types. The most common is nominally isolated from living spaces above it. Such crawlspaces are usually provided with screened vents in perimeter walls. The second crawlspace type is open to livable areas such as adjoining basements; it is not generally provided with vents.

Crawlspaces nominally isolated from living spaces and potentially ventilated with outdoor air are not generally decoupled from living spaces. The degree of decoupling depends on the presence of openings such as vents and cracks in the foundation wall, opening/closure condition of crawlspace vents, leakage potential between crawlspace and living spaces, and presence of leaky forced air heating ducts in the crawlspace (particularly cold air return ducts). Tracer gas studies have shown that crawlspaces are a significant source of air which infiltrates living spaces (on the order of 30 to 92%). Highest infiltration rates and radon flows are associated with closed vents and leaky air return ducts.

Assuming equal radon emanation potentials under dwellings with basement, slab-on-grade, and crawlspace substructures, highest indoor radon concentrations are expected to occur in those constructed on basements. This is so because basements have the highest surface area in contact with the ground and are more commonly constructed in regions with porous, well-drained soils. Basement radon concentrations are typically twice those in upstairs living spaces. Slab-on-grade dwellings are generally expected to have higher radon concentrations than those on crawlspaces because they have a greater surface area in direct contact with the ground. Crawlspaces without adequate ventilation, or with vents closed to conserve heat under cold climatic conditions, would have radon levels similar to those of slab-on-grade houses in which there are cracks sufficient to allow soil gas movement into the building interior.

The typical substructure in nonresidential buildings such as schools, commercial, and office buildings is slab-on-grade. Radon levels in such buildings are affected by pressure-driven flows through cracks/penetrations in the slab. These natural, pressure-driven flows are associated with meteo-

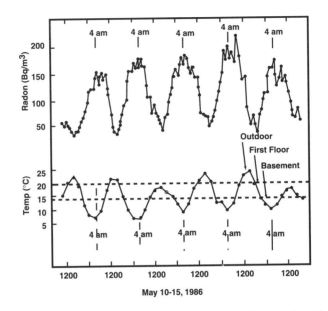

Figure 2.9 Relationship between indoor radon concentrations and outdoor temperatures (indoor/outdoor temperature differences). (From Kunz, Z., *Proc. 4th Internatl. Conf. Indoor Air Qual. Climate*, Berlin, 414, 1987.)

rological factors, the height and volume of the building, and operation of mechanical exhaust systems.

Radon transport is significantly enhanced when buildings are under significant negative pressure, particularly at floor level. In moderate to colder climates, most dwellings experience what is described as the "stack effect" (see Chapter 11). Indoor/outdoor temperature differences cause residential buildings to be under positive pressure near the roof and negative pressure near the floor. This phenomenon causes outdoor air to flow in at the base and out at the ceiling. The effect of temperature differences (responsible for the stack effect) on indoor radon levels can be seen in Figure 2.9. Note the strong diurnal variation in outdoor temperature and the corresponding variation in indoor radon concentration. Radon levels are at a maximum during the coolest part of the day when pressure differentials are greatest.

Meteorological variables such as outdoor temperature are not alone in affecting indoor radon concentration. Pressure-driven flows are also influenced by both the wind speed and direction. During moderate to high winds, the windward side of a building will be under positive pressure and the leeward side under high negative pressure. Indoor radon levels will be lower on the windward side and higher on the leeward side.

Other meteorological variables reportedly affect indoor radon levels. Drought conditions can cause cracks in otherwise impermeable clay soils, channeling radon upward from deeper soils and geological strata. Dry soil conditions also increase the volume of soil pores available for radon accumulation and transport. Heavy rains may also affect radon transport into buildings by causing a piston-like displacement of soil gas around building perimeters, forcing soil gas to flow inward. Frozen ground and concrete roadways may cause a similar phenomenon. In tropical climates, rainy weather contributes to elevated radon levels indoors.

Several investigators have proposed that transient atmospheric pressure changes associated with meteorological conditions can affect radon transport into buildings. Studies conducted in Florida have shown significantly increased indoor radon concentrations associated with semidiurnal atmospheric pressure changes in slab-on-grade houses built on low permeability soils. Peak concentrations occurred when other sources of house depressurization or pressurization were small; i.e., when houses as a whole are under neutral pressure relative to the outdoors.

In general, radon transport into buildings increases with increasing negative pressurization relative to the outdoors. Such negative pressurization is produced naturally by the stack effect and increased wind speeds. The stack effect increases with increasing temperature differentials between the inside and outside of buildings. As a consequence, highest radon concentrations in U.S. housing are thought to occur in the winter heating season when the thermal stack effect is the strongest. However, in some studies, highest concentrations were observed during summer months, ostensibly due to occupant operation of heating/cooling systems and strong seasonal variation in radon emanation potentials in the soil (soil gas radon concentrations were higher in summer than winter at several sites studied).

Natural forces are not the only cause of building depressurization. In mechanically ventilated schools and other large buildings, depressurization occurs when more air is exhausted from building spaces than is brought into the building through outdoor intakes. In dwellings, depressurization may result from the use of heating systems that require chimneys to exhaust flue gases from furnaces and fireplaces. Depressurization in basements can result from leaky furnace fan housings and cold air returns, and in crawlspaces from leaky air returns. Mechanical depressurization can significantly affect the flow of soil gas into buildings and, as a consequence, increase indoor radon levels.

B. Groundwater

Though most radon is transported into buildings through pressure-driven soil gas flows, groundwater serves as a limited, but sometimes significant, radon source in some geographical/geological regions. Radon has been reported to occur in well water, with concentrations ranging from a few 100 pCi/L to approximately 30,000 pCi/L. Highest concentrations are associated with drilled wells, particularly in areas with granitic bedrock. Radon is released from water when (1) temperature is increased, (2) pressure is increased, and (3) water is aerated. Optimum conditions for radon release and exposure occur during showering. Water with a radon concentration of

10^4 pCi/L can increase the indoor airborne radon concentration by 1 pCi/L under normal conditions of water use. Most groundwater supplies have radon concentrations that are <2000 pCi/L.

C. Building materials

Building materials derived from mineral substances often contain radioactive nuclides and thus have the potential for releasing radon as well as gamma radiation. These materials can be an important source of radon when radium contents are elevated (>1 pCi/g). Radon emanation depends on both radium content and porosity. Granite, having a relatively high radium content (and low porosity), often emits radon at lower levels than concrete with granitic sand.

D. Radon concentrations

Several different units are used to express radon concentrations in air, water, and soil. In the U.S., it is common to report radon concentrations in picocuries per liter (pCi/L). A picocurie is equal to 1×10^{-12} curies. A curie (Ci) is an amount of radioactive material that produces 10^{-10} nuclear disintegrations per second. A picocurie represents an amount of material that produces 2.2 radioactive decays per minute or 3.7×10^{-2} decays/sec. One picocurie is equivalent to a concentration of 7×10^{-7} parts per trillion (ppt).

In the International System of units (Systeme Internationale, SI), radon concentrations are expressed as becquerels per cubic meter (Bq/m^3). One Bq/m^3 is equal to one radioactive disintegration per second in a cubic meter of air. Thirty-seven Bq/m^3 equals 1 pCi/L or 37 disintegrations per second. Therefore one curie is equal to 3.7×10^{10} Bq.

Radon decay product concentrations are expressed in units called working levels (WL). A working level is a measure of the potential α-particle energy concentration (PAEC) associated with nuclear disintegrations involving radon and its decay products. The PAEC for one WL is 1.3×10^5 million electron volts (MEV) per liter. One WL would be equal to 100 pCi/L if the environment being measured is maintained at 100 pCi/L. In the real world, only a fraction of RDPs remain suspended in air and are therefore measurable. The ratio between measured and total RDPs is described as the equilibrium ratio (ER), where

$$ER = \frac{\text{measured RDPs}}{\text{total RDPs}} \tag{2.1}$$

A variety of factors affect the ER, including air circulation, use of electronic filters, suspended dust concentrations, and recent ventilation. Air circulation increases plateout, and electronic filters remove dust particles; thus, both decrease ER. Increased suspended particle concentrations decrease plateout and thus increase the ER. Ventilation decreases time available for indoor radon decay to take place. Equilibrium ratios typically range from 0.3 to 0.7. With an ER of 0.5, the actual radon concentration would be approximately 200 pCi/L. Working levels are used to quantify radon exposures in mines for purposes of determining compliance with occupational health standards and guidelines. An average exposure to 1 WL for 170 hours of a working month is described as a working level month (WLM).

All humans are exposed to radon and its radioactive decay products every moment of every day. Outdoor concentrations range from 0.1 to 30 pCi/L, with an average exposure concentration of 0.25 pCi/L. Radon concentrations in indoor air range from <1 to >3000 pCi/L, with a geometric mean of 1 to 2 pCi/L.

Radon levels and exposures in buildings vary considerably. This is especially true in residential buildings. In a single unusual case in Boyertown, Pennsylvania, radon levels as high as 2600 pCi/L (13 WL) were reported. This compares with USEPA's action guideline of 4 pCi/L annual average concentration, and the 8 pCi/L National Council on Radiation Protection (NCRP) unacceptable level. Radon guidelines used by a number of countries and organizations are given in Table 13.3.

It has been estimated that approximately 6% of U.S. residences have annual average radon levels ≥4 pCi/L. Based on short-term (one-week) measurements, approximately 30% of residences in midwestern states such as Indiana and Ohio have indoor radon levels ≥4 pCi/L. Because of building ventilation practices during warmer months, average annual concentrations are, in general, much lower.

The above guideline value (4 pCi/L) is based on an annual average. Unfortunately, most radon measurement values reported are based on short-term measurements (several days to a week); in some cases, extended measurements are made over a period of 3 to 6 months. Such measurements represent one-time sampling and generally do not reflect exposure to radon based on the actual guideline. Measured radon concentrations are not normally distributed. As a consequence, it is necessary to report median or geometric mean concentrations to reflect middle values. Geometric mean concentrations in 30,000 nonrandomly distributed homes, representing approximately 0.5% of the U.S. housing stock, are summarized by region in Table 2.5. These data, based on 3-month average measurements, had a geo-

Table 2.5 Geometric Mean Radon Levels (pCi/L) Measured in 30,000 U.S. Residences Summarized on a Regional Basis

Region	Geometric mean	% >4 pCi/L	% > 20 pCi/L
Northeast	3.43	44	8.9
Midwest	2.36	31	2.8
Northwest	0.64	4.6	0.2
Mountain states	2.90	36.5	5.4
Southeast	1.43	18.3	1.5
Total	1.74	29.3	5.5

Source: Data extracted from Alter, W.H. and Oswald, R.A., *JAPCA*, 37, 227, 1987.

metric mean value of 1.74 pCi/L. Five percent of the values were less than 0.2 pCi/L; 0.5% exceeded 100 pCi/L. The Northeast, Midwest, and Mountain States reported the highest geometric mean concentrations and potential exposures; the Northwest and Southeast the lowest. These data were based on purchased test kits and are not randomly representative of the U.S. housing stock. High regional geometric mean concentrations reflect, in part, both high regional interest in radon testing and localized areas of high radon concentrations. A geometric mean concentration of 0.9 pCi/L has been reported in a more random study of single-family dwellings from 32 areas of the U.S. (primarily urban areas).

Radon concentrations vary considerably from region to region and house to house. Houses located next to each other may have radon concentrations that differ by an order of magnitude or more. These differences reflect radon emission potentials under each structure as well as differences in radon transport into them.

E. Health effects

The major health concern associated with exposure to radon and its RDPs is their potential to cause cancer. Epidemiological studies of uranium and other deep miners have shown a very strong correlation between RDP exposure and lung cancer. Lung cancer risk is dependent on the total, or cumulative, dose (Figure 2.10).

The upper bronchial region of the lung is the primary tissue site exposed to α-particle emitters produced in radon decay. Other potential cancer sites include the buccal cavity (back of the mouth), pharynx (throat), upper portions of the gastrointestinal tract, and bone. There is also limited epidemiological evidence to suggest that radon exposure may be an important risk factor for acute myeloid leukemia, cancer of the kidney, melanoma, and certain childhood cancers.

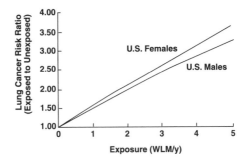

Figure 2.10 Cumulative radon RDP exposures and lung cancer rates. (From National Research Council, *Health Risks of Radon and Other Internally Deposited Alpha Emitters, BIER IV,* National Academy of Science Press, Washington, D.C., 1988. With permission.)

The main exposure/cancer risk is associated with inhalation of short-lived RDPs and, secondarily, radon itself. Since radon is inert, only a very small fraction of inhaled radon enters the bloodstream before it is exhaled, and only a very small fraction decays while in the body. Radon decay product exposures occur as a result of the inhalation of particle-attached and unattached RDPs. Deposition of attached RDPs depends on particle size. Large particles are deposited in the nose or pharynx; smaller particles enter the lung and are typically deposited at the bifurcations of the upper respiratory airways. Unattached RDPs usually settle on the surface of respiratory airways. Both attached and unattached RDPs can be absorbed into fluids in the respiratory airways or removed by respiratory defense mechanisms such as cilia and mucus (and subsequently swallowed, entering the gastrointestinal tract).

Radon-associated cancers in miners are bronchogenic (i.e., occur in the bronchi), indicating that the primary effective site of RDP deposition is the upper airways.

Though radon exposure and lung cancer have been strongly linked in epidemiological studies of miners, a similar link has yet to be established between RDP exposure in buildings (particularly dwellings) and lung cancer. The difficulty in establishing a quantitative relationship between building RDP exposures and lung cancer is due, in part, to the fact that lung cancer rates associated with all causes are high and vary among different populations and in different environments. Since building RDP exposure on a population basis is relatively low, an increase in lung cancer rates attributable to RDP exposure is likely to be low as well. Study populations must be large and exposures high in order to observe statistically significant differences in lung cancer rates.

A limited number of epidemiological studies have been conducted to determine whether exposure to indoor radon may be associated with increased lung cancer rates. Most have defined RDP exposure groups by geography and geology (i.e., ecologic studies). Of 15 such studies, seven showed positive statistical associations; six, no association; and two, negative associations. The power of these studies was limited because they did not control for confounding factors such as smoking or resident mobility; they were largely uninformative.

Studies based on the case-control approach can provide quantitative assessments of residential radon exposures and lung cancer rates with a much higher degree of statistical confidence than ecologic studies. With a case-control design, radon exposures of those with lung cancer are compared with appropriate controls who do not have lung cancer. To date, such studies have been based on very small population sizes and, not unexpectedly, report mixed results. In one of the largest case-control studies conducted to date (Sweden), lung cancer risk was found to increase with increasing cumulative exposure risk; the dose–response relationship was comparable to that observed in miners. The combined effect of RDP exposure and smoking was observed to be more than additive.

F. Risk assessment

The historical view among radiation biologists and radiation protection agencies has been that the induction of cancer from radiation exposure is a function of dose, and there is no threshold for effects. Based on this view, and limited epidemiological information on residential radon exposure and lung cancer, risk models have been developed to predict increases in lung cancer rates associated with different levels of radon exposures in dwellings.

These risk projection models employ assumptions associated with the time-dependent pattern of lung cancer occurrence and effects of such important cofactors as age at exposure, age at risk, and smoking.

The risk assessment models that have been widely used for indoor radon risk assessments are based on the 1988 National Research Council's Committee on Biological Effects of Ionizing Radiation report (BEIR IV). The BEIR IV committee concluded that exposure of target respiratory airway cells to α-particle energy in residences and mines yielded roughly equivalent dose responses. The BEIR IV model assumes that cigarette smoking multiplies effects of radon.

The manner in which exposures to RDPs and cigarette smoking are assumed to interact has a significant influence on risk projections. If interaction is multiplicative, the risk for smokers is multiplied by the risk for radon. If the interaction is additive, some excess risk is added to the background risk.

The problem of interaction is significant in making projections of risk associated with indoor radon exposure. A synergistic response is one in which the combined effect of two different exposure agents exceeds the sum of the individual effects. If the combined effect is the product of the independent risks, then the interaction is multiplicative. If the two agents interact synergistically, some cases of lung cancer can be attributed to each agent acting alone and some to the interaction. For radon exposure in residences, lung cancer associated with radon will occur in never smokers, past smokers, and current smokers. The potential effect of smoking (past and current) is predicted from annual lung cancer cases (Table 2.6). These risk estimates are based on the BEIR IV model and on an average annual radon exposure of 1.5 pCi/L, estimated to be the average exposure in the U.S.

Risk estimates in Table 2.6 indicate that approximately 16,000 lung cancer deaths per year may be attributed to radon (based on 1986 population data). The USEPA, using a modification of the BEIR IV risk model and adjusting for slightly lower doses in homes (compared to mines), estimated 13,300 radon-related lung cancer deaths per year, with a range for this estimate between 7000 and 30,000 deaths. Based on the BEIR IV model, exposure to an annual average radon concentration of 0.2 WLM/yr (approximately 1.5 pCi/L) would produce a radon-attributable lifetime risk of 0.7% for females.

These risks would be much higher in houses with high radon concentrations, such as those reported in the Reading Prong geological area, which runs through portions of Maryland, southeastern Pennsylvania, and central

Table 2.6 Lung Cancer Deaths in the U.S. Attributed to Radon Exposure

	Lung cancer death rate per year		
	Population (1000s)	All causes	Radon-attributable
1986 population			
Male			
Never smoked	63,900	1900	200
Former smoker	26,100	42,200	4700
Current, light smoker	18,400	21,800	2600
Current, heavy smoker	9000	22,900	3100
Female			
Never smoked	81,300	3100	300
Former smoker	17,100	14,800	1700
Current, light smoker	20,000	15,800	1900
Current, heavy smoker	5300	7900	1100
Male, total	117,400	88,800	10,700
Female, total	123,700	41,600	5000
Population, total	241,100	130,400	15,700

Source: From Nazaroff, W.W. and Teichmann, K., *Environ. Sci. Technol.*, 24, 774, 1990. With permission.

New Jersey, as well as in other areas of the U.S. In such cases, residential radon concentrations >100 pCi/L have been measured.

The increased lifetime risk of lung cancer attributable to radon can be seen in Figure 2.10, and the equivalency chart prepared by USEPA and presented in its citizens' radon information guide in Figure 2.11. Residing in a home with an exposure of 4 WLM/yr above background (equivalent to an annual exposure of 25 pCi/L) would be expected to triple the lung cancer rate. In the USEPA radon risk evaluation chart, this would be equivalent to smoking approximately 2 packs of cigarettes per day.

The lifetime risk estimates in the radon risk evaluation chart (Figure 2.11) and the chart itself were designed to educate consumers about the health hazard of radon exposure in their homes and the need to reduce exposures in houses where testing has shown radon to be elevated above the action level of 4 pCi/L annual average concentration. Risk estimates shown in this chart are generic, unlike those shown in Table 2.6. A logical consumer interpretation of the USEPA chart would suggest that the 13,000 to 15,000 annual predicted lung cancer deaths are due to radon exposure alone. For homeowners who have been the most concerned about radon (nonsmokers and never smokers), cancer risk is much smaller than the chart would suggest. Note that for never smokers, the estimated death rate is <1000 per year (Table 2.6). For never smokers, the risk associated with radon exposures in their home is very small, except in those houses that have very high (>10 pCi/L) annual average concentrations.

The action level used by USEPA of 4 pCi/L (annual average concentration) is based on mitigation studies of houses constructed on uranium mill tailings and phosphate mining wastes. After applying the best available

Indoor environmental quality

RADON RISK EVALUATION CHART

pCi/L	WL	Estimated number of LUNG CANCER DEATHS due to radon exposure (out of 1000)	Comparable exposure levels		Comparable risk
200	1	440–770	1000 times average outdoor level		More than 60 times non-smoker risk 4 pack-a-day smoker
100	0.5	270–630	100 times average indoor level		20,000 chest x-rays per year
40	0.2	120–380			2 pack-a-day smoker
20	0.1	60–210	100 times average outdoor level		1 pack-a-day smoker
10	0.05	30–120	10 times average indoor level		5 times non-smoker risk
4	0.02	13–50			200 chest x-rays per year
2	0.01	7–30	10 times average outdoor level		Non-smoker risk of dying from lung cancer
1	.005	3–13	Average outdoor level		20 chest x-rays per year
0.2	.001	1–3	Average indoor level		

Figure 2.11 USEPA radon risk estimate and equivalency chart. (From USEPA, *Citizens Guide to Radon*, EPA 86-004, 1986.)

control technology in high radon houses, it was concluded that reductions in radon levels below 4 pCi/L were not achievable. As a consequence, 4 pCi/L became, by default, the action level for remediation. This action level is not based on radiation biology assessments of safety. As such, it is not a magic number that determines whether the radon level in a particular dwelling is safe or not, as many homeowners (and in some cases radon professionals) are wont to believe. Since there appears to be no threshold, no amount of radon/RDPs indoors or outdoors is safe. Safety is a relative term. Exposures of 4 pCi/L annual average are less safe than 2 pCi/L; measured exposure concentrations of 4.05 and 3.9 pCi/L are not significantly different from each other and represent no difference in risk. For many homeowners, the difference is one of life and death or the difference between mitigating and not mitigating. For real estate transactions, it may be the difference between a deal maker or a deal breaker or the cost of installing a radon control system.

The risks of radon exposure in residences are real. Cancer risks are significantly greater than the 1 per 10^6 risk of cancer used by USEPA to regulate a number of cancer-causing substances. The risks are, however, different, in that radon is a naturally occurring substance and thus a risk attributable to nature. The risk is imperceptible to most humans and is long term. Thus it appears to cause less alarm than other cancer-causing substances.

III. Lead

Lead is a heavy metal which occurs naturally in soil, water, and to a more limited degree, ambient air (as mineral particles). It is found in the earth's crust in a number of different minerals. Metallic lead, characteristically bluish-gray in color, is extracted from relatively rich mineral ores such as galena (PbS), cerussite ($PbCO_3$), and angelsite ($PbSO_4$). In addition to its elemental form, it has two oxidation states, Pb(II) and Pb(IV).

Because of its relatively low melting point (327°C), malleability, and other desirable properties, lead is, and has been, the most widely used (based on mass) nonferrous metal. Humans have used it for over 8 millennia. As a result of the smelting of metal ores, combustion of lead-amended gasoline, and other uses, lead contamination of the environment is ubiquitous. Lead levels above background concentrations can be found in soil, drinking water, rivers and streams, plants, and polar ice. Indeed, ice core concentrations can be used to track the historical use of lead by humans.

In the past century, lead has been used as a pigment in paints; to manufacture storage batteries; and in the production of solders, munitions, plastics, and a variety of products. Until relatively recently, organic lead compounds (tetramethyl and tetraethyl lead) were widely used in North America to boost the octane rating of gasoline and serve as a scavenger for free radicals (antiknock agent).

Lead use has varied over the century. Because of health and environmental concerns, its use as a pigment in paint, gasoline fuel additive, and in soldering compounds has declined greatly, particularly in North America. Its use in storage battery production has, on the other hand, increased significantly.

Because of its widespread use, humans have been, and continue to be, exposed to lead concentrations well above background levels. Notable historical exposures have occurred from acidic liquids stored in lead containers (Roman and medieval times), inhalation of lead dusts from primary and secondary lead smelters, ingestion of food from lead-soldered cans, leaching from plumbing materials, inhalation of airborne lead dusts associated with the combustion of motor vehicle fuels, and ingestion of lead-based paint and contaminated dust and soil by children.

The combustion of lead-amended gasoline has likely been the single greatest source of lead contamination of both the environment and humans. In North America, the phase-out of leaded gasoline and implementation of strict regulatory limits on primary and secondary lead smelters has resulted in a decline of atmospheric lead emissions of over 99% in the past 3 decades.

Unlike a variety of other trace elements, lead is not required in the body to mediate any biochemical or physiological function. It is a very toxic substance, and increasingly, what once were considered to be low levels of exposure have become the subject of major public health concern. Total body burdens of lead in modern day humans are several hundred times greater than those of our preindustrial ancestors.

A. Lead in the indoor environment

Because of its widespread use and the nature of individual uses, it is not surprising that lead is a common contaminant of indoor spaces. Indoor lead contamination has been associated with lead aerosols produced from motor vehicle operation and lead and other nonferrous metal smelter emissions; the passive transport of lead dusts into vehicles and residences on the clothing and shoes of workers exposed to lead in lead-generating or lead-using industries (lead brought home) and other activities; home hobbies such as making stained-glass windows, munitions, fishing weights, etc.; deteriorating or damaged interior lead-based paint (LBP); and the passive transport of lead-contaminated residential soils indoors by people and pets. Significant lead contamination of household surface dusts is widespread in North American and European homes that have a history of LBP use.

1. Lead-based paint

Lead compounds such as white lead and lead chromate have been used as white pigments in paints for centuries. In addition to their pigment properties, lead compounds were valued because of their durability and resistance to weathering. Though used in interior and exterior paints, lead was more commonly used in exterior white paints. An old house with exterior lead-based paint is pictured in Figure 2.12. Lead compounds have also been used in the production of varnishes and primers. Red lead, or litharge (Pb_3O_4), has been widely used in the manufacture of weather-resistant metal coatings. Such coatings were used on bridges and other steel structures and occasionally on exterior residential metal products.

Lead poisoning of young children associated with LBP was initially recognized by public health authorities in the U.S. in the late 1940s and early 1950s. As a consequence, the paint industry voluntarily limited the lead content in paint to 1% solids by weight in 1955. Regulatory action subsequently limited this to 0.5% by weight in 1971 and 0.06% in 1978. Pre-1978

Figure 2.12   Old house with exterior lead-based paint.

housing in the U.S. is considered to have LBP and poses a potential lead exposure risk to children.

Direct and indirect exposures to LBP are considered to be the predominant contributors to elevated blood lead levels (≥10 μg/dL) in children in the U.S. Fifty million houses are estimated to contain LBP (defined as containing 1 mg/cm² or 0.5% by weight), with about 12 million children under the age of seven potentially exposed.

In the early history of LBP poisoning concerns, it was widely believed (in the public health community and regulatory agencies) that childhood lead exposures resulting in elevated blood lead levels (BLLs), and clinical symptoms were the result of the ingestion of LBP chips from deteriorated paint; ingestion was felt to be the result of compulsive mouthing behavior (described as pica) by infants and toddlers, and lack of supervision by parents, in low-income housing. In the 1980s, scientists concluded that the major source of pediatric lead exposure was house dust contaminated with lead from deteriorated LBP from interior and/or exterior sources. As a result of epidemiological studies which linked BLLs in children to house and hand dust, a consensus has emerged among research scientists and public health officials that the primary cause of elevated BLLs in young children is hand-to-mouth transfer of lead-contaminated house dust and soils through infant/toddler mouthing behaviors.

Based on studies at the University of Cincinnati, a simple exposure model was constructed to explain childhood lead exposures. This model was to serve as a tool for intervention efforts in reducing BLLs in children at risk. A modification of the Cincinnati model based on subsequent research studies is presented in Figure 2.13.

The model shows various relationships which have been statistically associated with elevated BLLs in children. Based on recent epidemiological studies, approximately 41% of the lead content of house dust can be linked to LBP, and another 18% with soil lead (passively transported indoors by

Lead Exposure Model for Young Children

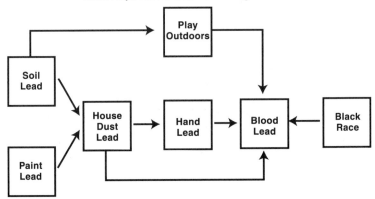

Figure 2.13 Childhood lead exposure model.

people and pets). Approximately 23% of lead on children's hands can be explained by exposure to house dust, and another 23% by exposure to soil. Both house dust and soil lead have been shown to be directly associated with children's BLLs. Hand lead is more weakly associated with BLLs than house dust lead, presumably because of the transient nature of lead contamination on children's hands.

The relatively weak association between hand lead and BLLs, compared to lead in house dust, may be, in part, due to other exposure pathways. Though not studied directly, some degree of lead dust exposure may occur by inhalation. Evidence for such potential exposures can only be inferred, although various lines of evidence are suggestive. These include: (1) studies showing that most lead-containing particles on children's hands have aerodynamic diameters <10 μm, the inhalable range; (2) chemical mass balance apportionment studies that indicate that 17 to 18% of house dust surface particles have a composition profile consistent with particles collected in area indoor samples (albeit indoor air lead levels were well below the USEPA ambient air quality standard); and (3) indoor particle monitoring studies which show that dust particle exposures (associated with individual activities) are on the order of 3 to 5 times higher than area samples collected nearby. This "personal dust cloud" may be particularly significant for a young child playing on dusty surfaces contaminated with lead. In addition, inhaled lead deposited in lung tissue may be more effectively extracted from dust particles than ingested lead. This is true in adults, but differences are likely to be smaller in children.

Investigators have attempted to characterize paint, dust, and soil lead levels associated with different housing types. Measures of environmental lead in populations of different residence types in Cincinnati are summarized in Table 2.7. Highest concentrations of lead in all tested parameters were observed in old housing, particularly 19th century housing in poor condition. Highest surface dust, dustfall dust, and outdoor soil lead concentrations

Table 2.7 Environmental Lead Levels (geometric mean values) Associated with Housing Types and Their Physical Condition in Cincinnati, OH

Lead measure	Post 1945, private, satisfactory condition	Public housing	Subsidized, rehabilitated housing	19th century, private, satisfactory condition	19th century, private, poor condition
Paint-XRF (mg/cm²)	1.2	1.7	1.2	7.3	10.5
Interior surface dust (ppm w/w)	332.0	490.0	622.0	1680.0	2360.0
Interior surface dust (µg/m²)	130.0	250.0	250.0	770.0	2100.0
Interior dustfall (ppm w/w)	176.0	179.0	221.0	464.0	563.0
Interior dustfall (ng/m²/30 days)	35.0	54.0	75.0	139.0	199.0
Soil (ppm w/w)	98.0	138.0	221.0	692.0	905.0
Hand lead (µg Pb/subject)	4.3	4.8	7.5	10.5	15.5

Source: From Clark, S. et al., *Chem. Speciation & Bioavailability,* 3, 163, 1991. With permission.

were associated with older houses and higher lead paint concentrations. The relationship between housing type and BLLs in young children can be seen in Figure 2.14. Note the increase in BLLs from age 3 to 18 months.

B. Blood lead levels

Blood lead is a measure of lead exposure. It is the fraction of the body burden that correlates most closely with recent environmental exposures. The half-life of lead in blood is approximately 36 ± 5 days. In the early phases of

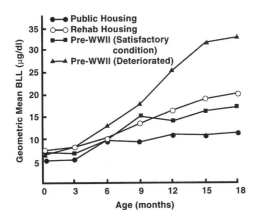

Figure 2.14 Relationship between housing type and blood lead levels in children. (From Clark, C.S. et al., *Environ. Res.,* 38, 46, 1985. With permission.)

Table 2.8 Blood Lead Levels (BLLs) in Different Population Groups in the U.S.

Population group	Period			
	1976–1980		1988–1991	
	Geometric mean (µg/dL)	95 Percentile (µg/dL)	Geometric mean (µg/dL)	95 Percentile (µg/dL)
All persons	12.8	25.0	2.8	9.4
Males	15.0	27.0	3.7	10.9
Females	11.1	20.0	2.1	7.4
Children ages 1–5	15.0	28.0	3.6	12.2

Source: Data extracted from Pirkle, J.A. et al., *JAMA*, 272, 284, 1994.

exposure, BLLs rise rapidly, and, as a consequence, are sensitive indicators of recent absorption. Blood lead represents only about 1% of the total body burden, with most lead stored in bones.

Blood lead levels are widely used to assess human exposure because they are a sensitive indicator of exposures that have been correlated with various health endpoints. Population BLL data for the U.S. has been collected and assessed for the periods 1976 to 1980 and 1988 to 1991 by the National Health and Nutrition Examination Survey (NHANES), a large-scale study periodically conducted by the National Center for Health Statistics of the Centers for Disease Control (CDC). Geometric mean BLLs for different population groups are summarized in Table 2.8. Notable in these data were the relatively high BLLs present in all population groups in the period 1976 to 1980 (average BLLs of circa 13 µg/deciliter [dL]), and the sharp decline in BLLs over the following decade. This decline was presumably due to the phase-down in use of leaded gasoline and other limitations on lead use. During the late 1970s, approximately 5% of the population had BLLs >25 µg/dL, with higher concentrations in males than females. Children in the age group 1 to 5 years had BLLs in the same range as adult males.

Children in the age range of 1 to 5 years are at particular risk of being adversely affected by lead because their brains are not fully developed (see *Health effects* below). Consequently, lead contamination of residences and subsequent exposures has become a major children's health issue.

The prevalence of elevated BLLs in selected populations of children in the U.S. is summarized in Table 2.9. Based on these 1988 to 1991 data, it is evident that approximately 9% of U.S. children ages 1 to 5 had BLLs that were ≥10 µg/dL, the CDC level of concern. Highest prevalence levels for elevated BLLs were found in children ages 1 to 2, and non-Hispanic black children. In the latter case, black children had BLLs twice as high as white and Mexican–American children. The increased risk for elevated BLLs among black children in the U.S. has been reported in a number of studies.

This racial disparity may be due to a variety of factors. Studies indicate that the major source of environmental lead exposure for black children is the indoor environment, whereas the outdoor environment is the major source for white children. Though painted surfaces in white children's homes

Table 2.9 Prevalence (%) of Elevated Blood Lead in Selected Populations
of Children in the U.S. in the Period 1988–1991

Population	Age, yrs.	≥20 μg/dL	≥15 μg/dL	≥10 μg/dL
All	1–5	1.1	2.7	8.9
Non-Hispanic white	1–2	0.8	2.1	8.5
	3–5	0.4	0.7	3.7
Non-Hispanic black	1–2	5.4	10.2	21.6
	3–5	2.9	6.0	20.0
Mexican-American	1–2	1.9	2.9	10.1
	3–5	0.7	1.4	6.8

Source: Data extracted from Brody, D.J. et al., *JAMA*, 272, 277, 1994.

have higher lead concentrations, they are reported to be in better condition, as are floor conditions. It is also believed that black children have a lower calcium intake than white children; calcium inhibits lead absorption.

Lead in house dust is considered to be the major risk factor for elevated BLLs in children. As such, it is important to know what levels are excessive and in need of remediation. The current Department of Housing and Urban Development (HUD) floor dust clearance level for lead abatement projects is 100 μg/ft² (1080 μg/m²). This value is also a guideline action level for residential risk assessment. A guideline value of 50 μg/ft² (540 μg/m²) has recently been proposed by USEPA as a lead hazard threshold. However, there is evidence that lead dust levels <10 μg/ft² (100 μg/m²) would be necessary to assure that 95% of the population of U.S. children would have BLLs below the national goal of 10 μg/dL. Other studies indicate that a mean lead floor dust level of 65 μg/ft² (702 μg/m²) is associated with a 10 μg/dL increase in BLLs in children.

C. Health effects

Lead is a very toxic substance affecting a variety of target organs and systems including the brain and the nervous, renal, reproductive, and cardiovascular systems. Effects are dose dependent. They may be acute (clinically obvious) or chronic (typically symptoms/effects are not easily diagnosed). Acute exposures, with BLLs > 60 μg/dL, may produce colic, shock, severe anemia, nervousness, kidney damage, irreversible brain damage, and even death. Acute responses may differ in adults and children since children under age five are more vulnerable to the neurotoxic effects of lead. Chronic exposure may result in a variety of symptoms, depending on the level of exposure. Acute and chronic exposure responses associated with different BLLs in children and adults are summarized in Table 2.10. The distinction between acute and chronic symptoms is not clear, especially for children; for adults, acute symptoms generally occur when BLLs are >60 μg/dL. At the low end of BLLs in Table 2.10, health endpoints may include neurodevelopmental effects in children, increased blood pressure and related cardiovascular effects in adults, and possibly cancer.

Table 2.10   Blood Lead Levels and Associated Health and Physiological Effects in
Children and Adults

Blood lead level ($\mu g/dL$)	Children	Adults
<10		Early signs of hypertension ALA-D inhibition

10–15	Crosses placenta
	Neuro-developmental effects
	ALA-D inhibition
	Impairment of IQ
	Increased erythrocyte
	  protoporphyrin
	Reduced gestational age and
	  birth weight	

15–20		Increased erythrocyte protoporphyrin

20–30	Altered CNS electrophysical
	  response
	Interference with Vitamin D
	  metabolism	

30–40	Reduced hemoglobin synthesis Peripheral nerve dysfunction	Systolic hypertension Altered testicular function Peripheral nerve dysfunction

40–50		Reduced hemoglobin
		  synthesis
		Overt subencephalopathic
		  neurologic symptoms

60	Peripheral neuropathy	Reproductive effects in females
70	Anemia	
80		Anemia Encephalopathy symptoms
80–100	Encephalopathy symptoms Chronic nephropathy Colic and other gastrointestinal symptoms	Chronic nephropathy Encephalopathy symptoms

Hematological changes (effects on blood chemistry and related physiological changes) are one of the earliest manifestations of chronic lead exposure in both adults and children. Lead interferes with heme synthesis. Heme
is a component of hemoglobin and cytochrome P-450 and other electron-
transferring cytochromes. As such, heme is essential to the proper functioning of the brain, kidney, liver, and blood-forming tissues. Interference with
hemoglobin synthesis results in mild to severe anemia. Inhibition of enzymes
involved in heme biosynthesis has been observed at BLLs as low as $10\,\mu g/dL$,
with an unknown threshold.

Lead decreases the production of vitamin D. Inhibition of vitamin D production in children has been observed at BLLs as low as 12 µg/dL.

In addition to lead's effect on blood, evidence from both epidemiological and animal studies indicates that low-level lead exposure is related to increases in blood pressure (hypertension). A relationship appears to exist between diastolic blood pressure in middle-aged males and a range of BLLs down to 7 µg/dL. Studies with females have shown a less consistent association. Recent studies which have focused on knee bone (patella) lead concentrations have shown a twofold increase of hypertension with an increase from the 10th to the 90th percentile lead concentration in middle-aged female nurses, with no association with BLLs.

The mechanism for this lead–hypertension association is unknown. It has been proposed that the effect may be due to the interaction of lead with membrane-bound enzyme systems responsible for transmembrane fluxes of ions such as sodium, potassium, and calcium. Lead directly affects the cellular transport of sodium and indirectly alters a number of calcium-regulating processes.

High blood pressure is a major risk factor for coronary artery disease, stroke, and renal (kidney) insufficiency. The effect of childhood lead exposures on subsequent risk of hypertension and cardiovascular disease in adulthood has not yet been evaluated; it should be of some public health concern.

Lead exposures have been associated with sperm abnormalities, reduced fertility, and altered testicular function in male industrial workers with BLLs in the range of 40 to 50 µg/dL. As more females enter the workforce in once male-dominated industries, an elevated lead body burden in pregnant females may pose a significant risk to developing fetuses. Because lead readily passes through the placenta, the fetus can be exposed through its mother. Blood lead levels in pregnant females appear to be higher than population averages, presumably due to the mobilization of bone lead along with calcium. Several epidemiological studies have reported associations between maternal BLLs and preterm delivery and low birth weight. An inverse relationship between maternal BLLs and birth weight with BLLs down to 15 µg/dL has been reported. Other studies have failed to observe such a relationship.

Lead is a potential carcinogen. It can cause mutations and cell transformation and interfere with DNA synthesis in mammalian cell cultures. Animal studies have shown that it can induce kidney tumors. Though not definitive, epidemiological studies suggest a causal relationship between lead exposures and cancer. Based on this evidence, USEPA has identified lead as a Group 2B human carcinogen (carcinogenicity has been confirmed in animal studies; human studies are inconclusive).

The nervous system appears to be the primary target organ system for lead exposure. Lead causes encephalopathy (brain damage) at BLLs of ≥ 80 µg/dL. Children with BLLs in this range may experience permanent brain damage characterized by severe mental retardation and recurrent convul-

Table 2.11   Centers for Disease Control Blood Lead Guidelines for
Childhood Lead Poisoning

Blood lead concentration (μg/dL)	Assessment
< 9	Child is not considered to have lead poisoning.
10–14	Child should have more frequent blood lead screening; community lead poisoning prevention programs should be implemented if many children are in this range.
15–19	Child should receive nutritional and educational interventions and more frequent screening; environmental investigation and interventions should be conducted if levels persist.
20–44	Child should receive medical evaluation, environmental evaluation and remediation; child may require pharmacologic treatment.
45–69	Child needs medical and environmental interventions including chelation therapy.
> 70	Medical emergency; child must receive immediate medical and environmental management.

Source: From CDC, *Preventing Lead Poisoning in Children*, DHHS, Washington, D.C., October, 1991.

sions. Lead may impair peripheral nerve conduction at BLLs as low as 20 to 30 μg/dL (the high end of present-day environmental exposures), with brain wave changes evident at 15 μg/dL (BLLs commonly reported in black children). There appears to be no apparent threshold for brain wave changes.

A number of prospective epidemiological studies have been conducted to evaluate the relationship between BLLs and neurodevelopmental effects in children. Taken together, these studies suggest that pre- and post-natal BLLs as low as 10 to 15 μg/dL up to the age of seven are associated with neurodevelopmental effects. These include decreased measures of intelligence, short-term memory loss, reading and spelling underachievement, impairment of visual motor function, poor perception integration, disruptive classroom behavior, and impaired reaction time. There is no known threshold for such effects.

The effect of lead exposure on children's learning potential as evaluated by IQ has been observed at blood lead levels as low as 10 μg/dL. In the BLL range of 10 to 20 μg/dL, lead effect on IQ is reported to be small, with some investigators estimating a 2–3 to 4–7 point decline in IQ with each 10 μg/dL increase in BLL.

The impact of childhood exposures on the learning required for successful participation in adult employment and the full societal benefits which accrue are unknown, as are any untoward social behaviors and demands on future social services. The potential for such impacts obviously exists.

Over the past half-century, the level of public health concern associated with BLLs has changed as our understanding of lead health risks has increased. In 1970, the CDC BLL level of concern was 60 μg/dL; in 1971,

40 µg/dL; in 1975, 30 µg/dL; and in 1985, 25 µg/dL. In 1991, CDC revised its guidelines to take neurodevelopmental effects into account. These revised guidelines, designed to be used by medical practitioners, public health agencies, and community action groups, are summarized in Table 2.11. The minimum level of concern is now 10 µg/dL. Table 2.11 recommends increasing levels of concern, with increasing public health, environmental, and medical intervention, with stepwise increases in BLL.

Readings

Beard, M. and Iske, E.D., Eds., *Lead in Paint, Soil, and Dust: Health Risks, Exposure Studies, Control Measures, Measurement Methods, and Quality Assurance*, American Society for Testing and Materials, Philadelphia, 1995.

Benarde, M.A., Ed., *Asbestos: The Hazardous Fiber*, CRC Press, Boca Raton, 1995.

Breen, J.J. and Stroup, C.R., Eds., *Lead Poisoning: Exposure, Abatement, Regulation*, Lewis Publishers, Boca Raton, 1995.

California Environmental Protection Agency Air Resources Board, *Proposed Identification of Lead as a Toxic Air Contaminant*, Executive Summary, 1997.

Centers for Disease Control, *Preventing Lead Poisoning in Children*, DDHS, Washington, D.C., October, 1991.

Costellino, N., Costellino, P., and Samola, M., *Inorganic Lead Exposure: Metabolism and Intoxication*, Lewis Publishers, Boca Raton, 1994.

Health Effects Institute-Asbestos Research (HEI-AR), *Asbestos in Pubic and Commercial Buildings: A Literature Review and Synthesis of Current Knowledge*, Health Effects Institute, Cambridge, MA, 1991.

Health Effects Institute–Asbestos Research (HEI-AR), Proceedings of workshop on asbestos operations and maintenance in buildings, *Appl. Indust. Hygiene*, Vol. 9, 1994.

Lamphear, B.P. and Roghmann, K.J., Pathways to lead exposure, *Environ. Res.*, 74, 67, 1997.

Maroni, M., Siefert, B., and Lindvall, T., Physical pollutants, *Indoor Air Quality. A Comprehensive Reference Book*, Elsevier, Amsterdam, 1995, chap. 3.

Maroni, M., Siefert, B., and Lindvall, T., Application of risk assessment: radon, *Indoor Air Quality. A Comprehensive Reference Book.* Elsevier, Amsterdam, 1995, chap. 16.

Millstone, E., *Lead and Public Health: The Dangers for Children*, Taylor & Francis, Washington, D.C., 1997.

National Research Council Committee on Health Effects Exposure to Radon, *Health Effects of Exposure to Radon: Time for Reassessment*, National Academy of Science Press, Washington, D.C., 1994.

National Research Council, *Health Risks of Radon and Other Internally Dispersed Alpha Emitters. BEIR IV,* National Academy of Science Press, Washington, D.C., 1988.

National Research Council Committee on Lead in the Environment, *Lead in the Human Environment: A Report.* National Academy of Science Press, Washington, D.C., 1980.

Nazaroff, W.W. and Teichmann, K., Indoor radon, *Environ. Sci. Technol.*, 24, 774, 1990.

Pirkle, J.L. et al., The decline in blood lead levels in the United States. The National Health and Nutrition Examination Surveys (NHANES), *JAMA*, 272, 284, 1994.

Samet, J.M., Radon and lung cancer revisited, in *Indoor Air and Human Health*, 2nd ed., Gammage, R.B. and Berven, B.A., Eds., CRC Press, Boca Raton, 1996, 325.

Samet, J.M., Radon, in *Indoor Air Pollution: A Public Health Perspective*, Samet, J.M. and Spengler, J.D., Eds., Johns Hopkins University Press, Baltimore, 1991, 323.

Schwartz, J. and Levin, R., The risk of lead toxicity in homes with lead paint hazard, *Environ. Res.*, 54, 1, 1991.

Selikoff, I.J. and Lee, D.H.K., *Asbestos and Disease*, Academic Press, New York, 1978.

Shakih, R.A., Asbestos exposures of building occupants and workers, in *Indoor Air and Human Health*, 2nd ed., Gammage, R.B. and Berven, B.A., Eds., CRC Press, Boca Raton, 1996, 341.

Spengler, J.D., Samet, J.M., and McCarthy, J.F., Eds., *Indoor Air Quality Handbook*, McGraw Hill, New York, 2000, chaps. 38–40.

U.S. Department of Energy, *Indoor Air Quality Environmental Information Handbook: Radon.* DOE/PE/72013-2, Washington, D.C., 1986.

USEPA, *Lead in Your Home: A Parents' Reference Guide*, EPA 747-B-98-002, Washington, D.C., 1998.

USEPA, *A Citizens Guide to Radon*, 2nd ed., EPA 402-K-92-001, Washington, D.C., 1994.

USEPA, *Radon — A Physician's Guide*, EPA 402-K-93-008, Washington, D.C., 1993.

USEPA, *Radon Measurement in Schools*, EPA 402-R-92-014, Washington, D.C., 1993.

USEPA, *Asbestos in Your Home*, Washington, D.C., 1990.

Questions

1. What is asbestos? How does it differ from other fibers?
2. Describe differences among different types of asbestos fibers used in commercial applications.
3. What makes asbestos so useful?
4. Describe major uses of asbestos fibers in buildings.
5. Where may you expect to find asbestos-containing products in residences?
6. Concentrations of airborne asbestos fibers are usually expressed in what units?
7. How are concentrations of airborne asbestos fibers determined?
8. How do indoor asbestos concentrations compare to those reported in workplace exposures?
9. In a building containing ACM, who is most at risk? Why?
10. Describe diseases associated with asbestos exposures.
11. What is the relationship between asbestos exposure, tobacco smoking, and disease?
12. In a building, what factors increase risks of exposure to the general building population?
13. What is radon? Describe its formation and decay.
14. Since radon is inert, why should it be considered to be potentially dangerous to human health?
15. What factors contribute to increased radon exposures in homes?
16. Radon concentrations may be expressed in pCi/L and working levels (WL). What is the distinction between these two measures?
17. Describe differences in radon concentrations in houses as they relate to substructure type.
18. What causes radon transport into buildings?
19. What are the relative contributions of groundwater and building materials to indoor radon concentrations?
20. What health risks are associated with exposure to elevated radon levels?

21. Describe lead, its sources and uses.
22. Lead is commonly found in indoor environments. What are its sources?
23. Define lead-based paint within its regulatory context.
24. What is the major source of lead exposure in children?
25. Describe the relationship between blood lead levels and exposures to lead-contaminated dust and soil.
26. Describe a model and its components for lead exposure and elevated blood lead levels in children.
27. Describe the health effects of lead associated with different blood lead levels.
28. Describe how blood lead levels have changed in both adults and children in the U.S., and why.
29. Describe individual risk factors for elevated blood lead levels in children.
30. How are children uniquely affected by lead exposures?
31. At the present time, what is considered a safe level of exposure (based on blood lead) to environmental lead?
32. What evidence is there to implicate inhalation as an exposure pathway for children to lead?

chapter three

Combustion-generated contaminants

Indoor spaces are commonly contaminated with substances that result from combustion. This has been the case since humans discovered the utility of fire and attempted to use it under various levels of control to cook food and provide warm living conditions in cold environments.

If fuels and materials used in combustion processes were free of contaminants and combustion were complete, emissions would be limited to carbon dioxide (CO_2), water vapor (H_2O), and high-temperature reaction products formed from atmospheric nitrogen and oxygen (NO_x). However, fuels and other combusted materials, e.g., tobacco, are never free of contaminants. Also, combustion conditions are rarely optimal; as a consequence, combustion is usually incomplete. When burned, fuels such as natural gas, propane, kerosene, fuel oil, coal, coke, charcoal, wood, and gasoline, and materials such as tobacco, candles, and incense, produce a wide variety of air contaminants. Some of these are generic to combustion while others are unique to materials being combusted. Substances produced in most combustion reactions include CO_2, H_2O, carbon monoxide (CO), nitrogen oxides (NO_x) such as nitric oxide (NO) and nitrogen dioxide (NO_2), respirable particles (RSP), aldehydes such as formaldehyde (HCHO) and acetaldehyde, and a variety of volatile organic compounds (VOCs); fuels and materials that contain sulfur will produce sulfur dioxide (SO_2). Particulate-phase emissions may include tar and nicotine from tobacco, creosote from wood, inorganic carbon, and polycyclic aromatic hydrocarbons (PAHs).

Sources of combustion-generated pollutants in indoor environments are many. In highly developed countries, they include emissions from: (1) a variety of vented and unvented combustion appliances, (2) motor vehicles (which may move from an outdoor [ambient] to an indoor environment), and (3) fuel-powered machinery such as floor burnishers, forklifts, and

Zambonis used in a number of indoor environments. They also include tobacco smoking and the increasingly popular activities of burning candles and incense. In developing countries they include indoor cooking fires which are not vented, or only poorly vented, to the outdoor environment.

I. Vented combustion appliances

Combustion of fuels such as wood and coal produces large quantities of smoke that humans in many advanced societies have for centuries found to be unacceptable in their domiciles. Chimneys and flues were developed and used to carry smoke away from cooking and heating fires. They exhaust by-products of fires while providing space heat with varying degrees of success (depending on how well they were designed and the effectiveness of the natural draft that carries exhaust up and outwards). Energy-inefficient fire-places were later replaced by well-vented stoves, which provided heat in local areas, and furnaces, whose energy could be used to heat an entire building.

Vented appliances are designed to provide a mechanism by which com-bustion by-products are carried through fluepipes or chimneys by natural or mechanical means. The effectiveness of these appliances varies, as would be expected. All vented combustion appliances will, from time to time, cause some degree of direct indoor contamination. With modern gas, propane, and oil-fired furnaces, indoor contamination is relatively limited except when a system malfunction occurs. Indoor air contamination from wood- or coal-burning appliances, such as fireplaces, stoves, and furnaces, is more common and varies with appliance, building design, and environmental factors.

For the past half century or more, residences in the colder regions of North America have been heated by natural gas, propane, or oil in well-designed furnaces with properly designed and installed flue/chimney sys-tems. Such furnaces produce little smoke but can produce significant CO emissions, which pose a potentially serious public health risk if they are not properly vented. Venting of flue gases is achieved by the use of natural or mechanical draft. In natural draft furnace systems, warm combustion gases rise by convection from the fire box (combustion chamber) and are carried upward by building air which flows into a draft hood on the side of the furnace where it joins and mixes with flue gases. The system is an open one. Should there be insufficient draft, flue gases will spill into the building environment surrounding the furnace and quickly be transported through-out the building. Such draft failures are not uncommon; in most cases they result in relatively limited flue-gas spillage and are of minor concern.

Mechanical draft systems which have a fan to exhaust flue gases have been used for many years, particularly in oil-fired furnaces. These systems are becoming the norm in North America with the development of medium-to high-efficiency (80 to 90%) gas and propane-fired furnaces. Because flue gases contain little heat to carry them upward, high-efficiency furnace sys-tems must be mechanically vented. Such venting is accomplished without chimneys. Since mechanical draft systems require no draft hood, the prob-

Table 3.1   Factors Contributing to Flue Gas Spillage and CO Poisoning in Residences

- Corroded, cracked heat exchangers
- Dislodged or damaged fluepipes
- Improperly installed fluepipes
- Changes in appliance venting (mechanical draft furnace combined with a natural draft hot water heater)
- Inadequate combustion air/tight building envelope
- Exhaust ventilation competes with furnace/fireplace for air
- Downdraft in chimney
- Blocked chimney

ability of flue-gas spillage by backdrafting is less than in systems which use natural draft.

A. Flue-gas spillage

Serious flue-gas spillage occurs in North American homes, with occasional deaths and, more commonly, sublethal CO poisoning. Flue-gas spillage has been reported with gas furnaces, gas water heaters, and wood-burning appliances. It occurs in residences with aging or poorly installed or maintained combustion/flue systems. Major causes and contributing factors of flue-gas spillage and reported CO poisonings are summarized in Table 3.1.

Flue-gas spillage occurs when upward airflow is too slow to exhaust all combustion products. Under circumstances such as chimney blockage, flue-gas flow is stalled, resulting in significant contamination of indoor spaces and a major CO exposure risk. In backdrafting, outdoor air flows down the chimney or flue and spills through the draft hood. Backdrafting can occur when a house is depressurized by competing exhaust systems (e.g., fireplace and furnace), when the chimney is cold, and under some meteorological conditions. Backdrafting can be a significant problem in energy-efficient houses where infiltration air is not adequate to supply the needs of mechanical exhaust systems, fireplaces, and furnace/hot water heating systems. This potential problem is being addressed in building codes which require that sufficient combustion and makeup air be provided by contractors.

B. Wood-burning appliances

Wood-burning appliances, such as fireplaces and stoves, have a long history in North America. The popularity of wood-burning appliances re-emerged in the 1980s in response to increased energy costs associated with the rising price of petroleum (which occurred as a result of military conflicts in the Middle East). An estimated 5 million wood-burning stoves were being used in the U.S. to provide supplementary space heating. Wood-burning furnaces were also being used to provide whole-house heating as well.

Wood-burning appliance use in the 1980s was based on the premise that energy costs could be reduced by setting back thermostats and spot heating occupied major living areas with a wood-burning stove. Such practices were anticipated to decrease overall space heating costs by reducing energy used and substituting what was perceived to be a lower-cost and environmentally friendly fuel.

As wood burning for space heating became popular, concerns were raised about the potential impact of wood-burning appliances on ambient air quality in communities where wood burning was common (e.g., Corvallis, OR; Butte, MT; Aspen, CO; and Watertown, NY). The impact of wood-burning stoves on ambient air quality was deemed so great that the USEPA promulgated a New Source Performance Standard (NSPS) for new wood-burning stoves that manufacturers must meet to reduce emissions and protect ambient air quality.

Wood-burning stoves vary in design. There are two basic types: conventional and airtight. Conventional stoves have relatively low combustion efficiencies (in the range of 25 to 50%) and tend to cause significantly more indoor and outdoor contamination that airtight stoves, which have combustion efficiencies >50%. Efficiencies of new stoves covered by the NSPS have, by necessity, increased in order to meet regulatory requirements. Wood stoves and furnaces that comply with the NSPS have significantly lower emissions of CO and particles to the atmosphere.

A variety of investigators have attempted to determine the impact of wood-burning appliances on indoor air quality (IAQ). Special attention has been given to contaminants such as CO, NO, NO_2, SO_2, RSP, and PAHs. Elevated indoor levels of NO, NO_2, and SO_2 have been reported in some studies but not others. Reports of elevated indoor CO and RSP levels associated with wood appliance operation have been more consistent. Carbon monoxide concentrations in houses with nonairtight stoves have been reported in the range of a few parts per million (ppmv) to 30 ppmv (the latter under worst-case operating conditions).

Wood-burning appliances produce smoke, which is a combination of particulate and gas-phase contaminants. The former gives wood-burning smoke its "visible" characteristics. Smoke tends to leak from nonairtight stoves during operation and from both airtight and nonairtight stoves during refueling.

The effect of wood-burning appliance operation on indoor RSP concentrations has been evaluated by investigators who have made measurements of RSP in both indoor and ambient environments and compared them by calculating indoor/outdoor ratios. Significantly higher indoor/outdoor (I/O) concentrations were always observed for residences during wood-burning appliance operations. Highest I/O ratios were reported for nonairtight stoves (4 to 7.5:1) and fireplaces (6.1 to 8.5:1); lowest ratios were in homes with airtight stoves (1.2 to 1.3:1). It must be noted that these ratios were likely biased (to lower values) by higher outdoor suspended particle concentrations due to the operation of wood-burning appliances themselves.

The particulate phase of wood-burning emissions includes a variety of substances, most notably PAHs, a group of compounds with considerable

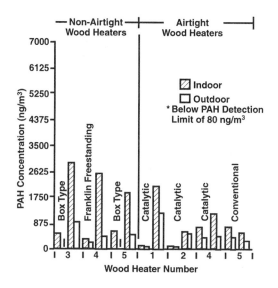

Figure 3.1 Polycyclic aromatic hydrocarbon concentrations associated with different wood-burning heating appliances. (From Knight, C.V., Humpreys, M.P., and Pennex, J.C., *Proc. Indoor Air for Health and Energy Conservation*, ASHRAE, Atlanta, 1986. With permission.)

carcinogenic potential. PAH concentrations associated with a variety of wood heater types, as well as outdoor concentrations, are indicated in Figure 3.1. Lowest indoor PAH concentrations and potential human exposures were associated with airtight wood stoves, particularly those equipped with catalytic systems. Catalytic systems are commonly used to meet the wood appliance NSPS.

II. Unvented combustion systems

A. Cooking stoves in developing countries

A majority of the world's households depend on biomass fuels, such as wood, animal dung, and crop residues, for their cooking and space heating needs, with wood being the principal fuel. Most biomass fuel use occurs in rural areas of developing countries (particularly the densely populated countries of Southeast Asia), although significant biomass fuel use also occurs in poor urban areas. Unprocessed biomass fuels are the primary cooking fuel in 75% of households in India, 90% of which use wood or dung. Biomass fuels are at the high end of the fuel ladder relative to pollutant emissions; not surprisingly they are at the low end of the ladder in terms of combustion efficiency and energy content.

In addition to unprocessed biomass, other fuel types used for cooking in developing countries include charcoal, kerosene, and coal. Unvented coal-burning cooking fires are increasingly being used in China where pop-

Table 3.2   Indoor Airborne Particulate Matter (TSP) Concentrations Associated with Biomass Cooking in Developing Countries

Location/report year	Measurement conditions	PM concentration ($\mu g/m^3$)
Papua New Guinea		
1968	Overnight/floor level	200–4900
1975	Overnight/sitting level	200–9000
India		
1982	Cooking with wood	15,800
	Cooking with dung	18,000
	Cooking with charcoal	5500
1988	Cooking, measured near ceiling	4000–21,000
Nepal		
1986	Cooking with wood	8800
China		
1987	Cooking with wood	2600
Gambia		
1988	24 hours	1000–2500
Kenya		
1987	24 hours	1200–1900

Source: From Smith, K.R., *Environment*, 30, 28, 1988. With permission.

ulation density is high, wood supplies are limited due to deforestation, and coal is abundant.

Most cooking stoves are simple pits, U-shaped structures made of mud, or consist of three pieces of brick. Most indoor cooking fires are not vented to the outdoor environment; only a small fraction have enclosed combustion chambers with flues.

Studies on exposure concentrations in households using biomass and other fuels for cooking in poorly ventilated environments have been conducted over the past three decades. Much of the focus of these studies has been the measurement of indoor concentrations of particulate matter, which is present in enormously high concentrations (based on North American, European, and Japanese ambient air quality standards). Indoor air concentrations of what is presumably total suspended particulate matter (TSP; particle size range of 0.1 to 100 μm) are summarized for rural households in a number of developing countries in Table 3.2. Personal daily exposures to particulate matter while using biomass cooking in India and Nepal are summarized in Table 3.3. What is notable in both cases is that daily indoor exposure to particulate matter in biomass-using households, particularly among adult women and young children, is significantly greater than that specified by ambient air quality standards for TSP (75 $\mu g/m^3$ annual geometric mean; 260 $\mu g/m^3$ 24-hour average not to be exceeded more than once per year) used in the U.S. until 1988 (since changed to a PM_{10} standard of 50 $\mu g/m^3$ annual arithmetic mean; 150 $\mu g/m^3$ 24-hour average) and World Health Organization (WHO) PM_{10} guidelines (40 to 60 $\mu g/m^3$ annual average; 100 to 150 $\mu g/m^3$ 24-hour average). In these households, daily indoor

Table 3.3 Personal Exposures to Airborne Particulate
Matter (TSP) during Biomass Cooking (2 to 5 hrs/day)
in Developing Countries

Country/year	Measurement conditions	PM concentration ($\mu g/m^3$)
India		
1983	In 4 villages	6800
1988	In 5 villages	4700
1988	In 2 villages	3600
1988	In 8 villages	3700
Nepal		
1986	In 2 villages	2000
1988	In 1 village	
	Traditional stove	8200
	Improved stove	300

Source: From Smith, K.R., *Environment*, 30, 28, 1988. With permission.

particulate matter exposure concentrations exceed standards for 24-hour average outdoor concentrations (which would not be permitted to be exceeded more than once per year in developed countries) by an order of 10 to 50+ times.

Inhalable particulate matter concentrations (PM_7) in suburban Mozambique homes using different fuel sources, including wood, are summarized in Table 3.4. Highest PM_7 concentrations, not unexpectedly, were associated with both wood and coal. Lower concentrations associated with wood fuels (when compared to those in Tables 3.2 and 3.3) may be due to the smaller cutoff diameter (7 μm) used for collected particles in Mozambique studies. Nevertheless, average PM concentrations were nearly 6 to 8 times higher than the U.S. 24-hour average PM_{10} air quality standard. Indeed, the U.S. 24-hour standard was exceeded even in the few homes using electricity and liquefied petroleum gas (LPG) for cooking. Such exposures are due to neighborhood ambient air pollution associated with the biomass cooking of others.

Table 3.4 Indoor Inhalable Particulate Matter (PM_7)
Concentrations in Mozambique Suburban Dwellings
using Different Cooking Fuels

Fuel	Average PM_7 concentrations ($\mu g/m^3 \pm SE$)	# of residences
Wood	1200 ± 131	114
Coal	940 ± 250	4
Kerosene	760 ± 270	10
Charcoal	540 ± 80	78
Electricity	380 ± 94	8
LPG	200 ± 110	3

Source: From Ellegard, A., *Environ. Health Perspect.* 104, 980, 1996.

Particulate matter associated with both biomass cooking and coal use has high PAH concentrations. In a study of 65 Indian households, benzo-α-pyrene concentrations measured indoors averaged 3900 ng/m^3, with a range of 62 to 19,284 ng/m^3, as compared with ambient concentrations of 230 and 107 to 410 ng/m^3, respectively. Indoor concentrations of this potent carcinogen can only be described as enormously high, while ambient concentrations are significantly lower but nevertheless high with reference to acceptable levels in developed countries.

As anticipated, significant concentrations of gas-phase substances are also associated with biomass, coal, and other cooking fuels. In the Mozambique studies, average indoor CO levels with wood and charcoal cooking were 42 and 37 ppmv, respectively. Elevated concentrations of CO in the range of 10 to 50 ppmv have been reported elsewhere. NO_2 and SO_2 levels in the range of ~0.07 to 0.16 ppmv and ~0.06 to 0.10 ppmv, respectively, have been reported in short-term (15-minute) measurements in Indian households using wood or dung for cooking. High aldehyde levels in the range of 0.67 to 1.2 ppmv have been reported in New Guinea households using biomass fuels.

B. *Gas and kerosene heating appliances*

A variety of unvented heating devices that burn natural gas, propane, or kerosene are used in the U.S. and other countries. These include gas and kerosene space heaters, water heating systems (primarily in European countries), gas stoves and ovens, ventless gas fireplaces, and gas clothes dryers. In each case, fuels are relatively clean-burning, i.e., they produce no visible smoke. They have also been designed to burn with relatively higher efficiencies. As a consequence, they pose little or no risk of CO poisoning associated with earlier appliances.

Unvented appliances used for home space heating have significant advantages. Costs are low compared to heating systems that include relatively expensive furnaces, ductwork, and flue/chimney systems. They can also be used to spot-heat a residence, or, as in New South Wales (Australia), classrooms. Heat is provided only where and when it is needed. Unvented space heaters can also be used in environments where vented combustion appliances may not make economic sense. These include vacation homes or cabins, recreational vehicles, detached garage workshops, and even tents.

The use of kerosene heaters to spot-heat residences in the U.S. became popular during the 1980s for the same reason that wood-burning appliances were popular at that time, i.e., to reduce energy costs. Kerosene heaters became popular only after low-CO-emitting devices became commercially available. More than 10 million kerosene heaters were sold in the U.S. by 1985.

There are three basic types of kerosene heaters: radiant, convective, and two-stage. They all utilize a cylindrical wick and operate at relatively high combustion temperatures. In the radiant type, flames from the wick extend up into a perforated baffle, which emits infrared energy (radiant heat). Such heaters operate at lower temperatures than convection heaters, which transfer heat from the wick by convection. In two-stage heaters, there is a second chamber above the radiant element that is designed to further oxidize CO and unburned or partially burned fuel components.

Laboratory studies of both unvented gas and kerosene space heaters indicate that they have the potential to emit significant quantities of CO_2, CO, NO, NO_2, RSP, SO_2, and aldehydes into indoor spaces. Based on the chemistry of combustion, they would also be expected to emit large quantities of water vapor. Emission potentials depend on heater type, operating and maintenance parameters, and the type of fuel (relative to SO_2 emissions) used. Radiant heaters produce CO at rates twice those of convection heaters and about 3 times those of two-stage heaters. Convection heaters have significantly higher emissions of NO and NO_2 compared to both radiant and two-stage heaters. Decreasing the wick height, a practice homeowners employ to decrease fuel usage, results in increased emissions of CO, NO_2, and formaldehyde (HCHO). Maltuned heaters have significantly higher emissions of CO and HCHO (as much as 20- to 30-fold). Emissions of SO_2 depend on the sulfur content of kerosene. Grade No. 1-K kerosene has a sulfur content of 0.04% by weight; grade No. 2-K may have a sulfur content as high as 0.30%. The latter has been more widely used than the former.

A variety of laboratory studies have been conducted to predict human exposures, and a few have attempted to measure contaminant levels in indoor spaces during heater operation. In one laboratory chamber study designed to simulate heater operation in a moderate-sized bedroom with an air exchange rate of one air change per hour (1 ACH), very high contaminant exposures were predicted (SO_2 levels >1 ppmv; NO_2 levels in the range of 0.5 to 5 ppmv; CO in the range of 5 to 50 ppmv; CO_2 in the range of 0.1 to 1%). Potential exposure concentrations under different ventilation conditions are illustrated in Figure 3.2 for radiant and convective kerosene heaters. Reference is also made in this figure to the National Ambient Air Quality Standards (NAAQS), Occupational Safety and Health Administration (OSHA) standards, and guidelines once recommended by the American Society of Heating, Refrigerating and Air-Conditioning Engineers (ASHRAE).

Unvented kerosene and gas-fueled space heaters have been used by homeowners and apartment dwellers under varying conditions of home air space volumes, ventilation rates, number of heaters used, and daily and seasonal hours of operation. As a consequence, exposure concentrations vary widely. In a study of 100 U.S. houses, NO_2 levels in homes operating one kerosene heater averaged ~20 ppbv; with two heaters, 37 ppbv; in control homes, NO_2 concentrations averaged ~4 ppbv. Over 49% of the residences had concentrations of NO_2 >50 ppbv during heater use, with approximately 8% exceeding 255 ppbv. Over 20% had average SO_2 levels >0.24 ppmv, the 24-hour ambient air quality standard. In other studies, carbon monoxide was reported in the range of 1 to 5 ppmv; there were also significant increases in RSP, in the range of 10 to 88 $\mu g/m^3$.

Kerosene heater usage has declined substantially from its peak in the mid-1980s. As a consequence, the number of individuals exposed has also

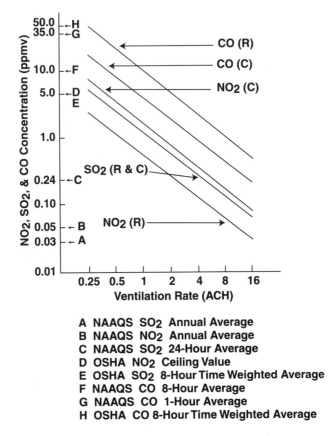

Figure 3.2 Contaminant levels associated with kerosene heater operation under controlled laboratory chamber conditions. (From Leaderer, B.P., *Science*, 218, 1113, 1982. With permission.)

decreased. Gas heater usage in southern states is, however, considerable. There is little scientific information available on combustion-generated contaminant levels in such residences and, as a consequence, little is known about potential public health risks associated with gas space heater operation.

C. Gas stoves and ovens

The use of natural gas and propane for cooking and baking is common in North America. Such appliances (outside the context of restaurants and cafeterias) are rarely provided with adequate local exhaust ventilation (if provided, such systems are rarely activated). Gas cooking stoves and ovens have been shown to be significant or potential sources of CO, CO_2, NO, NO_2, aldehydes, RSP, and VOCs.

Episodic increases in indoor CO levels in the range of 10 to 40 ppmv have been reported in residences. Peak levels of NO_x >0.5 ppmv may occur during the use of gas cooking appliances. Average concentrations are significantly lower. Indoor NO_2 levels in homes with gas cooking stoves have been reported to be in the range of 18 to 35 ppbv.

Gas cooking appliances are used intermittently so that exposures vary significantly. In low income areas of many northern U.S. cities, building occupants often use opened gas ovens as a continuous supplemental source of heat during cold winter conditions. Such use would be expected to result in higher exposure concentrations and longer exposure durations.

D. Gas fireplaces

Gas-burning fireplaces are widely used in North American homes for aesthetic reasons. Historically, emissions from these fireplaces were vented to the outdoor environment the same way as wood-burning fireplaces. In recent years, large numbers of ventless fireplaces have been sold and installed in new homes. Limited research studies have been conducted to determine the impact of their use on indoor contaminant levels. In one study, combustion by-products such as CO were reported to be low, while significantly higher concentrations were reported in higher-altitude Colorado studies. Carbon dioxide levels often exceeded 5000 ppmv, the OSHA 8-hour exposure limit.

Gas fireplace emissions are likely to be similar to those of gas heaters. Because of the intermittent use of ventless gas fireplaces, exposure to combustion by-products would, in most cases, be less than that associated with the use of gas heaters. Some users have reported moisture condensation and an oily film on windows, which they believe to be associated with ventless fireplace use. Along with elevated CO_2 levels, high production rates of water vapor may be expected.

III. Miscellaneous sources

Combustion sources described above are characterized by their use as home space heating or cooking appliances. There are a variety of other sources of combustion-generated substances which are common causes of indoor air contamination or cause contamination under relatively limited or unique circumstances. The most important of these is tobacco smoking. Other sources include candles, incense, and use of propane-fueled floor burnishers, propane-fueled forklifts and similar equipment, propane-fueled rink ice-making machines, arena events involving gasoline-powered vehicles, entrainment of motor vehicle emissions, and re-entry of flue gases.

A. Tobacco smoking

Approximately 24% of the adult population of the U.S. (~35 million individuals) smoke tobacco products daily. In nonresidential, nonindustrial buildings, smoking has either been banned or severely restricted. As a consequence, such buildings are unlikely to experience significant tobacco smoke-related contamination. However, significant indoor contamination continues

Table 3.5   Tobacco-Related Contaminant Levels in Buildings

Contaminant	Type of environment	Levels	Nonsmoking controls
CO	Room (18 smokers)	50 ppmv	0.0 ppmv
	15 restaurants	4 ppmv	2.5 ppmv
	Arena (11,806 people)	9 ppmv	3.0 ppmv
RSP	Bar and grill	589 $\mu g/m^3$	63 $\mu g/m^3$
	Bingo hall	1140 $\mu g/m^3$	40 $\mu g/m^3$
	Fast food restaurant	109 $\mu g/m^3$	24 $\mu g/m^3$
NO_2	Restaurant	63 ppbv	50 ppbv
	Bar	21 ppbv	48 ppbv
Nicotine	Room (18 smokers)	500 $\mu g/m^3$	—
	Restaurant	5.2 $\mu g/m^3$	—
Benzo-α-pyrene	Arena	9.9 ng/m^3	0.69 ng/m^3
Benzene	Room (18 smokers)	0.11 mg/m^3	—

Source: From Godish, T., *Sick Buildings: Definition, Diagnosis & Mitigation*, CRC Press/Lewis Publishers, Boca Raton, 1995.

to occur in residences, restaurants, and other environments not subject to smoking restrictions (Table 3.5).

Smokers subject both themselves and countless millions of nonsmokers to a large variety of gas and particulate-phase contaminants. Tobacco smoke reportedly contains several thousand different compounds, with approximately 400 quantitatively characterized. Some of the major air contaminants associated with tobacco smoke include RSP; nicotine; potent carcinogenic substances such as PAHs and nitrosamines; CO; CO_2; NO_X; and irritant aldehydes such as acrolein, HCHO, and acetaldehyde.

Exposure to tobacco smoke occurs from what is described as second-hand or environmental tobacco smoke (ETS). This smoke consists of exhaled mainstream smoke (MS) and sidestream smoke (SS); the latter is emitted from burning tobacco between puffs. Qualitatively, ETS consists of the same substances found in MS; however, quantitative differences exist between SS and MS.

On a mass basis, SS includes approximately 55% of total emissions from cigarettes. Sidestream smoke is produced at lower temperatures than MS and under strongly reducing (chemically) conditions. As a consequence, significant differences in production rates of various contaminants occur for SS and MS. These differences can be seen for a number of gas and particulate-phase substances in Table 3.6. Sidestream/mainstream smoke ratios (SS/MS) >1 indicate quantitatively higher concentrations in SS. Sidestream emissions may be fractionally to several times higher for some substances to an order of magnitude higher for others. As an example of the latter case, SS emissions of the suspected carcinogen *n*-nitrodimethylamine are on the order of 20 to 100 times greater than those in MS.

Environmental tobacco smoke undergoes chemical/physical changes as it ages. These changes include the conversion of NO to the more toxic NO_2

Table 3.6   Ratios of Selected Gas and Particulate-Phase Components in
SS and MS Tobacco Smoke

Vapor phase	SS/MS ratios	Particulate phase	SS/MS ratios
Carbon monoxide	2.5–4.7	Particulate matter	1.3–1.9
Carbon dioxide	8–11	Nicotine	2.6–3.3
Benzene[a]	10	Phenol	1.6–3.0
Acrolein	8–15	2-Naphthylamine[a]	30
Hydrogen cyanide	0.1–0.25	Benzo-α-anthracene[a]	2.0–4.0
Nitrogen oxides	4–10	Benzo-α-pyrene[a]	2.5–3.5
Hydrazine[a]	3	N-Nitrosodiethanolamine[a]	1.2
N-Nitrosodiethanolamine[a]	20–100	Cadmium	7.2
N-Nitrosopyrrolidine	6–30	Nickel[a]	13–30

[a] Animal, suspected, or human carcinogen.

Source: From U.S. Surgeon General, *The Health Consequences of Involuntary Smoking.* DHHS Pub.
No. (PHS) 87-8398, Washington, D.C. 1986.

and volatilization of substances from the particulate phase. Such volatilization reduces the mass median diameter of smoke particles and may, as a
consequence, increase their potential for pulmonary deposition in both nonsmokers and smokers.

Occupant exposures to components of ETS depend on several factors.
These include the type and number of cigarettes consumed per unit time,
the volume of building space available for dilution, building ventilation rate,
and proximity to smokers. Highest exposure concentrations would be
expected for those closest to smokers in small, poorly ventilated spaces
where a high rate of smoking is occurring (this is particularly the case in
some residences). The effect of tobacco smoke on levels of several contaminants generated in relatively high-smoking-density indoor environments has
been reported (Table 3.5). Levels in homes with smokers (where exposures
may be of considerable consequence) have not been reported. However, high
respirable particle (RSP) concentrations in indoor environments (as compared to those outdoors) have been suggested to be due to tobacco smoking,
believed to be the single most important contributor to RSP levels in residences and other buildings.

B. Candles and incense

Candles have been used as a source of illumination in buildings for thousands of years. Burning candles and incense has also been, and continues to
be, used as a part of religious worship and ritual. Soot-stained frescoes and
other paintings in European churches are a testament to building contamination associated with the long-term use of candles.

Burning candles and incense for aesthetic reasons in residences has
become increasingly popular. Several recent studies have attempted to characterize emissions from candles relative to potential health concerns and
building soiling potential. Burning candles can produce significant quantities

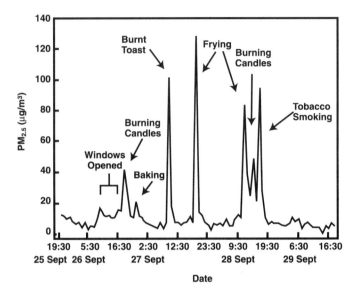

Figure 3.3 Effects of various indoor combustion sources/cooking activities on indoor $PM_{2.5}$ concentrations. (Courtesy of P. Koutraikis, Harvard University.)

of carbon particles which are predominantly in the respirable size range. Carbon or soot production varies with candle type, with scented candles reported to emit higher levels. Particularly high carbon emissions have been reported when candles are extinguished. The effect of candle-burning, as well as other combustion activities, on short-term $PM_{2.5}$ levels can be seen in Figure 3.3.

In addition to soot particles, candles can be expected to emit NO_x, CO, and aldehydes. Scented candles can be expected to emit a variety of odoriferous aldehydes, alcohols, and esters. Soot particles are likely to contain significant quantities of PAHs. Because of their small size and potential to contain potent carcinogens, carbon particles associated with significant candle-burning may be of public health concern.

At present, homeowners are primarily concerned with black deposits on indoor surfaces and appliances. Such deposits appear on surfaces where there are significant thermal differences. This phenomenon, known as "ghosting," occurs on wall surfaces near ceilings, electrical outlets, and stud areas on outside walls; near heating sources and light fixtures; etc. It is due to the thermophoretic deposition of airborne particles. Examination of such particles may reveal the presence of finely divided carbon particles or small burned candle fragments.

The "ghosting" phenomenon has also been reported in residences in which no candle burning has taken place. In one residential investigation, the author identified fragmented neoprene (which contains carbon black)-coated insulation in an air conditioning unit as the probable source of the apparent "carbon" deposits or, as some have described it, "black magic dust."

In some parts of the world (but not in the U.S.), candle manufacturers use lead-containing wicks. Such candles produce large quantities of very fine lead aerosol particles. These candles are imported into the U.S. and are estimated to comprise 3% of all candles sold.

C. *Propane-fueled burnishers*

Burnishers are used to polish tile floors in relatively small-to-large buildings. They may be powered electrically, or be propane-fueled. Propane-fueled burnishers can produce significant emissions of CO and other combustion by-products.

Research has shown that employees using propane-fueled burnishers are intermittently exposed to several hundred ppmv CO. Case investigations of their use in poorly ventilated buildings (e.g., a day-care center) have recorded CO concentrations >500 ppmv and high carboxyhemoglobin (COHb) levels (as high as 26% COHb) in those exposed.

Propane-fueled burnishers are widely used in large retail stores and other buildings by building staff or commercial cleaning companies. The actual extent of their use is not known and little is known about potential exposures in occupied buildings. Their use may explain, in part, news stories of illness complaints consistent with CO exposure in a number of retail establishments.

D. *Propane-fueled forklifts*

Forklifts powered by engines utilizing propane are widely used in American industry and warehouse operations. In such environments they are usually the major cause of elevated CO levels and worker exposures. Concentrations that average 10 to 50 ppmv are not uncommon, with higher levels in travel lanes.

Forklifts and similarly powered equipment pose IAQ problems when they are used in buildings with attached office spaces or other multipurpose functions. Significant contamination of office environments with CO and other combustion by-products can occur as a consequence. Office CO levels of 50 to 75 ppmv have been recorded. Office buildings attached to warehousing operations are at particular risk of CO and other combustion by-product exposures.

E. *Ice resurfacing machines*

Ice skating and hockey rinks in North America, northern Europe, and Japan are subject to considerable emissions and associated high concentrations of a variety of combustion-generated contaminants. Such emissions are associated with the use of fuel-powered ice resurfacing machines (commonly called Zambonis). Typically these are propane- or gasoline-fueled; less commonly, diesel fuels are used.

In the past three decades, a number of studies have been conducted to evaluate concentrations of CO and NO_2 in ice skating rinks. In one of the earliest studies conducted in the U.S., CO concentrations in the range of 12 to 250 ppmv were reported in six rinks (with a median of 100 ppmv). Highest concentrations were measured during resurfacing activities; when propane-fueled resurfacers were used, concentrations were in the range of 157 to 304 ppmv, while gasoline-fueled machines were associated with 80 to 170 ppmv. Concentrations of CO measured more recently are much lower, with breathing zone levels in the range of 10 to 30 ppmv.

Significant recent scientific attention has focused on NO_2 exposures in ice skating rinks. Short-term exposure concentrations >1 ppmv are commonly reported. In a survey study of 332 rinks in nine countries, 7-day NO_2 concentrations averaged 221 and 228 ppbv at breathing height near the ice surface and spectators' area, respectively, with highest concentrations of 2.68 and 3.18 ppmv. Seven-day average NO_2 concentrations of 360 ppbv in 70 rinks in the northeastern U.S. have been reported.

The World Health Organization (WHO) recommends a 1-hour NO_2 exposure guideline of 213 ppbv. This guideline was exceeded in 40% of the 332 rinks surveyed; 55% of the 70 U.S. rinks surveyed exceeded it.

Concentrations of NO_2 in ice skating rinks depend on a number of factors. The most important is fuel type. Highest NO_2 concentrations are associated with propane, followed by gasoline- and diesel-fueled machines, respectively. Fuel use varies from country to country. Propane-fueled resurfacers are used in approximately 70 to 75% of rinks in Canada and northern Europe, 60% in the U.S., and 0% in Japan. In the last case, gasoline-fueled resurfacers are used exclusively. Increased NO_2 levels have also been associated with increased numbers of resurfacing operations, lower ventilation rates, and smaller rink air volumes.

Compared to other exposure situations in developed countries, human exposures to CO and NO_2 in ice skating rinks are relatively high. For full-time employees, individuals who exercise significantly (hockey players and figure skaters), and those who practice for extended hours, exposures may be of particular concern. For other users and spectators, exposures are relatively transient.

F. Arena events

A variety of events involving motor vehicle operation are held in indoor arenas. These include tractor pulls, monster truck shows, and motorcross (motorcycle) events. Mean instantaneous arena CO concentrations in the range of 68 to 436 ppmv have been reported for Canadian tractor pulls. Mean CO concentrations for three different arena competitions (tractor pulls and monster truck shows) of 79, 106, and 140 ppmv have been reported for U.S. arenas. Average 5-hour personal exposure concentrations of 38 ppmv and a peak (3 minutes) of 226 ppmv CO have been reported in Canadian motorcross competitions.

G. Entrainment

Combustion-generated substances enter and contaminate buildings when they are produced by outdoor sources. Such sources include motor vehicles, nearby incinerators and boilers, residential heating appliances, etc.

All buildings are ventilated to some degree by mechanical systems, natural forces (infiltration), or both. Contaminants produced outdoors become readily entrained in air brought into buildings by either mechanical or natural forces.

Outdoor air is commonly contaminated by motor vehicle emissions, the dominant source of ambient air pollution in the U.S. Because of atmospheric mixing, combustion by-products and their derivatives associated with motor vehicles have a limited impact on IAQ, particularly in regard to health concerns.

Problems arise when the emission source is close to a building. Entrainment of motor vehicle emissions occurs near loading docks where truck emissions (often diesel) become entrained in air drawn through open loading dock doorways and/or outdoor air intakes located nearby. Loading docks on the lowest level of a tall building may be a major pathway for a significant stack effect (see Chapter 11) and entrainment of motor vehicle emissions. Complaints of motor vehicle odors are common in school buildings where idling motor vehicles are in close proximity to outdoor air intakes on unit ventilators. They are also common in buildings served by diesel- and gasoline-powered trucks idling near loading docks or outdoor air intakes.

One of the most common entrainment phenomena occurs in the fall of the year when millions of homeowners burn leaves (where there are no bans on this activity). Entrainment occurs through open windows and by infiltration in residences whose occupants wish to protect themselves from neighbors' leaf-burning smoke.

In residential neighborhoods, entrainment of combustion by-products from the gas heating and wood-burning systems of nearby households is common. Such entrainment occurs by infiltration.

The nature and significance of the entrainment of combustion-generated contaminants in residences and other buildings have not been systematically studied. In some field investigations there are suggestions that CO levels may become elevated by several ppmv. For the most part, entrainment is categorized as a nuisance. Occupants complain of "diesel odors," "gas odors," etc. Such odors, however, often serve as a focus of broad IAQ concerns which occur in problem buildings.

H. Re-entry of flue gases

The contamination of indoor air by combustion by-products vented to the outdoors through flues and chimneys often occurs indirectly as a result of re-entry. In residences it is common for fireplace wood smoke to be drawn indoors after it has been exhausted through chimney flues. This re-entry problem occurs during the heating season when houses are under strong negative pressure at their base. Indeed, this negative pressure is enhanced as a result of the operation of combustion appliances such as furnaces and fireplaces. Re-entry occurs by infiltration (see Chapter 11) caused by pressure differences associated with indoor/outdoor temperature differences and wind speed. Wood smoke odor can re-enter houses through bathroom exhaust systems and chimneys with more than one flue liner.

Because of its unique odor, it is easy to detect re-entry phenomenon associated with wood smoke. Since combustion by-products associated with natural gas, propane, and oil are more difficult to detect by odor, little is known about the re-entry of flue gases from these sources. They may be inferred to be similar to those from wood smoke by analogy and from tracer gas studies which have been conducted to characterize re-entry problems in buildings.

Re-entry commonly occurs in large buildings where flue gas discharges are located close to outdoor air intakes. This is especially the case when intakes are located downwind of short boiler chimneys. Elevated levels of CO (several ppmv) have been reported in buildings associated with flue gas re-entry phenomenon.

IV. Health concerns and health effects

Combustion-generated contaminant exposures can pose a variety of acute and chronic health risks. These may include acute, potentially lethal, and sublethal effects from exposure to CO; irritation-type symptoms of the eyes and mucous membranes of the upper respiratory system; irritation of the upper airways that may cause asthmatic symptoms in individuals with asthma or contribute to the development of asthma; initiation of pulmonary symptoms and pulmonary function changes; development of lung and respiratory system cancers; and development of heart disease. Health risks depend on the nature of exposures, contaminants and their levels, and the duration of exposure.

A. Carbon monoxide

Carbon monoxide is emitted from all combustion sources. Exposures vary considerably with source type, both in terms of peak and average exposure concentrations, as well as duration. There are two major concerns: short-term exposure to relatively high concentrations that have the potential to cause death or acute illness, and chronic exposures to relatively low concentrations that may be associated with unvented combustion appliances and other circumstances.

Carbon monoxide poisoning episodes associated with malfunctioning combustion appliances and systems are not uncommon in North America. Many dozens of deaths due to CO poisoning are reported in North America each year, as are numerous other cases of near-fatal occurrences. In addi-

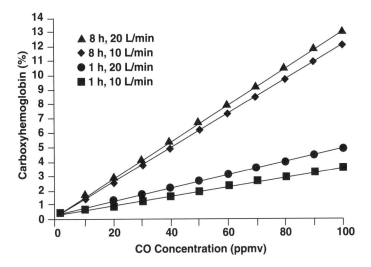

Figure 3.4 Carboxyhemoglobin levels in human blood as a function of exposure duration. (From USEPA, *EPA/600/8-90/045f*, 1991.)

tion, many more cases associated with sublethal exposures occur over a broad range of concentrations (30 to several 100 ppmv). Under these exposure conditions, individuals may experience a variety of central nervous system symptoms which are dose dependent. These may include, at relatively low concentrations (40 to 60 ppmv), headache and low levels of fatigue, and, at higher concentrations (75 to 200 ppmv), nausea, vomiting, and sleepiness.

The major immediate effect of CO is to chemically bond with hemoglobin in the blood to form carboxyhemoglobin (COHb). Carbon monoxide competes with oxygen (O_2) for hemoglobin binding sites, with an affinity for hemoglobin that is 200 times greater than that of O_2. Because it competes with O_2 for hemoglobin and is more tightly bound to it, CO can significantly reduce the amount of O_2 transported to body tissues. It also binds to intracellular proteins such as myoglobin, cytochrome oxidase, tryptophan oxidase, and dopamine hydroxylase. Such binding may cause extravascular effects.

Because brain tissue is very sensitive to changes in O_2 availability, it responds relatively quickly to diminished O_2 supplies associated with CO exposures. As a consequence, central nervous system-type effects are associated with CO exposures.

Exposure to CO can be determined by measurement of COHb in blood. This is usually <1% for individuals not exposed to CO. Among cigarette smokers, COHb concentrations vary from 3 to 8%. Blood COHb levels associated with different CO concentrations, durations, and breathing rates are illustrated in Figure 3.4. At the Occupational Safety and Health Administration's (OSHA) permissible exposure limit (PEL) of 50 ppmv 8-hour time-weighted average (TWA), an individual would have approximately 5% of his/her hemoglobin bound as COHb; at 100 ppmv it would be over 10%.

Table 3.7   Human Responses Associated with Different Blood COHb Levels

Blood COHb level	Effect
0–1%	None
2.5%	Impairment of time interval discrimination in nonsmokers.
2.8%	Onset of angina pectoris pain shortened in exercising patients; duration of pain lengthened.
3.0%	Changes in relative brightness thresholds.
4.5%	Increased reaction time to visual stimuli.
10%	Changes in performance in driving simulation.
10–20%	Headache, fatigue, dizziness, loss of coordination.

Source: From NACPA, USDHEW, *Publication No. AP-62*, Washington, D.C., 1970.

Well-documented human responses to short-term CO exposures expressed as % COHb saturation are summarized in Table 3.7.

Carbon monoxide exposures can affect individuals with cardiovascular disease. The lowest observed physiological effect level for patients with exercise-induced ischemia (tissue O_2 deficiency) is somewhere between 3 and 4% COHb. Exposure to enough CO to produce 6% COHb appears to be sufficient to cause arrhythmia in exercising patients with coronary artery disease. There is suggestive evidence that CO exposure may increase the risk of sudden death from arrhythmia in such patients. In addition, epidemiological studies indicate that a significant relationship may exist between increased CO exposure and increased mortality and cardiovascular system complaints.

Several studies have been conducted in hospital emergency rooms to evaluate the incidence of elevated COHb levels ($\geq 10\%$) in patients with influenza-like or neurologic symptoms. In the first case, 24% of patients with flu-like symptoms had blood COHb levels $\geq 10\%$. In the second case, 3% of patients reporting neurologic symptoms had elevated COHb; incidence increased to 12% when only individuals with gas-fueled heating systems were considered. These studies indicate that subacute CO poisoning is commonly misdiagnosed.

The effects of chronic exposures to relatively low levels of CO are unknown. There is evidence from animal studies to suggest that CO exposures may contribute to atherosclerosis.

The effects of CO are likely to differ among those exposed. Subpopulations at higher risk would include the unborn, infants, the elderly, individuals with preexisting disease that decreases O_2 availability, and individuals using certain medications and recreational drugs.

B. Irritants

Combustion-generated contaminants include a number of mucous membrane and upper respiratory system irritants. Most notable are aldehydes such as HCHO, and in some cases acrolein, RSP, and SO_2. Aldehydes are common irritants of the eyes, nose, throat, and sinuses. Respirable particles vary in composition. Their primary effect would be to irritate the upper respiratory passages and bronchi. Because of its solubility in tissue fluids, SO_2 (associated with some combustion appliances) can cause bronchial irritation. Irritants are described in a later section on ETS.

C. Nitrogen oxides

The nitrogen oxides include NO and NO_2. The former is a relatively nontoxic gas produced in the high-temperature reaction of nitrogen (N_2) and O_2 in combustion. Its importance lies in its relatively rapid oxidation to NO_2, a substance with considerably greater toxicity. Nitrogen dioxide is relatively nonsoluble in tissue fluids. As a consequence, it enters the lungs where it may expose lower airways and alveolar tissue.

Animal studies at 0.5 ppmv NO_2 have indicated that such exposures can cause a variety of pathological changes including the destruction of cilia lining respiratory airways, obstruction of bronchioles, and disruption of alveoli. Other animal exposure studies indicate that NO_2 exposure may inhibit respiratory defense mechanisms, allowing bacteria to invade and multiply in lung tissue.

A causal relationship between low-level NO_2 exposures and respiratory symptoms or disease is suggested from epidemiological studies that have attempted to evaluate potential health effects associated with gas cooking stoves and kerosene heaters. Results of studies with gas cooking stoves have been mixed. Some studies indicate a significantly increased risk of experiencing one or more pulmonary symptoms for children under the age of seven in homes with a gas stove; exposure to NO_2 has been suggested as the cause. However, follow-up studies have been unable to demonstrate a dose–response relationship between respiratory symptoms and NO_2 levels. In all cases, NO_2 was measured with relatively long averaging times (circa 7 days) and may have not been indicative of potential effects associated with short-term peak exposures. Many epidemiological studies have been conducted to evaluate the relationship between gas cooking (and, to a limited extent, associated NO_2 levels) and respiratory illness. They have not shown a consistent pattern between such exposures and respiratory symptoms or lung function changes in either children or adults.

However, in a relatively recent Australian study, the presence of a gas stove and exposure to NO_2 were observed to be significant risk factors for both asthma and respiratory symptoms in children. The presence of a gas stove in the household increased the likelihood of asthma by threefold and respiratory symptoms by twofold. This association between gas stoves and asthma was significant even after data were adjusted for NO_2 (NO_2 was significantly associated with asthma when the presence of a gas stove was not taken into account). These results suggest that an additional risk may be associated with gas stoves that is separate from 4-day average NO_2 exposures.

A few studies have focused on kerosene heaters. In a U.S. study, children under the age of seven exposed to average NO_2 levels >16 ppbv were reported to have more than a twofold higher risk of lower respiratory symptoms and illness (fever, chest pain, productive cough, wheeze, chest cold, bronchitis, pneumonia, or asthma) than children at lower exposures in non-kerosene stove-using homes. Increased risks of upper respiratory symptoms (sore throat, nasal congestion, dry cough, croup, or head cold) were also reported.

D. Carcinogens and cancer

By-products of combustion other than those associated with tobacco smoke may contain a variety of carcinogenic substances. Most notably these include PAHs. Many PAHs are potent carcinogens, particularly benzo-α-pyrene. Benzo-α-pyrene exposure in ambient (outdoor) air has been implicated as a risk factor for lung cancer in European epidemiological studies. Since PAH concentrations are often higher in wood smoke, exposures associated with wood-burning appliances may pose a cancer risk.

Few studies have attempted to evaluate cancer risks associated with exposure to nontobacco-related combustion by-products. Interestingly, a U.S. study designed to evaluate potential risk factors for childhood leukemia (see Chapter 4) observed a significant increase in childhood leukemia in households where incense was burned ≥ once a week during pregnancy, with a significant increase in this risk when incense was burned regularly by the mother. These risks were observed even after data were adjusted for confounding variables. Burning incense has been reported to produce benzo-α-pyrene, several other PAHs, and the nasal carcinogen, sinapaldehyde.

E. Environmental tobacco smoke

Numerous studies have been conducted to evaluate the potential health effects of exposure to ETS on nonsmokers. These have included surveys of nonsmoker complaints of irritant symptoms; controlled laboratory chamber studies of human exposures, which have focused on irritation symptoms; cross-sectional epidemiological studies attempting to evaluate risk factors for sick building (SBS)-type symptoms; epidemiological evaluations of respiratory health risks including pulmonary function changes, asthma, and lung cancer; and epidemiological studies focusing on a variety of nonrespiratory health risks.

Several survey studies have attempted to evaluate irritant effects associated with exposure to ETS. As can be seen in Table 3.8, groups of allergic and nonallergic nonsmokers reported relatively high prevalence rates for symptoms such as eye and nasal irritation, headache, and cough. Individuals with allergy reported a significantly higher prevalence of respiratory symptoms (nasal symptoms, cough, wheezing).

Studies of human exposures to ETS, for the most part, have demonstrated dose–response relationships between tobacco smoke complaints and symptoms of nasal and eye irritation. Annoyance with air quality was observed to increase with higher ETS levels. Factors apparently responsible for irritation and annoyance included acrolein, HCHO, and RSP.

Table 3.8 Symptoms Prevalence (%) in
Nonallergic and Allergic Nonsmokers
Associated with Reported Exposure to ETS

Symptom	Nonallergic	Allergic
Eye irritation	69	73
Nasal symptoms	29	67
Headache	32	46
Cough	25	46
Wheezing	4	23
Nausea	9	15
Hoarseness	4	16
Dizziness	6	5

Source: Data extracted from Speer, F., *Arch. Environ. Health*, 16, 443, 1968.

A number of cross-sectional epidemiological studies have been conducted to evaluate sick building-type symptom risk factors in office buildings. Studies that have evaluated passive smoking as a risk factor for such symptoms have consistently shown a relationship between office worker perceptions of ETS exposure and symptom prevalence. Other studies attempting to evaluate objective measurements and symptom prevalence have reported weaker relationships. Exposure to ETS may have an indirect effect: because of the subjective annoyance expressed by nonsmokers to the presence of ETS in office environments, it is likely that such annoyance may increase SBS-type symptom reporting rates.

Among the most significant concerns associated with exposure to ETS are potential respiratory and other effects associated with chronic exposures. Numerous epidemiological studies have been conducted to evaluate such health risks, particularly lung cancer.

Reports published by the Surgeon General and the National Research Council in 1986 comprehensively reviewed the scientific literature on human exposure to ETS. Both concluded that such exposure was a cause of disease in nonsmokers.

A number of epidemiological studies have linked exposure to ETS with an increased occurrence of lower respiratory tract illness during infancy and childhood. The USEPA estimates that between 150,000 and 300,000 of such cases in infants and young children annually are due to ETS exposure. Of these, between 7500 and 15,000 children are hospitalized. Among the reported illnesses are bronchitis and pneumonia, which are infectious in nature. It appears that an effect of ETS exposure is to predispose infants and children to respiratory infections (likely due to its irritant properties). The risk of respiratory illness appears to be greater in the first year of life and is closely linked with maternal smoking. Greater numbers of smokers and higher smoking rates increase the risk of respiratory illness in young children. For school-age children, parental smoking increases the risk of chest illness, the prevalence of which increases with the number of smoking parents. In addition to respiratory illness, a number of studies have shown a relationship between otitis media, an ear infection that can lead to varying degrees of hearing loss, and exposure to ETS. ETS exposure appears to increase the incidence of fluid in the middle ear, an indication of chronic middle ear disease.

Numerous studies have reported higher prevalence of common respiratory symptoms (cough, phlegm production, and wheeze) in children of smoking parents. In a study of 10,000 school children conducted in the U.S., parental smoking was associated with a 30% increase in the prevalence of persistent cough. In other studies, prevalence of chronic wheeze increased significantly as the number of smokers in the household increased.

Although wheeze is a common symptom of asthma, the relationship between exposure to ETS and asthma has been less well-established. In some studies, parental smoking has been shown to be a strong predictor of asthma in children, whereas others have failed to demonstrate a relationship. Nevertheless, USEPA has concluded that ETS is a risk factor for new cases of asthma.

Though exposure to ETS has not been established as a cause of asthma in children, there is evidence to indicate that parental smoking increases the clinical severity of asthma. These include symptom frequency, changes in lung function, and emergency room visits. The USEPA estimates that 200,000 to 1 million asthmatic children in the U.S. have their asthmatic condition worsened by exposure to ETS. Cessation of smoking by parents, or not smoking indoors, has been reported to result in a significant decline in asthmatic symptoms in children.

Lung function studies have been conducted on adults. In some studies, decreased lung function (determined by spirometry) was associated with reported ETS exposure in the home and workplace. Lung function decreased with increased smoking rates and years of exposure. Other studies, however, have reported no apparent relationship between presumed ETS exposure and adult lung function.

Numerous epidemiological studies have been conducted to evaluate a potential causal relationship between exposure to ETS and lung cancer. These have varied in design and population size evaluated. Evidence from cohort and case control studies does not uniformly establish a strong relationship between increased lung cancer risk in individuals exposed to ETS; however, most studies indicate increased risk among nonsmokers married to smokers.

Based on the biological plausibility of the association between exposure to ETS and lung cancer and supporting epidemiological evidence, the International Agency for Research on Cancer (IARC), USEPA, and OSHA have concluded that ETS is a Class A carcinogen, i.e., a proven human carcinogen.

The extent of the lung cancer hazard associated with ETS is uncertain. In 1992, USEPA published findings from a meta-analysis of 31 published studies which predicted a 19% greater risk of a nonsmoker contracting lung cancer when living with a smoker. A subsequent evaluation, which included additional studies by the California Environmental Protection Agency (Cal

EPA), came to a similar conclusion. The USEPA has projected that ETS is associated with approximately 3000 cases of lung cancer among nonsmokers in the U.S. each year.

A variety of other cancers and exposure to ETS have been studied. These include breast, cervical, and nasal cancers. Studies on both breast and cervical cancer are inconclusive, but do have a degree of biological plausibility. Based on several studies, there is an increased risk of nasal cancer in nonsmokers exposed to ETS that varies from 1.7 to 3.0 times greater than in those not exposed. This appears to be inconsistent with the apparent fact that there is no increase in nasal cancer risk in smokers.

Epidemiological evidence suggests that exposure to ETS increases the risk of cardiovascular disease in never-smokers. Studies that have controlled for confounding risk factors, such as blood cholesterol, blood pressure, and obesity, have reported an increased relative risk ranging from 1.59 to 2.01 associated with exposure to ETS. Both clinical and animal studies indicate that exposure to ETS can increase blood COHb concentrations, increase blood platelet aggregation, impair platelet function, lower high-density lipoprotein cholesterol concentration, and increase fibrinogen concentration. The projected increase of deaths from ischemic heart disease among never- and former smokers of 35,000 to 40,000 deaths per year is approximately an order of magnitude higher than that for lung cancer projections. The lifetime risk of a male never-smoker living with a current or former smoker dying from heart disease by the age of 74 has been estimated at approximately 10%; the lifetime risk for a male never-smoker living with a nonsmoker is estimated at 7%. Corresponding risks for female never-smokers were 6% and 5%, respectively.

F. Biomass cooking

Given the high exposures to combustion-generated contaminants reported above, significant adverse health effects can be expected. Indeed, the WHO estimates that biomass cooking is responsible for approximately 2.2 to 2.5 million premature deaths a year.

Exposure to cooking smoke is a known risk factor for a number of respiratory diseases. These include increased risk of acute respiratory infection, chronic obstructive lung disease, cor pulmonale, and lung cancer.

Acute respiratory illness includes infections by a number of bacteria and viruses. In developing countries, respiratory infections in children often proceed to pneumonia and death. Pneumonia is the chief cause of death in children worldwide (approximately 4.3 million deaths per year). High incidence rates occur in developing countries such as India. Based on epidemiological studies, exposure to pollutants from biomass cooking increases the risk (on the order of 2 to 3 times) of respiratory infections resulting in pneumonia.

Chronic obstructive lung disease (COLD) is commonly associated with tobacco smoking. Studies in countries such as New Guinea, India, and Nepal have led investigators to conclude that biomass cooking is a substantial risk factor for COLD, on the order of 2 to 4 times greater than for those not exposed. In Nepal, 15% of nonsmoking women who are 20 years old or older are reported to experience chronic bronchitis, a chronic obstructive lung disease. In China, COLD is associated with long-term exposure to coal smoke from cooking fires.

Cor pulmonale, a form of heart disease that develops secondary to chronic lung disease, occurs more frequently among biomass cooking fuel-using, nonsmoking women in India and Nepal. It is also reported to occur at an earlier age than is normally the case.

Exposure to biomass fuel emissions is reported to be equivalent to smoking several packs of cigarettes per day. Because of its high benzo-α-pyrene and other PAH concentrations, those exposed to high levels of biomass fuel smoke would be expected to be at high risk of developing lung cancer. Extrapolations from animal studies also suggest that lung cancer should be common among exposed individuals. Despite these facts, lung cancer among nonsmokers in biomass fuel-using areas is rare. On the other hand, greater lung cancer risk among women in China using coal for cooking has been well established.

Several studies have suggested a link between biomass cooking and increased risk of developing tuberculosis. This link is suggested to be related to suppression of pulmonary immune defense mechanisms by components of wood smoke such as benzo-α-pyrene.

Readings

Coultas, D.B. and Lambert, W.E., Carbon monoxide, in *Indoor Air Pollution: A Health Perspective*, Samet, J.M. and Spengler, J.D., Eds., Johns Hopkins University Press, Baltimore, 1991, 187.

Department of Energy, *Indoor Air Quality Environmental Information Handbook: Combustion Sources*, DOE/EV/10450-1, Washington, D.C., 1985.

Dockery, D.W., Environmental tobacco smoke and lung cancer: Environmental smoke screen, in *Indoor Air and Human Health*, 2nd ed., Gammage, R.B. and Berven, B.A., CRC Press/Lewis Publishers, Boca Raton, 1996, 309.

Marbury, M.C., Wood smoke, in *Indoor Air Pollution: A Health Perspective*, Samet, J.M. and Spengler, J.D., Eds., Johns Hopkins University Press, Baltimore, 1991, 209.

Maroni, M., Siefert, B., and Lindvall, T., *Indoor Air Quality. A Comprehensive Reference Book*, Elsevier, Amsterdam, 1995, chaps. 1, 4, 17.

National Research Council, *Environmental Tobacco Smoke: Measuring Exposures and Assessing Health Effects*, National Academy Press, Washington, D.C., 1986.

Samet, J., Nitrogen dioxide, in *Indoor Air Pollution: A Health Perspective*, Samet, J.M. and Spengler, J.D., Eds., Johns Hopkins University Press, Baltimore, 1991, 170.

Samet, J.M., Cain, W.S., and Leaderer, B.P., Environmental tobacco smoke, in *Indoor Air Pollution: A Health Perspective*, Samet, J.M. and Spengler, J.D., Eds., Johns Hopkins University Press, Baltimore, 1991, 131.

Smith, K.R., Fuel combustion, air pollution exposure, and health: the situation in developing countries, *Ann. Rev. Energy Environ.*, 18, 529, 1993.

Smith, K.R., *Biofuels, Air Pollution and Health: A Global Review*, Plenum Publishing Co., New York, 1987.

Spengler, J.D., Samet, J.M., and McCarthy, J.F., Eds., *Indoor Air Quality Handbook*, McGraw-Hill Publishers, New York, 2000, chaps. 29, 30.

USDHHS, *The Health Consequences of Involuntary Smoking. A Report of the Surgeon General*, DHHS Pub. No. (PHS) 87-8398, Washington, D.C., 1986.

USEPA, *Air Quality Criteria for Particulate Matter*, Vol. III, EPA/600/AP-95-001C, Washington, D.C., 1995.

USEPA, *Respiratory Health Effects of Passive Smoking: Lung Cancer and Other Disorders*, EPA/600/6-90/006B, Washington, D.C., 1993.

USEPA, *Air Quality Criteria for Carbon Monoxide*, EPA/600/8-90/045F, Washington, D.C., 1991.

Questions

1. What are the major contaminants produced as a result of combustion?
2. What health concerns are associated with flue gas spillage?
3. What is the major exposure concern associated with the use of wood-burning appliances?
4. How do exposures from biomass cooking compare with exposures from unvented combustion appliances?
5. What contamination/exposure concerns are associated with regular candle burning?
6. How may one be exposed to significant SO_2 levels associated with indoor combustion systems?
7. Describe major health concerns associated with exposures to environmental tobacco smoke.
8. Identify sources of polycyclic aromatic hydrocarbon (PAH) exposures indoors. What health risks may be associated with exposures to PAHs?
9. Describe emission characteristics of kerosene heaters and potential health risks associated with emission exposures.
10. Describe the nature of health concerns associated with the use of gas cooking stoves.
11. Exposure to tobacco smoke reportedly causes mucous membrane irritation. What are the likely contaminants responsible?
12. Describe exposure concerns associated with floor burnishers, forklifts, and ice-resurfacing machines.
13. What contaminant exposures are unique to environmental tobacco smoke?
14. What health risks may be associated with exposures to nitrogen oxides?
15. Significant levels of combustion-generated contaminants are measured in a building despite the fact that no apparent combustion source is present. Explain how this could be the case.
16. Describe how re-entry/entrainment phenomena affect combustion-related indoor air quality.
17. Carbon monoxide exposures are a concern with vented combustion appliances and are generally not a concern with unvented appliances. Explain why.
18. How are asthma and ETS exposure related?
19. Who is at greatest risk of harm from exposure to environmental tobacco smoke? Why?

20. Who is at greatest risk of exposure to smoke from biomass cooking? Why?
21. Irritation is a major symptom associated with combustion-generated contaminants. Describe potential causes.
22. Describe cancer and cardiovascular risks associated with exposure to ETS.
23. Describe exposure and health risks associated with biomass cooking.
24. Describe the nature of combustion-generated RSP and potential health risks.

chapter four

Organic contaminants

A large variety of natural and synthetic organic compounds can be found in indoor environments. These include very volatile organic compounds (VVOCs) which have boiling points ranging from <0°C to 50–100°C, volatile organic compounds (VOCs) with boiling points ranging from 50–100°C to 240–260°C, semivolatile organic compounds (SVOCs) with boiling points ranging from 240–260°C to 380–400°C, and solid organic compounds (POMs) with boiling points in excess of 380°C. In the last case, POMs may be components of airborne or surface dusts.

Organic compounds reported to contaminate indoor environments include a large variety of aliphatic hydrocarbons, which may be straight, branch-chained, or cyclic (contain single bonds [alkanes] or one or more double bonds [alkenes]); aromatic hydrocarbons (contain one or more benzene rings); oxygenated hydrocarbons, such as aldehydes, alcohols, ethers, ketones, esters, and acids; and halogenated hydrocarbons (primarily chlorine and fluorine containing). These may be emitted from a number of sources including building materials and furnishings, consumer products, building cleaning and maintenance materials, pest control and disinfection products, humans, office equipment, tobacco smoking, and other combustion sources.

Organic compounds which are seen as relatively distinct indoor contamination problems include the aldehydes, VOCs/SVOCs which include a large number of volatile as well as less volatile compounds, and pesticides and biocides which are, for the most part, SVOCs.

I. Aldehydes

Aldehydes belong to a class of compounds called carbonyls. Carbonyls, which include aldehydes and ketones, have the functional group

$$\overset{\displaystyle H}{\underset{\displaystyle }{-\overset{|}{C}=O}}$$

in their chemical structure. The carbonyl is in a terminal position in aldehydes. A compound is described as an aldehyde if it has one terminal carbonyl, a dialdehyde if it has two, and a trialdehyde if it has three carbonyls.

Aldehydes include saturated (single bonds) aliphatic, unsaturated (one or more double bonds) aliphatic, and aromatic or cyclic compounds. Saturated aliphatic aldehydes include formaldehyde (one carbon), acetalaldehyde (two carbons), propionaldehyde (three carbons), butryaldehyde (four carbons), valeraldehyde, glutaraldehyde, etc. As can be seen in Table 4.1, glutaraldehyde is a dialdehyde with carbonyls on both ends of the molecule. Unsaturated aliphatic aldehydes contain a carbon–carbon double bond (Table 4.1). They include acrolein (acryaldehyde), methacrolein, and crotonaldehyde. Methacrolein is commonly used to produce methyl methacrylate, an eye-irritating ester used as an adhesive in many industrial applications. Aromatic aldehydes include compounds such as benzaldehyde and cinnamaldehyde.

Table 4.1 Chemical Structures and Properties of Common Aldehydes

Compound	Structure	M. W.	Solubility (g/L)
Formaldehyde	$\overset{O}{\overset{\|\|}{H-C-H}}$	30.03	560
Acetaldehyde	$\overset{O}{\overset{\|\|}{CH_3-C-H}}$	44.05	200
Glutaraldehyde	$\overset{O\qquad\quad O}{\overset{\|\|\qquad\quad\|\|}{H-C-(CH_2)_3-C-H}}$	100.12	Miscible
Acrolein	$\overset{O}{\overset{\|\|}{CH_2=CH-C-H}}$	56.06	210
Crotonaldehyde	$\overset{O}{\overset{\|\|}{CH_3-CH=CH-C-H}}$	76.09	181
Benzaldehyde	$\bigcirc\!\!-\overset{O}{\overset{\|\|}{C}}-H$	106.11	3

Source: From Leikauf, G.D., in *Environmental Toxicants: Human Exposures and their Health Effects*, Lippman, M., Ed., Van Nostrand Reinhold (John Wiley & Sons), New York, 1992, chap. 2. With permission.

Individual aldehydes differ in their molecular structure, solubility, chemical reactivity, and toxicity. Only a relative few have industrial and commercial applications which may result in significant indoor exposures, are byproducts of other processes, or have biological activities that have the potential for posing major public health concerns. Those which, at present, are known to cause either significant indoor air contamination and/or adverse health effects include formaldehyde (HCHO), acetaldehyde, acrolein, and glutaraldehyde. Many aldehydes are potent sensory (mucous membrane) irritants; some are skin sensitizers; and there is limited evidence that several aldehydes may be human carcinogens.

A. Sensory irritation

Because of their solubility in aqueous media and their high chemical activity, aldehydes as a group are potent mucous membrane irritants (affecting eyes and mucous membranes of the upper respiratory tract). This irritation is associated with maxillary and ophthalmic divisions of trigeminal nerves in nasal and other mucosa which respond to chemical/physical stimuli. These serve as respiratory defense mechanisms through the perception of pain or irritation and reduced contaminant inhalation.

Measured decreases in respiratory rates in rats and mice on exposure to irritant chemicals have been used to evaluate the irritation potential of aldehydes and other substances using a standard mouse bioassay. Doses required to cause a reduction of breathing rates by 50% (RD_{50}) for selected aldehydes are summarized for mouse bioassays in Table 4.2. As can be seen, RD_{50} values range by more than three orders of magnitude. Formaldehyde and the unsat-

Table 4.2 RD_{50} Values for Swiss-Webster Mice Exposed to Aldehydes

Chemical	RD_{50} value (ppmv)
Formaldehyde	3.2
Acetaldehyde	2845
Propionaldehyde	2052
Acrolein	1.03
Butryaldehyde	1015
Isobutryaldehyde	4167
Crotonaldehyde	3.53
Valeraldehyde	1121
Isovaleraldehyde	1008
Caproaldehyde	1029
2-Ethybutryaldehyde	843
2-Furaldehyde	287
Cyclohexane carboxaldehyde	186
3-Cyclohexane-1-carboxaldehyde	95
Benzaldehyde	333

Source: From Steinhagen, W.H. and Barrow, C., *Toxicol. Appl. Pharmacol.*, 72, 495, 1982. With permission.

urated aldehydes, acrolein and crotonaldehyde, were the most potent sensory irritants; the saturated aldehydes, acetaldehyde and propionaldehyde, were least potent. These data, which do not include glutaraldehyde (a potent irritant in rats), indicate that only a few aldehydes have the potential to be significant mucous membrane irritants at the relatively low concentrations that occur in indoor environments, such as residences and nonresidential, nonindustrial buildings. Of these HCHO, acrolein, and glutaraldehyde are the most notable. Though a relatively weak sensory irritant, acetaldehyde is a common contaminant of both indoor and ambient (outdoor) air.

B. Formaldehyde

Formaldehyde is molecularly the smallest and simplest aldehyde. It is unique because the carbonyl is attached directly to two hydrogen atoms (Table 4.1). Due to its molecular structure, HCHO is highly reactive chemically and photochemically. It has good thermal stability relative to other carbonyls and has the ability to undergo a variety of chemical reactions, which makes it useful in industrial and commercial processes. As a consequence it is among the top 10 organic chemical feedstocks used in the U.S.

Formaldehyde is a colorless, gaseous substance with a strong, pungent odor. On condensing, it forms a liquid with a high vapor pressure (boiling at $-19°C$). Because of its high reactivity, it rapidly polymerizes with itself to form paraformaldehyde. As a consequence, liquid HCHO must be held at low temperature or mixed with a stabilizer (such as methanol) to prevent/minimize polymerization.

1. Uses/sources

Formaldehyde is commercially available as paraformaldehyde, which contains varying lengths of polymerized HCHO molecules. It is a colorless solid that slowly decomposes and vaporizes as monomeric HCHO at room temperature. It has been used in a variety of deodorizing commercial products, such as lavatory and carpet preparations.

Formaldehyde is also commercially available as formalin, an aqueous solution that typically contains 37 to 38% HCHO by weight and 6 to 15% methanol. Because of HCHO's volatility, formalin also has a strong, pungent odor. In solution it is present as methylene glycol ($CH_2(OH)_2$); in concentrated solutions it is in the form of polyoxymethylene glycol ($HO-CH_2O)_n$-H).

As a chemical feedstock, HCHO is used in many different chemical processes. Of particular significance to indoor environments is its use to produce urea and phenol–formaldehyde resins (50% of HCHO consumed annually).

Urea–formaldehyde (UF) copolymeric resins are used as wood adhesives in the manufacture of pressed-wood products such as particle board, medium-density fiber board (MDF) and hardwood plywood, finish coatings (acid-cured), textile treatments (permanent-press finishes), and in the production of urea–formaldehyde foam insulation (UFFI). Urea–formaldehyde wood adhesives are colorless and provide excellent bonding performance. They are, however, somewhat chemically unstable, releasing monomeric HCHO on hydrolysis of methylol end groups and, less commonly, methylene bridges. Their decomposition is sensitive to product moisture levels as well as relative humidity. Because of resin sensitivity to moisture, UF-bonded wood products are intended only for indoor use. Historically UF-based adhesives were formulated with relatively high HCHO to urea ratios (F:U 1.5:1) to enhance performance by ensuring that there was sufficient HCHO present to achieve cross-linking of all primary and secondary amino groups. Because of this excess HCHO associated with the resin, UF-based wood adhesives emitted significant levels of free HCHO into indoor environments, particularly in the first months or so in the life of a product. Because of high HCHO emissions, indoor concentrations, and health complaints, UF-bonded wood products are presently manufactured with low F:U ratios (e.g., 1.05:1) and thus emit much less HCHO. Though HCHO emissions from UF-bonded wood products are substantially lower than those of two decades ago (<10%), they continue to be a significant source of indoor HCHO concentrations. Most emissions are associated with the hydrolytic decomposition of the resin copolymer.

Phenol–formaldehyde resins receive significant use as exterior-grade adhesives in the manufacture of softwood plywood and oriented-strand board (OSB) products that are widely used in new home construction. Phenol–formaldehyde (PF)-bonded wood products have historically had low HCHO emissions compared to UF-bonded wood products. Emissions from the latter were once 1000 times greater than from PF-bonded products.

Formaldehyde is produced in the thermal oxidation of a variety of organic materials. As a consequence, it is found in the emissions of motor vehicles, combustion appliances, wood fires, and tobacco smoke. It is also produced in the atmosphere as a consequence of photochemical reactions and hydrocarbon scavenging processes, and in indoor air as a result of chemical reactions.

2. Exposures

Formaldehyde is omnipresent in both ambient and indoor environments. Ambient concentrations are usually <10 ppbv in urban/suburban locations but may reach peak levels of 50 ppbv or more in urban areas subject to significant atmospheric photochemistry (e.g., south coast of California). Formaldehyde levels in indoor environments are on average significantly higher (order of magnitude or more) in residential, institutional, and commercial buildings than background ambient levels. Concentrations vary from structure to structure, depending on the nature of sources present and environmental factors which may affect emissions and indoor concentrations.

Historically, the major sources of HCHO emissions have been wood products bonded with UF resins, UF-based acid-cured finishes, and in houses retrofit insulated (in the 1970s and early 1980s) with UFFI. Formaldehyde

Table 4.3   Formaldehyde Emissions from Construction Materials,
Furnishings, and Consumer Products

Product	Emission rate range ($\mu g/m^2/day$)
Medium-density fiberboard	17,600–55,000
Hardwood plywood paneling	1500–34,000
Particle board	2000–25,000
UFFI	1200–19,200
Softwood plywood	240–720
Paper products	260–280
Fiberglass products	400–470
Clothing	35–570

Source: Data extracted from Pickrell, J.A. et al., *Environ. Sci. Tech.*, 17, 753, 1983; Matthews, T.G., *CPSC-IAG-84-1103*, Consumer Product Safety Commission, Washington, D.C., 1984; and Grot, R.A. et al., *NBSIR 85-3225*, National Bureau of Standards, Washington, D.C., 1985.

emission rates from a variety of construction materials and consumer products available in the early 1980s marketplace are summarized in Table 4.3.

Pressed wood products have been the major source of HCHO contamination in indoor environments. Particle board has been used as underlayment in conventional homes; floor decking in manufactured homes; components of cabinetry, furniture, and a variety of consumer products; and as a decorative wall paneling. Because of marketplace changes, it is now little used as underlayment in conventional houses, and fewer than 50% of new manufactured homes are constructed with particle board floor decking. Hardwood plywood has been used as a decorative wall covering and as a component in cabinets, furniture, and wood doors. Medium-density fiber board has been used in cabinet, furniture, and wood door manufacture. Acid-cured finishes, which often contain a mixture of urea and melamine–formaldehyde resins, are used as finish coatings on exterior wood cabinet components, fine wood furniture, and hardwood flooring.

Urea–formaldehyde foam insulation or similar products are occasionally used to retrofit insulate houses in North America and are commonly used in the United Kingdom. Prior to a ban by the Consumer Product Safety Commission (CPSC) in the U.S. (subsequently voided in a federal appeals court) and a ban by the Canadian government in the early 1980s, UFFI was applied in over 500,000 U.S. residences and 80,000 in Canada.

Formaldehyde concentrations in U.S. residences based on data collected in the late '70s to mid '80s are summarized in Table 4.4. As indicated previously, significant improvements (reduced emission rates) in UF-bonded wood products have occurred since the mid '80s, and there have been changes in products used in construction. As a consequence, HCHO levels in building environments (particularly residences) are significantly lower in houses built since 1990 than in those constructed previously. Formaldehyde levels in new mobile homes are rarely >0.20 ppmv, and are more likely to be in the range of 0.05 to 0.15 ppmv. In other new residential buildings

Table 4.4 Formaldehyde Concentrations in U.S. Houses
Measured in the Period 1978–1989

Study	N	Concentration (ppmv)		
		Range	Mean	Median
Urea-Formaldehyde Foam Insulated Houses				
New Hampshire	71	0.01–0.17	—	0.05
Consumer Product				
Safety Commission	636	0.01–3.4	0.12	—
Manufactured Houses				
Washington	74	0.03–2.54	—	0.35
Wisconsin	137	<0.10–2.84	—	0.39
California	663	—	0.09	0.07
Indiana	54	0.02–0.75	0.18	0.15
Conventional Houses				
Texas	45	0.0–0.14	0.05	—
Minnesota	489	0.01–5.52	0.14	—
Indiana (particle board underlayment)	30	0.01–0.46	0.11	0.09
California	51	0.01–0.04	0.04	—

Source: From Godish, T.J., *Indoor Air Pollution Control*, 1st ed., Lewis Publishers, Chelsea, MI, 1989. With permission.

constructed in the U.S. and Canada, concentrations are unlikely to exceed 0.10 ppmv, with concentrations <0.05 ppmv the norm. In office buildings, HCHO levels are rarely >0.05 ppmv, with concentrations in the range of 0.02 to 0.03 more common.

3. Factors affecting formaldehyde levels

Formaldehyde levels in building environments are affected by a number of factors. These include the potency of formaldehyde-emitting products present, the loading factor (m^2/m^3), which is described by the surface area (m^2) of formaldehyde-emitting materials relative to the volume (m^3) of interior spaces, environmental factors, materials/product age, interaction effects, and ventilation conditions.

As indicated in Table 4.3, formaldehyde-emitting materials have historically differed in their emission potential. These differences have decreased with product improvements. Medium-density fiber board and acid-cured finishes have been among the most potent formaldehyde-emitting materials.

a. Loading factor. Mobile homes have had the highest reported concentrations of HCHO. This has been the case in good measure because of the high loading rate of formaldehyde-emitting wood products. In the past, mobile homes were constructed using particle board floor decking, Luan plywood wall covering, and wood cabinets (made from various combina-

Table 4.5 Effect of Temperature and Relative Humidity on Formaldehyde Levels in a Mobile Home Under Controlled Environmental Conditions

Temperature (°C)	Relative humidity (%)	Concentration (ppmv)	% maximum value
30	70	0.36	100
25	70	0.29	81
30	50	0.28	78
30	30	0.23	64
25	50	0.17	47
25	30	0.14	39
20	70	0.12	33
20	50	0.09	25
20	30	0.07	19

tions of particle board, MDF, Luan plywood, and hardwood). As a consequence, they had relatively high surface/volume ratios (expressed as m²/m³) of formaldehyde-emitting wood products.

   b. Temperature and relative humidity. Environmental factors such as temperature and humidity have significant effects on HCHO levels in buildings where UF-bonded wood products are major HCHO sources. The effects of temperature on indoor HCHO concentrations is exponential, whereas the effect of relative humidity is linear. Combined effects of various temperature and humidity regimes on HCHO levels in a mobile home can be seen in Table 4.5. Note that the highest combination of temperature and humidity (30°C, 70% RH) resulted in indoor concentrations that were 5 times greater than the lowest combination (20°C, 30% RH).

   Experimentally derived relationships between HCHO levels and temperature and HCHO levels and relative humidity have been used to develop equations to "correct" (or more appropriately, standardize) HCHO levels determined under different environmental conditions to temperature and humidity conditions such as 25°C and 50% RH. The Berge equation is widely used to standardize HCHO concentrations. It has the following form:

$$C_X = \frac{C}{[1 + A(H-H_O)e^{-R(1/T-1/T_O)}]} \tag{4.1}$$

where C_X = standardized concentration (ppmv)

   C = measured concentration (ppmv)

   e = natural log base 2.7818

   R = coefficient of temperature (9799)

   T = temperature at test (°K)

   T_O = standardized temperature (°K)

   A = coefficient of humidity (0.0175)

   H = relative humidity at test (%)

   H_O = standardized relative humidity (%)

The Berge equation is a relatively good predictor of HCHO concentration at standard conditions when measured under a variety of environmental conditions. It has been reported to have a standard error of ±12% within a 95% confidence level.

   c. Decrease in formaldehyde levels with time. Formaldehyde levels decrease significantly with time. A generalized relationship between HCHO levels and product or home age with time can be seen in Figure 4.1. Rapid reductions of HCHO levels can be seen to occur in the early life of formal-dehyde-contaminated residences or emitting products. After an initial rapid decline, HCHO levels decrease at a much slower rate, with relatively elevated levels continuing for years.

   Several investigators have attempted to model changes in HCHO levels with time, using exponential model equations as well as statistical analyses. Exponential models that describe the decay of radioactive isotopes as well as first-order chemical reactions predict a constant half-life. Studies of field data indicate that HCHO decreases with time are exponential only in part, with half-lives that increase in time. Statistical and graphical evaluations of Wisconsin mobile homes tested for HCHO in the early 1980s indicated half-life values of 3, 5, 12, and 72 months.

   Because HCHO levels depend on a variety of source and environmental variables, it is unlikely that a model equation could be developed that would reliably predict the decay of HCHO levels under the many source and environmental conditions that have and continue to exist in North American residences. However, double exponential models have been shown to be relatively good predictors of changing HCHO levels with time (see Chapter 9).

   The decay rate is dependent on emission rates that are affected by temperature, relative humidity, and interaction effects between formaldehyde-emitting materials, as well as ventilation rates (increasing temperature, humidity, and ventilation rates increase emission rates and, as a consequence, decay rate). Therefore, half-lives would be expected to be shortened

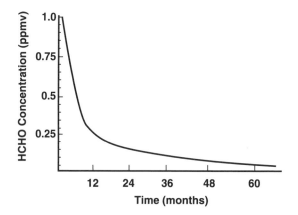

Figure 4.1   Generalized decrease of formaldehyde with time.

(HCHO levels would decline more rapidly) with increasing temperature, relative humidity, and ventilation. Interaction effects described below would, if all other factors were standardized, likely increase the time period required for a 50% reduction in HCHO levels.

d. Interaction effects. In building environments that contain multiple UF-based emission sources, measured concentrations are typically similar or are slightly above sources with the highest emission potential present (most potent source) rather than the sum of emissions/emission potentials of all formaldehyde-emitting sources. Such interaction effects are due to a vapor pressure phenomenon. High vapor concentrations associated with emissions from potent sources suppress emissions from less potent sources.

e. Ventilation. Ventilation associated with infiltration, opening windows, and mechanical induction can affect indoor concentrations of HCHO as well as emission rates. Formaldehyde levels decrease with increasing ventilation rates. The relationship is not linear because a doubling in the ventilation rate is associated with only a 30 to 35% decline in HCHO levels (due to increased emission rates). Natural ventilation associated with infiltration appears to have a significant effect on HCHO levels. Under controlled conditions, HCHO levels reach their maximum values when indoor/outdoor temperature differences are small. Lowest HCHO levels in northern climates are observed during the cold season, especially on cold winter days (Figure 4.2) when indoor/outdoor temperature differences are large.

f. Tobacco smoke. Since HCHO is a by-product of combustion processes, smokers, as can be expected, are exposed to high HCHO levels (on the order of 40 to 250 ppmv in a single puff). Formaldehyde emissions from burning cigarettes are indicated in Table 4.6. Nonsmokers are exposed to significantly lower levels from environmental tobacco smoke (ETS) because of the significant dilution effects that occur. Because of interaction effects

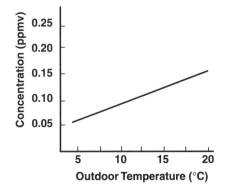

Figure 4.2 Relationship between indoor formaldehyde levels and outdoor temperatures.

Table 4.6 Aldehydes in Cigarette Smoke

Aldehyde	Emission (mg/pk)		
	Mainstream	Sidestream	ETS[a]
Formaldehyde	3.4	14.5	1.3
Acetaldehyde	12.5	84.7	3.2
Acrolein	1.5	25.2	0.6

[a] Environmental tobacco smoke integrated over 2 hours.

Source: From Leikauf, G.D., in *Environmental Toxicants: Human Exposures and Their Health Effects*, Lippmann, M., Ed., Van Nostrand Reinhold (John Wiley & Sons), New York, 1992, chap. 10. With permission.

between ETS, HCHO, and UF-based HCHO, the effect of tobacco smoke on indoor concentrations would be small (because of the modulating effect of UF-based sources).

4. Health effects

a. Mucous membrane irritation and neurotoxic effects. The effects of HCHO on human health have been extensively investigated in human exposure studies and field and epidemiological studies of workplace and residential buildings. Major health concerns have included mucous membrane irritation, neurological-type symptoms, potential sensitization, and upper respiratory system cancers.

The ability of HCHO to cause irritation or inflammatory-type symptoms and symptoms of the central nervous system (e.g., headache, fatigue) is known from controlled animal studies, reports of occupational exposures, field investigations, and epidemiological studies of humans exposed to HCHO in residential environments.

Controlled human exposure studies at concentrations in the range of those reported in residential environments (≥ 0.03 ≤ 1.0 ppmv) have been shown to cause significant eye and nose irritation among healthy volunteers in 90-minute exposures compared to unexposed controls. In 5-hour human exposure studies, significant decreases in nasal mucus flow were observed at 0.25 ppmv in healthy volunteers, with slight to significant discomfort and eye and throat dryness increasing in frequency with HCHO concentrations above 0.25 ppmv.

Elevated prevalence rates of mucous membrane symptoms, central nervous system symptoms (headache, fatigue, sleeplessness), as well as nausea, diarrhea, unnatural thirst, and menstrual irregularities have been reported in field investigations of HCHO exposures in residential and nonindustrial workplaces. The author observed significant dose–response relationships between HCHO levels and symptom severity for 14 symptoms/health problems including eye irritation, dry/sore throat, runny nose, bloody nose, sinus irritation, sinus infection, cough, headache, fatigue, depression, difficulty sleeping, nausea, diarrhea, chest and abdominal pain, and rashes for a range of concentrations with a median value of 0.09 ppmv in a study of mobile homes and homes with particle board underlayment. Studies in Canada and California indicate significant dose-dependent symptoms associated with HCHO exposure concentrations of <0.10 ppmv.

  b. Asthma.   Asthmatic reactions to HCHO and asthmatic/pulmonary-type symptoms have been reported in field investigations of occupational and residential HCHO exposures. These studies suggest that HCHO may be a pulmonary irritant and may be capable of inducing asthmatic attacks by specific sensitization reactions. Controlled bronchial challenge studies (which are the standard for confirming the induction of asthmatic symptoms on exposure to a substance) of asthmatic patients in the U.S. and Canada have been unable to identify any changes in pulmonary function at exposure concentrations of 2 to 3 ppmv. Less well-controlled bronchoprovocation tests in Europe indicate that formaldehyde-induced pulmonary function changes in exposed workers occur but that such responses are rare. Significantly increased prevalence rates of asthma and chronic bronchitis among children in homes with HCHO levels in the range of 0.06 to 0.12 ppmv (especially in homes with ETS) compared to those less exposed were observed in an epidemiological study of HCHO concentrations, respiratory symptoms, and pulmonary function conducted in Arizona. A linear decrease in peak expiratory flow rates with HCHO levels was observed with an estimated decrement of 22% at 0.06 ppmv. This study is significant since it apparently shows a strong dose–response relationship between pulmonary symptoms and HCHO concentrations in the range of ≥0.06 to 0.12 ppmv; levels which are still likely to occur in U.S. mobile homes. Also notable is that pulmonary function decrements were observed in both asthmatic and nonasthmatic children

  The induction of asthma or asthmatic symptoms may occur as a consequence of a specific hypersensitivity reaction or a response to a nonspecific irritant. The former suggests that an immunological mechanism is responsible. Though specific IgE antibodies for HCHO have not been demonstrated, several investigators have shown that some formaldehyde-exposed individuals develop antibodies against formaldehyde–human serum albumin conjugates. The clinical significance of these findings has not been established.

  c. Cancer.   Formaldehyde has been shown to cause a variety of genotoxic effects in cell culture and *in vitro* assays. These include DNA-protein cross-links, sister chromatid exchange, mutations, single-strand breaks, and aberrations in chromosomes. These results indicate that HCHO is both genotoxic and mutagenic and is therefore likely to be carcinogenic. Several high-concentration, chronic animal exposure studies have demonstrated that HCHO can cause squamous cell carcinomas in the nasal passages of rats and mice.

  A number of epidemiological studies have attempted to evaluate the potential relationship between HCHO exposure and upper respiratory system cancer. These studies have not provided conclusive evidence of a causal relationship. Formaldehyde exposures have been reported to be associated with slight to moderate increases in risk for cancers of the buccal cavity, nasopharynx, oropharynx, and lung. Studies of residents of mobile homes have shown a significant increase in the risk of nasopharyngeal cancer in individuals living in mobile homes (at exposure concentrations ≥0.10 ppmv) for more than 10 years (compared to a randomly chosen population). Based on available evidence, the USEPA, Occupational Safety and Health Administration (OSHA), and International Agency for Research on Cancer (IARC) have listed formaldehyde as a Class 2A (suspected human) carcinogen.

C. Acetaldehyde

Acetaldehyde is a two-carbon aliphatic aldehyde with a pungent, fruity odor. Though it is used in a variety of industrial processes, its presence in ambient and indoor air is almost always associated with the combustion oxidation of fuels and products such as tobacco (Table 4.6). It is a major constituent of automobile exhaust gases and is the predominant aldehyde found in tobacco smoke.

Exposures to acetaldehyde (in indoor environments) are likely to occur from the infiltration of ambient air, ETS, combustion by-products from unvented gas and kerosene appliances, flue gas spillage, leakage from wood-burning appliances, and, in developing countries, unvented by-products of wood, charcoal, and kerosene cooking fuels.

Exposures to acetaldehyde in indoor environments have not been characterized. Compared to HCHO, it is a relatively mild irritant of the eye and upper respiratory system. It is unlikely to cause irritant symptoms in most indoor environment situations because of anticipated low exposure concentrations. The exception to this may be cooking fires in developing countries. Acetaldehyde is a proven animal carcinogen and, as such, is a potential carcinogen in humans.

D. Acrolein

Acrolein is a three-carbon aldehyde with one double bond. It is highly volatile and has an unpleasant choking odor. It is used in the production of a number of compounds and products. It is released into the environment as a combustion oxidation product from oils and fats (containing glycerol), wood, tobacco, and automobile/diesel fuels.

There are few published reports of acrolein in indoor air. In an apparently unusual case, the author has observed significant acrolein levels (>0.1 ppm, the OSHA PEL) in school administrative office spaces associated with the application of a polyurethane insulation/rubberized roofing material.

Acrolein emissions and potential human exposures have been reported for tobacco smoke (Table 4.6). They also occur in the ambient and indoor environments with open windows, as a result of motor vehicle emissions and atmospheric photochemistry. Acrolein levels, like other aldehydes, peak at midday (on sunny days). Acrolein is present in wood smoke in significant quantities and, as a consequence, exposures may result from leaking wood-burning appliances and from cooking fires in developing countries.

Acrolein is a potent eye irritant causing lacrimation (tearing) at relatively low exposure concentrations. It is widely believed that acrolein exposures are the primary cause of eye irritation associated with tobacco smoke. It may also be the major cause of eye irritation associated with wood smoke. Though acrolein has been reported to be both geno- and cytotoxic, it has not been shown to be carcinogenic in animal studies, nor has there been any epidemiological link between acrolein exposure and human cancer.

E. Glutaraldehyde

Glutaraldehyde is a five-carbon dialdehyde. It is a liquid with a sharp, fruity odor. Compared to the aldehydes described above, it has a low vapor pressure (0.20 mm Hg at 20°C) and thus volatilizes slowly.

Glutaraldehyde is the active ingredient in disinfectant formulations widely used in medical and dental practice. Significant or incidental exposures have been reported among hospital, medical service, dental, veterinary, and funeral service staff. Exposure concentrations in hospital environments have been reported to range from 0.001 to 0.49 ppmv. It has also been reported to occur in some carbonless copy papers and is used as a biocide in duct-cleaning services.

Potential human health effects associated with glutaraldehyde exposures have been reported (case reports, field investigations, limited epidemiological studies, challenge, and animal studies). These have included irritant symptoms of the nose and throat and other symptoms such as nausea and headache. They have also included pulmonary symptoms such as chest tightening and asthma. Other exposure concerns include reproductive effects and cancer. Limited studies suggest that exposure to glutaraldehyde increases the risk of spontaneous abortion among pregnant females. Cancer risks have been inferred from chemical reactivities and mutagenic activity of similar compounds. However, there is no direct evidence that glutaraldehyde is either an animal or human carcinogen.

II. VOCs/SVOCs

A number of studies have been conducted to both identify and quantify VOCs in indoor environments. Despite inherent difficulties in sampling and identifying VOCs present in mixtures at low concentrations, available evidence indicates that a large and variable number of VOCs are present in indoor air. VOC compounds vary widely from building to building, depending on sources present as well as human activities. Studies conducted to date have reported the presence of many different aliphatic and aromatic hydrocarbons and hydrocarbon derivatives such as oxygenated and halogenated hydrocarbons.

The aldehydes are a group of VOCs that have received special attention in indoor environments because of their irritant effects at relatively low concentrations. The aldehydes, however, represent only a fraction of a larger group of organic compounds found to contaminate air in residential and nonresidential indoor environments. Concentrations of individual VOCs and SVOCs as described below are very low and, therefore, it is generally believed that exposure to any individual VOC (other than HCHO) poses little or no risk of causing acute symptoms in building occupants. Nevertheless, because of the large number of VOCs, and to a lesser extent SVOCs, present, health concerns have been expressed relative to the collective effect (i.e., additive, synergistic) of exposures to a large number of substances, as well as substances which are potentially carcinogenic in humans.

A. VOCs in residential buildings

VOCs detected and quantified in U.S. residences include aromatic hydrocarbons such as toluene, ethyl benzene, *m*- and *p*- isomers of xylene, naphthalene, and methylnapthalene, alkanes such as nonane, decane, undecane, dodecane, tridecane, tetradecane, pentadecane, and hexadecane, as well as compounds such as mesithylene, cumene, limonene, and benzaldehyde. Toluene was observed in the highest concentrations (45 to 160 μg/m³). With the exceptions of toluene, xylene, and benzaldehyde, average concentrations of other individual VOCs were <20 μg/m³.

Volatile organic compounds and their levels have been characterized in hundreds of randomly selected German houses. The 57 individual VOCs identified included straight (*n*-), branch-chained, and cyclic alkanes, aromatic hydrocarbons, chlorinated hydrocarbons, terpenes, carbonyls, and alcohols. Concentration ranges of individual compounds varied by three orders of magnitude. With the exception of toluene, mean concentrations were <25 μg/m³, with most <10 μg/m³ (low ppbv range expressed as mixing ratios). The average sum of identified VOCs was approximately 0.4 mg/m³, with a range of 0.07 to 2.67 mg/m³. In a study of new and older occupied apartments, Danish investigators reported summed VOC concentrations in newer apartment units to average 6.2 mg/m³ and 0.4 mg/m³ in older units.

B. VOCs in nonresidential buildings

A number of investigations have been conducted to characterize VOCs in nonresidential buildings such as offices, schools, and other institutional environments. Concentrations of individual VOCs in California office buildings have been reported to range from 3 to 319 μg/m³. Major identified compounds included straight, branch-chained, and cyclic alkanes, followed by a variety of aromatic hydrocarbons, with toluene the most common and abundant VOC. Chlorinated hydrocarbons, such as tetrachloroethylene, 1,1,1-trichloroethane, and trichloroethylene, were also commonly measured.

Table 4.7   Frequency of Detected VOCs/SVOCs in the USEPA BASE Study
of 56 Randomly Selected Office Buildings

VOCs		
Frequency = 81 to 100%		
Acetone	2-Butanone	Toluene
Styrene	*m*- and *p*-Xylene	TXIB
n-Undecane	4-Ethyltoluene	*n*-Dodecane
2-Butoxyethanol	Nononal	2-Ethyl-1-hexanol
n-Decane	Nonane	*o*-Xylene
Octane	α-Limonene	Butyl acetate
Benzene	*n*-Hexane	1,1,1-Trichloroethane
Pentanal	Hexanal	1,3,5-Trimethylbenzene
Ethylbenzene	α-Pinene	1,2,4-Trimethylbenzene
Texanol 1 and 3	Tetrachloroethane	4-Methyl-2-pentanone
Phenol	Naphthalene	
	Frequency = 61 to 80%	
1,4-Dichlorobenzene		
	Frequency = 41 to 60%	
3-Methyl pentane	Trichloroethane	
	Frequency = 21 to 40%	
Methylene chloride	Trichlorofluoromethane	*t*-Butyl methyl ester
	Frequency = 1 to 20%	
Trichloro-trifluoroester	Chloroform	Carbon tetrachloride
4-Phenycyclohexene	Carbon disulfide	Chlorobenzene
1,2,4-Trichlorobenzene	1,2-Dichlorobenzene	

Source: From Girman, J.R. et al., *Proc. Indoor Air '99*, Edinburgh, 2, 460, 1999.

In qualitative and quantitative analyses of air samples collected from public access buildings, USEPA identified hundreds of aliphatic hydrocarbons as well as 200 or so other organic compounds. These included aromatic hydrocarbons, halogenated hydrocarbons, esters, alcohols, phenols, ethers, ketones, aldehydes, and epoxides. Concentrations in new buildings were generally much higher than in established or older buildings, often by an order of magnitude or more.

As a part of its BASE (Building Assessment Survey Evaluation) study, USEPA conducted VOC/SVOC measurements in 56 randomly selected private office buildings. The frequency of detected VOCs/SVOCs is summarized in Table 4.7. Thirty-four VOCs were detected in 81+% of samples collected. Concentrations of 12 VOCs observed at the highest concentrations are summarized in Table 4.8. The highest median concentrations were observed for acetone, toluene, d-limonene, xylenes, 2-butoxyethanol, and *n*-undecane. With the exception of d-limonene (used as a fragrance), these substances are commonly used as solvents and tend to be in the highest concentration range.

Table 4.8 Median and Maximum Concentrations of 12 VOCs Observed at
the Highest Concentrations in the USEPA BASE Study of 56 Randomly
Selected Office Buildings

VOC	Median concentration ($\mu g/m^3$)	Maximum concentration ($\mu g/m^3$)
Acetone	29.0	220.0
Toluene	9.0	360.0
d-Limonene	>0.1	140.0
m- and *p*-Xylene	5.2	96.0
2-Butoxyethanol	4.5	78.0
n-Undecane	3.7	58.0
Benzene	3.7	17.0
1,1,1-Trichloromethane	3.6	450.0
n-Dodecane	3.5	72.0
Hexanal	3.2	12.0
Nonanal	3.1	7.9
n-Hexane	2.9	21.0

Source: From Girman, J.R. et al., *Proc. Indoor Air '99*, Edinburgh, 2, 460, 1999.

Mean (and median) concentrations of VOCs of individual compounds
in nonresidential buildings tend to be below 50 $\mu g/m^3$, with most below
5 $\mu g/m^3$. Volatile organic compounds in nonresidential buildings are fewer
in number than those found in residences but are qualitatively similar. Concentrations tend to be lower (by as much as 50% or more) both on an
individual basis and TVOC (total volatile organic compounds) levels.

C. Sources/emissions

Investigators in North America and Europe have attempted to characterize
emissions of VOCs from building materials and consumer products. VOC
emissions from such products/materials vary with the product as well as
with its manufacturer. Major VOCs emitted from a sample of products are
identified in Table 4.9.

Most compounds listed in Table 4.9 are nonpolar substances. Polar substances, on the other hand, are commonly used in a variety of personal and
home-care products such as colognes, perfumes, deodorants, soaps, detergents, shampoos, air fresheners, and cosmetics. In a study of these materials,
USEPA observed that polar compounds such as ethanol, benzaldehyde, α-
terpineol, and benzyl acetate were found in over 50% of 31 products tested.
In an intensive study of five products (two colognes, one perfume, one soap,
and one air freshener), USEPA confirmed the presence of the above four
compounds as well as α-pinene, β-pinene, camphene, myacene, limonene,
diethylene glycol monomethyl ether, linalool, β-phenethyl alcohol, benzyl
alcohol, estragole, and menethyl acetate in one or more products, with
limonene and linalool found in all five products tested.

Table 4.9 VOCs Emitted from Building Materials and Consumer Products

Material/product	Major VOCs identified
Latex caulk	Methylethylketone, Butyl propionate, 2-Butoxyethanol, Butanol, Benzene, Toluene
Floor adhesive	Nonane, Decane, Undecane, Dimethyloctane, 2-Methylnonane, Dimethylbenzene
Particle board	Formaldehyde, Acetone, Hexanal, Propanol, Butanone, Benzaldehyde, Benzene
Moth crystals	p-Dichlorobenzene
Floor wax	Nonane, Decane, Undecane, Dimethyloctane, Trimethylcyclohexane, Ethylmethylbenzene
Wood stain	Nonane, Decane, Undecane, Methyloctane, Dimethylnonane, Trimethylbenzene
Latex paint	2-Propanol, Butanone, Ethylbenzene, Propylbenzene, 1,1-Oxybisbutane, Butylpropionate, Toluene
Furniture polish	Trimethylpentane, Dimethylhexane, Trimethylhexane, Trimethylheptane, Ethylbenzene, Limonene
Polyurethane floor finish	Nonane, Decane, Undecane, Butanone, Ethylbenzene, Dimethylbenzene
Room freshener	Nonane, Decane, Undecane, Ethylheptane, Limonene, substituted aromatics (fragrances)

Source: From Tichenor, B.A., *Proc. 4th Internatl. Conf. Indoor Air Qual. Climate*, Berlin, 1, 8, 1987.

As is the case with HCHO, source emissions decrease rapidly with time. Higher temperature and ventilation rates significantly increase emission rates and thus accelerate decay rates. Decay rates of individual compounds decrease very rapidly with time, as can be seen with a wet (freshly applied finish coating) and dry (carpeting) product (Figures 4.3 and 4.4).

Time-dependent decreases in VOC concentrations have been reported in source emission studies. In newly built Swedish preschool buildings, more than 50% of 160 different compounds initially identified were undetectable within a 6-month period. Similar declines in VOC concentrations have also been observed by USEPA.

D. Polyvalent alcohols and their derivatives

Manufacturers have significantly reduced solvent levels in paints, varnishes, and adhesives during the past decade. These water-based products emit low levels of low-volatility VOC compounds which are being used both as solvents and dispersants. Most commonly they are highly polar ether or ester derivatives of polyvalent alcohols, with boiling points in the range of 125 to 255°C.

These compounds include glycol, glycol–ethers, and associated acetates (esters formed by reaction of an alcohol with acetic acid). They are chemically stable, colorless, and inflammable liquids with an ethery to sweetish odor. They are miscible with water and organic solvents.

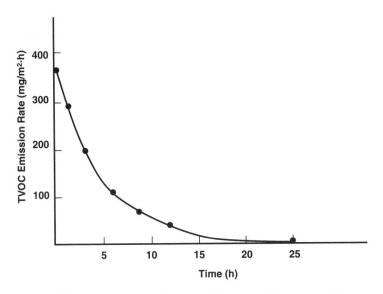

Figure 4.3   TVOC emission decay of a wet (oil-based paint) product. (Data extracted from Guo, H., Ph.D. thesis, Murdoch University, Perth, Australia, 1999.)

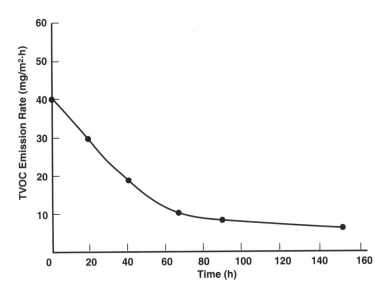

Figure 4.4   TVOC emission decay of a dry (carpeting) product. (Data extracted from Guo, H., Ph.D. thesis, Murdoch University, Perth, Australia, 1999.)

Table 4.10   Glycols/Glycol Derivatives Measured in 200 German Apartments

Compound	Mean concentration ($\mu g/m^3$)	Maximum concentration ($\mu g/m^3$)
Ethylene glycol monomethyl ether	0.8	20.1
Ethylene glycol monoethyl ether	8.1	259.7
Ethylene glycol monophenyl ether	2.3	108
Ethylene glycol monobutyl ether acetate	0.1	6.5
Diethylene glycol monomethyl ether	0.6	164
Diethylene glycol monoethyl ether	1.4	108
Diethylene glycol monobutyl ether	3.0	158
Diethylene glycol monobutyl ether acetate	1.7	302
1,2-propylene glycol	8.4	98
1,2-propylene glycol monomethyl ether	9.9	835
1,2-propylene glycol monobutyl ether	6.3	419
1,2-propylene glycol monophenyl ether	0.7	110

Source: From Pheninger, P. and Marchl, D., *Proc. Indoor Air '99*, Edinburgh, 4, 171, 1999.

Ethylene glycol, diethylene glycol, propylene glycol, and their derivatives are commonly used in paints, varnishes, adhesives, and other products. Concentrations of polyvalent alcohol compounds measured at detectable levels in German apartments are summarized in Table 4.10. Total glycol compound concentrations higher than 100 $\mu g/m^3$ were found in 30 (half had been recently renovated) out of 200 apartments.

Concentrations of polyvalent alcohols and their derivatives in indoor air are relatively low (because of their low vapor pressures). However, emissions continue for longer periods of time than for more volatile VOCs. Ethylene glycol emissions from latex paint do not show the rapid declines observed for higher boiling point VOCs (Figure 4.5). As a consequence, product emissions of polyvalent alcohols and their derivatives may occur for months or years.

E. SVOCs

A number of organic compounds present in indoor environments have boiling points in the range of 240 to 260°C to 380 to 400°C with vapor pressures ranging from 10^{-1} to 10^{-7} mm Hg. Such compounds are described as being semivolatile organic compounds (SVOCs). They exist in gas and condensed phases (adsorbed to particles). They include a variety of compound types including solvents, linear alkanes, aldehydes and acids, pesticides (discussed in the following section), polycyclic aromatic hydrocarbons (PAHs), polychlorinated biphenyls (PCBs), and plasticizers. Concentrations of selected

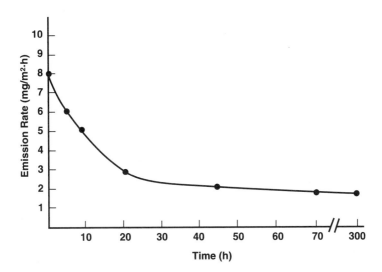

Figure 4.5 Ethylene glycol emissions from latex-based paint. (Data extracted from Roache, N.F. et al., in *Characterizing Sources of Indoor Air Pollution and Related Sink Effects*, Tichenor, B.A., Ed., ASTM, Conshohocken, PA, 98, 1996.)

SVOCs identified in nonresidential Danish buildings are summarized in Table 4.11. Plasticizers were the most common SVOCs measured.

1. Plasticizers

Substances used as plasticizers are ubiquitous contaminants of indoor air and settled dust. They are used in vinyl products to make them soft and relatively

Table 4.11 Concentrations of Selected SVOCs Measured in Four Danish Office Buildings, a Day Care Center, and School Classroom During Winter Time

Compound	Concentration (ng/m³)				Day care center	School classroom
	Office 1	Office 2	Office 3	Office 4		
Tridecane	1128	890	1070	980	1144	4599
Texanol	2055	4378	1910	3871	5492	1451
Tetradecane	1273	1239	1409	1301	1515	3192
Pentadecane	1003	863	1173	1226	1306	1488
Dodecanoic acid	1144	987	2174	2146	3040	1987
TXIB	957	710	658	1282	8341	7803
Hexadecane	934	746	1146	1178	1308	1223
Heptadecane	993	718	1039	965	952	692
Octadecane	862	364	478	475	491	502
Diisobutyl phthalate	1393	977	891	578	1240	1279
Dibutyl phthalate	789	710	1078	840	1346	1195
Di-2-Ethylhexyl phthalate	201	278	241	404	1053	111

Source: From Clausen, P.A., Wolkoff, P., and Svensmark, B., *Proc. Indoor Air '99*, Edinburgh, 2, 434, 1999.

Table 4.12   Concentrations of Phthalate Compounds Measured in 125
California Residences

Compound	Concentration (ng/m³)		Indoor/outdoor ratio
	Median	90th percentile	
Diethylphthalate	340	840	4.3
Di-*n*-butylphthalate	420	1300	14.0
Butylbenzyl phthalate	34	240	4.3
Di-2-ethylhexyl phthalate	110	240	1.7
Di-*n*-octyl phthalate	BLD	9.7	1.4

Source: From Sheldon, L. et al., *Proc. Indoor Air '93*, Helsinki, 3, 109, 1993.

flexible and typically comprise 25 to 50% of the weight of products such as vinyl floor covering. As plasticizers slowly vaporize with time, vinyl products lose their desirable soft properties, become hard and brittle, and crack.

Commonly used plasticizers include phthalic acid esters such as diethylhexyl phthalate (DEHP), dibutyl phthalate (DBP), diisobutyl phthalate (DiBP), benzyl butyl phthalate (BBP), and diethyl phthalate (DEP). A commonly used nonphthalate plasticizer is TXIB (2,2,4-trimethyl-1,3-pentanediol diisobutryate).

Because of their low vapor pressures, concentrations of individual vapor-phase plasticizer compounds in indoor air are generally very low. Nevertheless, significant vapor-phase concentrations may be associated with the installation of new building products such as vinyl floor covering or wall paper where the loading factor is high (high surface to volume ratio). Such high loading factors also occur in new automobiles where the volatilization and subsequent condensation/deposition of plasticizers is reportedly responsible for the "greasy window" phenomenon which is exacerbated by greenhouse-type heating of automobile interiors on sunny days.

Measurable levels of both vapor- and particulate-phase phthalic acid esters have been reported in buildings. Concentrations of DEHP as high as 110 to 230 μg/m³ have been associated with emissions from vinyl wall covering, with mean concentrations of DBP and DiBP in the range of 2 to 16 μg/m³. Air concentrations (vapor and aerosol phase) of phthalate compounds in 125 California residences are summarized in Table 4.12. These concentrations are approximately half of those reported for Danish nonresidential buildings (Table 4.11).

Phthalates associated with airborne and settled dust have been reported for Norwegian residences. Total phthalate concentrations in suspended particulate matter ranged from 450 to 2260 ppm (w/w) with an average concentration of 1180 ppm. The settled dust concentration of phthalates was 960 ppm, with DEHP comprising approximately two thirds of this level. Suspended and settled dust levels were strongly correlated, particularly for DEHP and DBP. As such, resuspended settled dust appears to be the major route of human exposure.

TXIB, a nonphthalate plasticizer, has been reported in the concentration range 100 to 1000 μg/m³ in problem buildings. It was one of the most common substances detected in nonresidential U.S. buildings (frequency 81 to 100% in 56 BASE office buildings), with concentrations in the range of 0.2 to 2.8 µg/m3; in Danish buildings, concentrations were 1.45 to 5.9 µg/m^3 (Table 4.11).

2. *Nonplasticizer SVOCs*

A variety of nonplasticizer SVOCs are found in indoor air in measurable concentrations (Table 4.11). Texanol, found in the highest concentration, is a colorless liquid mixture, with 2,2,4-trimethyl-1,3-pentanediol monoisobutyrate, a compound closely related to TXIB, the dominant compound present. It is commonly added to latex-based paints and has a vapor pressure of 0.013 mm Hg.

3. *PCBs*

Polychlorinated biphenyls (PCBs) are a group of synthetic chemical compounds (209) characterized by the attachment of one to 10 chlorine atoms to a biphenyl moiety. They were produced in the U.S. from approximately 1930 to the mid-1970s and marketed under the trade name Arochlor™. Polychlorinated biphenyls were used as: dielectric fluids in transformers and capacitors, heat transfer and hydraulic fluids, lubricating and cutting oils, and plasticizers in adhesives. They were also used in paints, pesticides, inks, caulking, and sealants. As a consequence, PCBs are ubiquitous contaminants of many pre-1975 buildings.

Polychlorinated biphenyls represent a number of building contamination concerns. These include fluorescent light ballast leakage and failures which result in air and surface contamination. Polychlorinated biphenyl levels in buildings with PCB-containing transformers (191–888 ng/m^3) have been observed to be twice as high as those without such transformers.

Polychlorinated biphenyl contamination of building surfaces, as well as air, is a particularly significant problem when a building has experienced a structural fire. Fire can destroy dozens, if not hundreds, of PCB-containing ballasts. Resulting surface contamination poses a significant challenge to building owners and restoration personnel in their efforts to restore the fire-damaged/contaminated areas and materials. Air and surface PCB levels must be below guideline values to make the building acceptable for reoccupancy.

In one building subjected to a limited structural fire, elevated air and surface PCB levels were reported to be associated with (1) PCB-containing adhesive used to adhere fiberglass insulation to the exterior of metal supply air ducts and (2) the adhesive of fiberglass duct board. The burned building was also contaminated with polychlorinated dibenzofurans, polychlorinated dibenzo-*p*-dioxins, and combustion products of PCBs.

4. *Floor dust*

Low-volatility, high-potency organic compounds and those with high polarity can be expected to partition more to particles than to the vapor phase in air. SVOCs and particulate organic matter (POM) may be partially in the vapor phase and partially adsorbed to suspended particles or settled dust, or to indoor surfaces.

House dust can serve as a reservoir for SVOCs and POM where exposure may occur as a result of resuspension, ingestion, or dermal contact. Over 230 organic compounds (other than pesticides and polycyclic aromatic hydrocarbons) have been identified in house dust samples collected in southern Europe. The most commonly detected compounds were fatty acids, phthalic acid esters, and C_{20}–C_{35} n-alkanes. Concentrations were typically >200 μg/g house dust.

Floor dust has also been shown to have significant concentrations of linear alkylbenzene–sulfonates (LAS), substances widely used as surfactants in cleaning agents such as soap and detergents. LAS concentrations in the floor dust of seven Danish public buildings ranged from 34 to 1500 μg/g dust, with clothing being the most important source of LAS.

F. Health effects

It is widely believed in the scientific community that exposures to individual VOCs or SVOCs in indoor environments are not likely to be responsible for acute symptoms reported by individuals in buildings. Concentrations are typically two or more orders of magnitude lower than OSHA permissible exposure limits (PELs) and American Conference of Governmental Industrial Hygienists (ACGIH) threshold limit values (TLVs), which have been developed to protect most workers from adverse health effects (including irritation) associated with airborne exposures.

VOCs nevertheless have been and continue to be suspected as a cause of (or a risk factor for) sick building syndrome (SBS)-type symptoms (see Chapter 7). Reasons for this include: (1) many VOCs have the potential to cause sensory irritation and central nervous system symptoms characteristic of SBS, (2) concentrations, though low, are significantly higher indoors than in the ambient atmosphere, and (3) the potential for the many VOCs present to cause symptoms as a result of both additive and multiplicative (synergistic) effects.

In response to these concerns, Danish scientists have developed and tested what is described as the TVOC theory. In the TVOC theory, sensory irritation and even neurotoxic-type symptoms are thought to be due to the combined effects of exposure to VOCs. The biological mechanism for this phenomenon appears to be stimulation of the trigeminal nerve system, described as the common chemical sense.

1. Common chemical sense

The common chemical sense is one of two olfactory mechanisms by which humans perceive and respond to odor. The chemical sense organ consists of trigeminal nerves in the nasal cavity and eyes, as well as facial skin areas. They can be stimulated by a variety of chemicals as well as physical agents such as touch and temperature. Stimulation of trigeminal nerves produces irritation described as burning, stinging, or smarting. Other effects include changes in heart rate and respiratory frequency, as well as coughing, sneezing, and tearing. The most important outcome of trigeminal nerve stimulation (relative to explaining SBS-type symptoms) is the production of similar qualitative symptom responses by many different chemical exposures.

2. Exposure studies

Human exposure studies using VOC mixtures have been conducted to test the TVOC theory. In initial studies, Danish scientists exposed healthy volunteers to a mixture of 22 VOCs weighted in concentration to those observed in VOC measurements in a Danish residence in the early '80s. These VOCs and their weighted proportions are listed in Table 4.13. Groups of subjects were exposed to 5 and 25 mg/m³ TVOC concentrations (toluene equivalent) in double-blind experiments for approximately 2.75 hours. Subjects exposed to VOC mixtures at both concentrations reported significantly greater irritation than those in clean air as well as diminished performance on neurobehavioral tests. These results were subsequently confirmed in additional Dan-

Table 4.13 VOC Mixture Used in Danish Human Exposure Studies

Compound	Weight ratio
n-Hexane	1.0
n-Nonane	1.0
n-Decane	1.0
n-Undecane	0.1
1-Octene	0.01
1-Decene	1.0
Cyclohexane	0.1
3-Xylene	10.0
Ethylbenzene	1.0
1,2,4-Trimethylbenzene	0.1
n-Propylbenzene	0.1
α-Pinene	1.0
n-Pentanal	0.1
n-Hexanal	1.0
Isopropanol	0.1
2-Butanol	1.0
2-Butanone	0.1
3-Methyl-3-butanone	0.1
4-Methyl-2-pentanone	0.1
n-Butylacetate	10.0
Ethoxyethylacetate	1.0
1,2-Dichloroethane	1.0

Source: From Molhave, L. et al., *Atmos. Environ.*, 25A, 1283, 1991. With permission.

Table 4.14 Dose–Response Model Relationship Between TVOC Exposures and
Discomfort/Health Effects

TVOC concentration (mg/m^3)	Response	Exposure range
<0.20	No effects	Comfort range
0.20–3.0	Irritation/discomfort possible	Multifactorial exposure range
3.0–25.0	Irritation and discomfort; headache possible	Discomfort range
>25.0	Neurotoxic effects	Toxic range

Source: From Molhave, L., *Proc. 5th Internatl. Conf. Indoor Air Qual. & Climate*, Toronto, 5, 15, 1990.

ish studies, and in part, by USEPA studies at 25 mg/m³. Though USEPA scientists observed a significant increase in headache and a feeling of discomfort, they were unable to confirm neurobehavioral changes. Concentrations of 5 mg/m³ and 25 mg/m³ would be approximately 1 and 5 ppmv toluene-equivalent concentrations. The threshold for acute effects on exposure to submixtures of six to nine of the original 22 VOCs has been reported to be <1.7 mg/m³.

3. *TVOC theory application*

Danish scientists have proposed a dose–response model to relate exposures to TVOCs and health effects. These model relationships are summarized in Table 4.14. At TVOC exposure concentrations <0.2 mg/m³, no acute health effects are expected, whereas irritant symptoms would be expected in all exposure cases >3 mg/m³. Neurotoxic effects other than headache would be expected above an exposure concentration of 25 mg/m³. In the exposure range of 0.2 to 3.0 mg/m³, both irritation and discomfort would be expected to occur if other factors, such as temperature and other contaminant exposures, interacted with TVOC exposures. Because of the limited number and unique nature of contaminants used in their study (buildings vary widely in VOC types and concentrations), Danish scientists have cautioned that the TVOC concept should not be used as a generic indicator of potential health risks in buildings. At best, TVOC concentrations are an indicator of the risk of nonspecific sensory irritation to relatively unreactive VOCs with a limited range of vapor pressures.

4. *Systematic building studies*

A number of investigators have attempted to determine whether SBS-type symptoms are associated with exposure to TVOCs. No apparent relationships were observed in non-problem Danish municipal and California office buildings. However, in problem Swedish school buildings, significant positive correlations were observed between SBS-type symptom prevalence and the logarithmic value of TVOC concentrations (Figure 4.6). Significant

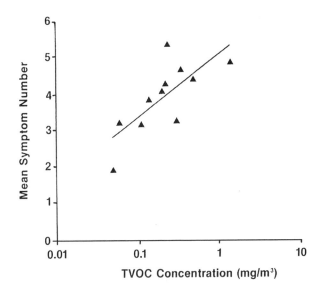

Figure 4.6   Dose–response relationships between building TVOC concentrations and symptom prevalence. (From Norback, D., Michel, I, and Winstrom, J., *Scand. J. Work Environ. Health*, 16, 121, 1990. With permission.)

log–linear relationships were also observed between symptom prevalence rates and concentrations of terpenes, *n*-alkanes (C_8–C_{11}), and butanols in investigations of problem Swedish office buildings. TVOC concentrations varied from 0.07 to 0.18 mg/m³ in problem schools and 0.05 to 1.38 mg/m³ in problem office buildings. In the former case, TVOC levels would have been at the no-effects level of the Danish model (Table 4.14).

Elevated SBS-type symptom prevalence in office building investigations conducted by other Swedish scientists appeared to have been associated with reduced TVOC levels in office environments. On further data analysis, a log–linear dose–response relationship was observed between "lost" TVOCs between intake air and air in office spaces (Figure 4.7). Symptoms including fatigue, heavy-headedness, hoarse/dry throat, and dry facial skin were collinearly related to low level (up to 40 ppbv) increases in room HCHO concentrations.

An epidemiological study has been recently conducted on a cohort of 170 pregnant women in the United Kingdom in order to evaluate potential health effects among them and their subsequently newborn children relative to exposures to VOCs as well as VOC-containing products. The use of air fresheners was observed to significantly increase the risk of ear infection and diarrhea in infants and depression and headaches in mothers. Animal studies indicate that air fresheners may be potent respiratory irritants. Elevated TVOC levels were associated with both the use of air fresheners and aerosols. Aerosol use appeared to increase the risk of diarrhea and vomiting in infants and headaches in mothers.

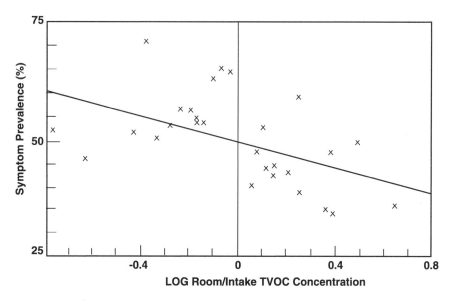

Figure 4.7 Dose–response relationships between symptom prevalence and "lost" TVOCs. (From Sundell, J. et al., *Indoor Air*, 3, 82, 1993. With permission.)

5. Asthma

There is some scientific evidence to indicate that exposures to plasticizer chemicals such as DEHP may be partly responsible for the significant increase in asthma in developed countries in the last two decades. Mono-2-ethylhexyl phthalate (MEHP), the hydrolysis product of DEHP, has been shown to cause bronchial hyperreactivity in rats. An association between plastic interior surfaces and bronchial obstruction in young children has been reported by Finnish scientists who suggest DEHP is the potential causal agent.

There is a structural similarity between DEHP, MEHP, and prostaglandins. As a result, these compounds may mimic the effect of prostaglandins that mediate inflammatory responses such as those that cause asthmatic attacks. High local exposure to DEHP can occur on deposition sites in respiratory airways with subsequent local health effects.

6. Cancer

A number of VOCs commonly found in indoor VOC samples are either mutagenic or carcinogenic (e.g., benzene, styrene, tetrachloroethylene, 1,1,1-trichloroethane, trichloroethylene, dichlorobenzene, methylene chloride, and chloroform). This suggests that VOC exposures in indoor environments have the potential (albeit at very low individual risk) for causing cancer in humans. The nature of potential health risks associated with multiple potential carcinogens at very low levels of exposure is unknown.

Significant relationships between parental occupation and cancer in children have been reported in over a dozen different epidemiological studies. It has been suggested that paternal exposure to VOCs during employment as well as maternal VOC exposures during pregnancy are responsible for increased risks of childhood leukemia. A significant increase in the risk of childhood leukemia has been associated with paternal exposure to chlorinated solvents, methylethyl-ketone, spray paint, dyes and pigments, and cutting oils. Infant VOC exposures or maternal exposures during pregnancy may result from passive transport on clothing or body surfaces from the work environment. As such, this reflects a "toxics brought home" phenomenon.

G. Indoor air chemistry

The dose–response relationship with "lost" VOCs and symptom prevalence and an increase in HCHO levels suggest that health complaints in buildings may be associated with exposures to new VOCs/aldehydes that are produced as a consequence of chemical reactions between airborne substances, most notably VOCs.

There is increasing evidence that chemical reactions between ozone (O_3) and unsaturated hydrocarbons, such as d-limonene, α-pinene, α-terpene, styrene, isoprene, etc., at levels commonly found in indoor air can produce aldehydes that may affect air quality and human health. This may be associated with reaction pathways involving hydroxyl (OH^-) or nitrate (NO_3) radicals produced as a result of O_3-initiated reactions.

Simple aldehydes such as HCHO, hexanal, nononal, and decanal, and aromatic aldehydes such as benzaldehyde and tolualdehyde may be produced by such reactions. Other reaction products may include one or more carbonyls, or a carbonyl with a double bond (e.g., 2-nonenal). Reactions involving O_3, isoprene, and OH^- can result in the production of methacrolein (CH_2=$C(CH_2)CHO$). Isoprene is emitted by plants and humans. Ozone reactions with building furnishings such as carpeting have been observed to produce measurable levels of 2-nonenal, *n*-nonenal, and C_1–C_3 and C_6–C_8 aldehydes.

Aldehydes produced by indoor chemical reactions also participate in other chemical reactions. Aldehyde reactions with NO_3 radicals can produce peroxy nitrates. The formation of peroxyacyl nitrate (PAN) begins with the reaction of either NO_3 or OH^- with acetaldehyde to form the acetyl radical (CH_3CO), which reacts with nitrogen dioxide (NO_2) to form PAN ($CH_3C(O)NO_2$). Hydroxyl radical reactions with propionaldehyde and benzaldehyde produce peroxy propionyl nitrate (PPN) and peroxy benzoyl nitrate (PBN), respectively.

Peroxyacyl nitrate, PPN, and PBN are all potent eye irritants, causing symptoms at relatively low concentrations (circa 20 ppbv). Aldehydes produced as a result of indoor chemical reactions have lower odor thresholds and may cause more sensory irritation than their precursors. Odor thresholds are reported as 1.9 pptv for *cis*-2-nonenal and *trans*-6-nonenal, 17 pptv for 8-nonenal, and 24 pptv for *trans*-3-nonenal and *cis*-3-nonenal.

Indoor chemistry involving production of aldehydes, and aldehydes as reactants, has only recently emerged as a potential explanation for IAQ complaints in buildings. Such research has been limited, and considerable additional effort is required to elucidate the potential role of indoor chemistry and its by-products in contributing to symptoms and health complaints in indoor environments.

III. Pesticides

Because of their known toxicity, pesticides are used to control one or more pest organisms. As such they are unique contaminants of indoor environments. There is a limited level of understanding by users that such products represent a potential toxic hazard to them and other building occupants, albeit the toxic risk to users is perceived to be very low. Pesticide contamination of indoor spaces can also occur inadvertently by passive transport (track in) from lawn, garden, and agricultural applications.

Based on a 1990 National House and Garden Pesticide Use Survey, over 85% of the 66.8 million American households use pesticides, with 20% commercially treated for indoor pests such as cockroaches, ants, and fleas. Approximately 18 million households use pesticides on lawns, 8 million on gardens, and 14 million on ornamental plants.

There are approximately 20,000 different household pesticide products, which include 300 active ingredients and 1700 so-called inert (having no pesticidal activity) ingredients. Such an extensive presence and use in U.S. households, not surprisingly, results in unintended exposures and misapplications. In 1993, Poison Control Centers reported over 140,000 cases of likely pesticide exposure, with approximately 93% associated with home use. Twenty-five percent of these had frank symptoms of pesticide poisoning; half occurred in children under the age of six.

Pesticide products used indoors include a large variety of substances and applications. These include biocides used by manufacturers to prevent the biodegradation of products and by homeowners/tenants to disinfect surfaces such as toilet bowls, nursery surfaces, etc. They also include insecticides to control cockroaches, flies, ants, spiders, and moths; termiticides; flea and tick sprays and shampoos for pets; and insecticides for both indoor and outdoor plants (lawn, garden, and ornamentals). Fungicides are used primarily in lawn treatments and garden sprays and, except for some biocidal applications, are not commonly used indoors. Herbicides are used exclusively for lawn care, brush control, and agricultural applications.

A. Biocides

Biocides are pesticidal compounds/formulations that are used to control the growth of microorganisms such as bacteria (which include actinomycetes) and fungi. They control microorganisms by killing them directly or limiting their growth. A biocide that kills bacteria behaves like an antibiotic; it is either bactericidal, bacteriostatic, or both.

Biocides are used in building environments as disinfectants in cleaning agents for sinks, counters, bathtubs, showers, toilet bowls, baby nurseries, diaper pails, and some soap products. They are also used in products such as furnace and air handling system filters, carpeting, and paints. They are used to control the growth of microbial slimes on exterior elements of cooling coil condensate drip pans and in sumps of cooling towers.

Antimicrobial compounds commonly used in residential applications include sodium/calcium hypochlorite, ethanol, isopropyl alcohol, pine oil, o-phenylphenol, 2-benzyl-4-chlorophenol (BCP), glycolic acid, and phosphoric acid. Quaternary ammonium compounds are used for more heavy-duty applications in hospitals, large buildings, etc. *Ortho*-phenylphenol has been the principal ingredient in a widely advertised commercial product used in nurseries and in the disinfectant treatment of toilet bowls in residences, schools, and hotel/motel rooms. It has a characteristic strong chemical odor. Its use has been declining in recent years.

A variety of compounds are used as biocides in carpeting products. These include such apparently nonvolatile compounds as zincomedine (zinc complex of pyrethione) and organosilane compounds such as octadecy-lamino-trimethylsilylpropyl ammonium chloride. Biocides have been historically used in carpet manufacture to reduce potential microbial growth associated with accidental wetting and wetting associated with shampooing. Shampoos themselves may contain biocides.

Biocides are used in paint formulations to prevent microbial degradation of the product or to protect substrate materials coated with paints (mold-inhibiting paints are widely available).

Mercury biocides have been used extensively to control the growth of mold and bacteria during production and storage of interior, water-based, latex paints and to control mold growth in exterior latex paints on a long-term basis. Use of mercury biocides in latex paint was approved by USEPA despite their known adverse human health effects. Approximately 25 to 30% of interior latex paints manufactured in the U.S. contained organic mercury compounds before their use was discontinued in a voluntary agreement between manufacturers and USEPA in 1990.

Substances used in biocidal treatment of furnace/air handling unit filters include a commercial product called Intercept; cooling tower water is treated with quaternary ammonium compounds, bis (tri-*n*-butyl tin) oxide, and chlorine/hypochlorite.

Duct cleaning has been widely promoted by service providers in North America to improve IAQ. After cleaning, ducts are commonly treated with one or more biocides. A variety of formulations are used, including some that contain glutaraldehyde.

Biocides may be used to control problem infestations of bacteria or mold. Chlorine bleach is widely used by homeowners and building service personnel to control fungal growth on surfaces, and fungicidal formulations containing thiocarbamate compounds are used to control infestations on flooring joists and in post-flood rehabilitations.

B. Fungicides

Ortho-phenylphenol and pentachlorophenol (PCP) are the most common fungicides used in indoor environments. *Ortho*-phenylphenol is used as a biocidal disinfectant, as described previously. The predominant (80%) historical use of PCP, or penta, has been to preserve wood from fungal decay and insect borers. Other uses have included controlling cooling tower algae and fungi; waterproofing plywood and fiber board; and as a preservative in paint, leather, rope, twine, burlap, and cable. It has also been used in herbicidal formulations and as a disinfectant. Because it is a teratogen and possible human carcinogen, use has been restricted since 1986, a result of an agreement by PCP manufacturers and USEPA. Approved uses include preservation of wood products designed to be used outdoors, such as telephone poles and other pressure-treated lumber. It was once used in the U.S. to preserve timbers intended for log home construction and in Europe as a wood preservative for a variety of indoor applications.

C. Insecticides

Insects, and occasionally arachnids (mites, spiders, ticks), represent major pest control concerns in buildings. Pesticidal formulations used to control insects and arachnids are described as insecticides and acaricides, respectively. Active ingredients in insecticidal formulations used in or around building perimeters have included diazinon, *p*-dichlorobenzene, chlordane, heptachlor, chlorpyrifos, ronnel, dichlorvos, malathion, lindane, aldrin, dieldrin, bendiocarb, methoxychlor, propoxur, methyl demeton, naphthalene, and pyrethrins. Para-dichlorobenzene is the active ingredient in moth balls; naphthalene is the active ingredient in moth flakes. Major insecticidal classes or groups include chlorinated hydrocarbons, organophosphates, carbamates, and pyrethrins.

Acaricides have been developed to control dust mite populations in residences. The active ingredients are benzoic acid esters. They are applied as a powder to fabric surfaces that harbor dust mites, such as carpeting and upholstery. Dust mite acaricides have not been approved for use in the U.S.

Insecticides used indoors may be applied in a variety of ways: as an emulsion spray, fogging device (bug bomb), poison bait, or pet wash, or impregnated in pest strips and pet flea collars, etc. They may be: aerosolized (fly spray, bug bombs) or slowly vaporized (pest strips) into indoor air, applied directly to interior surfaces (cockroach control) or to the soil surrounding crawlspace and basement substructures and within the slab substructures (termiticides), and used as solids near or on surfaces to be protected.

Many insecticidal formulations consist of one or more active ingredients and so-called inert ingredients. Inert ingredients are proprietary compounds and usually are not disclosed. They often include a mixture of relatively low-volatility solvent oils which help to dissolve and suspend insecticidal compounds applied as emulsion sprays.

1. Organochlorines

Chlordane, heptachlor, lindane, aldrin, dieldrin, and *p*-dichlorobenzene are chlorinated hydrocarbons. Their use and availability has been sharply restricted in North America and Europe because of their suspected human carcinogenicity. Therefore, indoor exposures are unlikely to be associated with recent applications. However, because they are persistent, exposures may continue for several decades after a single building-associated application. *Para*-dichlorobenzene is one of the few chlorinated hydrocarbons still widely used in residences and other buildings (moth balls, cakes, and flakes, and as a deodorizing substance).

2. Organophosphates

Organophosphate insecticides used indoors have included dichlorvos, chlorpyrifos, diazinon, and malathion. Dichlorvos has been widely used in bug bombs (because of its relatively high volatility, 10^{-2} mm Hg), slow-release insecticidal strips, and flea collars for dogs and cats. Because of health concerns, its use has dramatically declined in the U.S. since 1988. Chlorpyrifos has been one of the most widely used indoor pesticides. It was used in most termiticidal treatments, to control cockroaches, and to control household flea infestations. These uses were phased out in the year 2000 as a consequence of a USEPA/industry agreement.

3. Carbamates

The carbamates are widely used insecticides. They are N-substituted esters of carbamic acid

$$(NH_2-C=O)$$
$$\overset{|}{O}H$$

Commonly used carbamate insecticides include carbaryl, aldicarb, carbofuran, propoxur, and bendiocarb. Carbaryl, also called Sevin™, is the most widely used of these compounds. Both carbaryl and propoxur have been detected in indoor environment samples.

4. Pyrethrins and pyrethroids

Pyrethrins and pyrethroids represent a class of related insecticidal chemicals. Pyrethrins are components of a natural extract of a *Chrysanthemum* species. The six main pyrethrin components are cyclopropane esters. Because of their sensitivity to photochemical degradation, use has been limited to indoor applications such as fly sprays. In the last two decades, however, synthetic derivatives (pyrethroids) have been developed which have greater environmental stability and higher biological activity. Pyrethroids have very low vapor pressures and are classified as nonvolatile chemicals. Pyrethroids include permethrin, deltamethrin, cypermethrin, cyflutrin, resmethrin, teliomethrin, and alletrin. Of these, permethrin is the most widely used. The pyrethroids are often formulated with piperonyl butoxide (PBO) which acts as a synergist to enhance their biological activity. Synergists reduce the rate of pyrethroid degradation after uptake by insects.

Pyrethroids are increasingly being used indoors for a variety of insect pest applications, including flea and cockroach control. They are replacing such compounds as chlorpyrifos, diazinon, and dichlorvos. Their use is expected to increase significantly as a result of the regulatory phaseout of chlorpyrifos applications indoors and for termiticidal treatments.

D. Indoor exposures and levels

1. Biocides

In the late 1980s, staff from the Centers for Disease Control (CDC) conducted an investigation of a 4-year-old child who was diagnosed with acrodynia (a relatively rare form of childhood mercury poisoning) after the child's home was painted with 12 gallons of interior latex paint containing 930 to 950 mg/L mercury. CDC staff concluded that the child had been poisoned by inhalation of mercury vapors from phenylmercuric acetate (PMA), one of the most commonly used mercury biocides in latex paint.

In response to this incident, CDC and other investigators measured elemental mercury levels in 21 residences that had been painted 3 months earlier with a mercury-containing paint and 16 control residences (not recently painted). The median air concentration in exposed homes was 0.3 $\mu g/m^3$, with a range from below the limit of detection (BLD) to 1.5 $\mu g/m^3$; unexposed homes had a BLD median level, with a range of BLD to 0.3 $\mu g/m^3$. Six of the exposed homes had mercury levels >0.5 $\mu g/m^3$, the guideline limit for acceptable indoor air levels recommended by the Agency for Toxic Substances and Disease Registry (ATSDR).

Mercury emission studies of 13 paints containing mercury biocides such as PMA, diphenyl mercury, dodecenyl mercury succinate, diphenyl mercury, phenyl mercuric oleate, and chloromethoxy propyl mercuric acetate, indicate that though biocides are organic mercury compounds, emissions are in the form of elemental mercury which, because of its high vapor pressure, readily becomes airborne.

Airborne mercury sorbs on household dust and materials such as carpet. As would be expected, higher dust mercury levels are found in homes recently painted with mercury-containing paints. Medians and ranges of mercury levels in house dust in a sample of exposed and "unexposed" U.S. homes were 3.76 mg/kg, 0.93 to 11.91 mg/kg, and 1.78 mg/kg, 1.0 to 4.8 mg/kg, respectively. It appears that mercury contamination of floor dust in U.S. homes is common; it may pose an exposure risk to small children.

Table 4.15 Detection Frequency and Mean Concentrations of Airborne Pesticides in Springfield/Chicopee, MA, Residences

Pesticide	Spring % households	Spring Mean concentration (ng/m³)	Winter % households	Winter Mean concentration (ng/m³)
o-Phenylphenol	90	44.5	72	22.8
Chlordane	50	199.0	83	34.8
Heptachlor	50	31.3	70	3.6
Propoxur	49	26.7	38	17.0
Chlorpyrifos	29	9.8	30	5.1
Diazinon	16	48.4	10	2.5
Dieldrin	12	1.0	34	4.2
Lindane (γ-HCH)	10	0.5	21	9.5
Dichlorvos	2	4.3	1	1.5
Malathion	2	5.0	0	0.0

Source: From Whitmore, R.W. et al., *Arch. Environ. Contam. Toxicol.*, 26, 47, 1994. With permission.

Ortho-phenylphenol was detected in the air of 70 to 90% of U.S. residences monitored in two states during 1987–88 in USEPA's NOPES (Non-Occupational Pesticide Exposure Study) research program (Tables 4.15 and 4.16), with mean concentrations of approximately 45 and 23 ng/m³, respectively (range of 0.03 to 1 µg/m³). It was also detected in 100% of floor dust samples collected in Washington, D.C. residences.

Table 4.16 Detection Frequency and Mean Concentrations of Airborne Pesticides in Jacksonville, FL, Residences

Pesticide	Spring % households	Spring Mean concentration (ng/m³)	Winter % households	Winter Mean concentration (ng/m³)
Chlorpyrifos	100	370.0	29	120.0
Propoxur	85	528.0	95	162.0
o-Phenylphenol	85	96.0	79	59.0
Diazinon	83	421.0	83	85.7
Dieldrin	79	1470	62	7.2
Chlordane	61	324.0	94	220.0
Heptachlor	58	163.0	92	72.2
Lindane (γ-HCH)	34	20.2	68	6.0
Dichlorvos	33	134.0	10	24.5
Malathion	27	20.8	17	20.4
α-Hexachlorocyclohexane	25	1.2	22	1.1
Aldrin	21	31.3	31	6.9
Bendiocarb	23	85.7	20	3.4
Carbaryl	17	68.1	0	0.0

Source: From Whitmore, R.W. et al., *Arch. Environ. Contam. Toxicol.*, 26, 47, 1994. With permission.

2. Insecticides and wood preservatives

A number of investigators have evaluated pesticide concentrations and potential exposures in building environments, particularly residences. These have included surveys associated with misapplications or alleged misapplications of pesticides, contamination concerns associated with wood preservatives, application-related illness complaints, nonoccupational exposures, and exposures among farm families and farm workers.

a. Air/surface dust concentrations. Highest air concentrations of insecticides have been reported for home fogging devices. These devices (described as bug bombs) are pressurized canisters used to control fleas, flies, mosquitoes, ants, spiders, moths, and cockroaches. A variety of active ingredients have been, and continue to be, used in bug bombs. The organophosphate, dichlorvos, was the most widely used insecticide in bug bombs in the 1970s and 1980s. Because of its relatively high vapor pressure, concentrations within the first few hours after application were often very high (in the range of 4 to 6 mg/m^3), decreasing rapidly thereafter. Dichlorvos use in pest strips has been shown to result in potentially high exposure levels under laboratory chamber conditions simulating a typical bedroom. Concentrations increased from an initial 200 μg/m^3 to 1300 μg/m^3 on the 7th day, decaying to <100 μg/m^3 after 56 days. Based on the WHO acceptable daily intake (ADI) guideline value of 4 μg/kg body weight, dichlorvos exposures of 200 μg/m^3 would have exceeded the WHO ADI by 8 to 16 times and, in worst case situations, approximately 100 times.

In response to occupant complaints associated with retrofit sub-slab termiticidal treatments, chlordane and heptachlor concentrations were measured in a number of Army and Air Force bases in the 1970s. Termiticides had been, in many cases, inadvertently injected into heating ducts located within or below the slab. Chlordane levels in approximately 14% of 500 Air Force residences were >3.5 μg/m^3. Approximately 50% of 157 private U.S. homes investigated in another study had concentrations >5 μg/m^3, an interim guideline recommended by a committee of the National Academy of Science. These relatively high chlordane levels were usually associated with improper applications.

Chlordane formulations have been available as products containing 72% chlordane:7 to 13% heptachlor, and 39.2% chlordane:19.6% heptachlor. Consequently, residences contaminated with chlordane are also contaminated with heptachlor. Because of the higher vapor pressures associated with heptachlor, chlordane and heptachlor concentrations are often in the same range, and, in some cases, heptachlor concentrations are higher than those of chlordane.

In the late 1980s, USEPA conducted a large-scale study of nonoccupational pesticide exposure (NOPES) in two communities in Massachusetts and Florida. The former was selected as a relatively high-use site; the latter, low-use. Prevalence rates and mean concentrations for the most commonly detected pesticides are summarized for two seasons in Tables 4.15 and 4.16.

Table 4.17 Prevalence and Concentrations of Pesticide
Residues Measured in Carpet Samples in Nine
Jacksonville, FL, Residences

Pesticide	# carpets	Mean concentration ($\mu g/g$)
Heptachlor	10	1.3
Chlorpyrifos	11	5.8
Aldrin	10	0.4
Dieldrin	10	2.2
Chlordane	10	14.9
DDT + metabolites	9	1.2
o-Phenylphenol	10	0.8
Propoxur	9	1.6
Diazinon	9	1.7
Carbaryl	5	1.4
Atrazine	2	0.7

Source: From Whitmore, R.W. et al., *Arch. Environ. Contam. Toxicol.*, 26, 47, 1994. With permission.

The seven most prevalent pesticides in both tables made up >90% of the total mass concentration of all monitored pesticides. These included four general-purpose insecticides (chlorpyrifos, propoxur, diazinon, and dichlorvos), two discontinued termiticides (chlordane and heptachlor), and the biocide/disinfectant, o-phenylphenol. As expected, concentrations were higher in Florida, with highest concentrations reported for summer months.

The moth agent and deodorizer, p-dichlorobenzene, has been commonly observed as a contaminant of residential indoor air; naphthalene less so. Measurable levels of p-dichlorobenzene were detected in over 50% of approximately 7850 residences investigated by USEPA, with mean concentrations in the range of 20 to 50 $\mu g/m^3$. Chamber emission studies confirm that use of moth cakes indoors can cause significant indoor levels. They also indicate that sink effects which result in the revaporization and deposition of p-dichlorobenzene may cause prolonged exposures.

Pentachlorophenol and lindane were widely used in the past to treat foundation timber, structural timbers exposed to weather, and in coatings such as paints, wood stains, and sealers. Pentachlorophenol was also used on timbers intended for log home construction. Airborne levels have been reported in complaint log homes in the range of 10 to 30 $\mu g/m^3$; in an office building with PCP-treated timbers they were 27.2 to 30.7 $\mu g/m^3$.

Carpet dust sampling for pesticides was conducted in nine Florida homes in the NOPES study. Frequency and mean concentrations of targeted pesticides found in 11 carpets are summarized in Table 4.17. The average number of pesticides found in carpet dust was 12, compared to 7.5 in air in the same residence. Thirteen pesticides were found in carpet dust that were not observed in air samples, presumably because of their low volatility. Highest concentrations were reported for chlordane, a discontinued pesticide once used for termiticidal treatment.

The pyrethroids are very persistent when used indoors. As much as 10% of the initial concentration of permethrin and deltamethrin is found in house dust 2 years after application. Pyrethroids are commonly observed in dust samples collected in buildings with prior pest control treatment. Dust concentrations in the range of 2 to 320 mg/kg permethrin and 2 to 12 mg/kg deltamethrin have been reported. Though less persistent, the catalyst PBO can still be detected 2 years after application. Pyrethroids are only observed in the gas phase immediately after application. At that time, high concentrations are associated with suspended particles, decreasing by 90+% 2 days later and relatively slowly thereafter.

b. Biological levels.   Concentrations of pesticides in human tissues and body fluids such as blood and urine may be an important indicator of exposure. Population exposures to organochlorines, organophosphates, and chlorophenoxy pesticides or their metabolites were evaluated in urine collected from a sample of U.S. citizens in the second National Health and Nutritional Examination Survey (NHANES) conducted during 1976–1980. Pentachlorophenol (PCP) was the most frequently occurring (72% of the sample population) pesticide residue observed in quantifiable concentrations (geometric mean of 6.3 ng/ml). An estimated 119 million individuals age 12 to 74 years old had been exposed to PCP at the time of the study. PCP was used as a wood preservative and insecticide. It may be present in biological samples as a metabolite of the insecticide lindane and the fungicide hexachlorobenzene.

Other quantified pesticide residues included 3,5,6-trichloro-2-pyrindol (a metabolite of chlorpyrifos, 5.8%), 2,4,5-trichlorophenol (3.4%), *p*-nitrophenol (2.4%), dicambra (1.4%), malathion dicarboxylic acid (0.5%), malathion α-monocarboxylic acid (1.1%), and 2,4-D (0.3%). Dicambra and 2,4-D are used as herbicides; 2,4,5-trichlorophenol as an antimicrobial biocide. Exposures were likely to have been inadvertent as a result of their direct use, indoor contamination, or with the possible exception of chlorpyrifos, residues on food.

Exposure to airborne PCP in U.S. PCP-treated log homes has been reported to result in elevated blood serum PCP levels, ranging from 69 to 1340 ppb (w/v), with a geometric mean of 330 ppb. Geometric mean concentrations were significantly higher in children aged 2 to 7 years (510 ppb). Serum PCP levels in a control population, on the other hand, were considerably lower, ranging from 15 to 75 ppb, with a geometric mean of 37 ppb.

Elevated cyclodiene organochlorine insecticides (dieldrin, aldrin, heptachlor, chlordane) in human breast milk have been reportedly associated with termiticidal applications on soil around home perimeters. The greatest contributor was heptachlor, presumably because of its higher volatility.

3. Lawn care products

Lawn care chemicals are used by approximately 18 million U.S. households to control weed species, insect pests, and fungal pathogens. They may be formulated to control a single pest problem, or combined to control weeds, insects, and fungal pathogens. Indoor exposure to individual pesticides may result from drift during application, or by passive transport on the shoes and clothing of applicators, on the shoes of residents or their children, and on pets. Once indoors, lawn care pesticides are protected from sunlight, rain, elevated temperatures, and microbial action. As a consequence, they persist on indoor dust and other surfaces for months or years as opposed to days for the same chemicals outdoors. House dust on which pesticides accumulate may serve as a source of exposure on resuspension. Lawn care chemicals used exclusively outdoors, such as the herbicide 2,4-D (2,4-dichlorophenoxy acetic acid and salts), the insecticide carbaryl (Sevin™, 1-napthylmethyl carbamate), and the fungicide chlorothalonil (tetrachloroiso-phthalonitrile), have all been identified in indoor environments.

a. Herbicides. The herbicide chemicals 2,4-D, dicambra (3,6-dichloro-2-methoxybenzoic acid, 3,6-dichloro-ortho-anesic acid and salts), MCCP (mecroprop-2,4-chloro-2-methylphenoxy propionic acid and salts), and glycophosphate (Roundup™ — isopropylamine salt of *n*-phosphonomethyl glycine) are widely used in lawn care formulations. They are applied as post-emergent products to kill weeds selectively or for total weed (and grass) control. Other products are used as pre-emergent chemicals to control pest species such as crabgrass. These include Benifen, Bensulide, DCPA, Pendimethalin, and Siduron.

Acid herbicides such as 2,4-D and glycophosphate are typically applied as amine salts and have extremely low vapor pressures. They are found in indoor air at very low concentrations and are almost always in the particulate phase.

2,4-D was detected in the air of 64 of 82 homes whose lawns had been commercially treated. Approximately 65% of particulate phase 2,4-D mass was associated with inhalable particles (≤ 10 µg); 25 to 30% with particles <1 µm in diameter. Detectable 2,4-D residues were reported on all indoor surfaces sampled a week after application. Track-in is the most important route of transport indoors. For high-activity homes, 58% of 2,4-D track-in has been estimated to be due to an indoor-outdoor dog; 25%, the applicator's shoes; and 8%, children. Spray-drift and postapplication air intrusion appear to be minor sources of indoor 2,4-D (≤ 1%). Once indoors, 2,4-D is subsequently transported to tabletops, window sills, and other horizontal surfaces by resuspension.

4. Agricultural areas

Pesticide contamination of residences and subsequent exposure of children and other residents may be particularly high in agricultural regions, even if residents are not involved in farm activities. Exposure may result from contaminated soil, dust, clothing, food, or water, or drift from spraying operations. Residential areas in many agricultural communities are surrounded by fields or orchards.

Studies conducted in fruit-growing regions of Washington have shown significant contamination of indoor dusts by organophosphorous compounds such as azinphosmethyl, chlorpyrifos, parathion, and phosmet. Azinphosmethyl and parathion are highly toxic and registered exclusively for agricultural use.

Organophosphate pesticide concentrations measured in house dust indicated that houses of farm families had significantly higher concentrations of azinphosmethyl, phosmet, and chlorpyrifos than reference or control families.

In general, household pesticide levels are higher in homes of farm workers compared to the homes of farmers. Pesticide levels in household dusts are inversely related to distance from commercial orchards.

5. *Misapplications*

Misapplication of pesticides in residences by homeowners and in some cases by personnel of private firms, particularly pest control services, are not uncommon. In most instances they are sporadic, occurring randomly around the country. In several recent cases, however, major "epidemics" of pesticide misapplications (and illegal use) have occurred, requiring significant USEPA response and mitigation efforts. These cases involved the highly toxic, restricted-use compound, methyl parathion. Stocks intended for agricultural use were surplused by owners to unknowing individuals who sold or applied them for cockroach and other insect control purposes. Such misapplications/illegal treatments were conducted in 232 houses in Ohio in 1994 (cleanup cost over $20 million), 1100 homes in Mississippi in 1996, and 100 homes in Illinois in 1997. A highly concentrated and illegal preparation of the carbamate insecticide aldicarb has also been used in inner-city communities.

6. *Children's exposures*

Because of their relatively high intake of food, water, and air per unit body mass, young children have unique exposure risks to pesticides and other toxicants compared to adults. Young children, because of their developmental immaturity, may be at higher risk of adverse health effects. Recognizing this, a committee of the National Research Council recommended that USEPA re-evaluate its decision-making process for setting acceptable pesticide exposures to reflect the unique exposure and health concerns of children. In 1996, Congress, in P.L. 104-170, the Food Quality Protection Act, directed USEPA to re-evaluate pesticide exposure tolerances in food and to consider all other sources of pesticide exposures in children, especially compounds with similar mechanisms of toxicity.

In response to such concerns, a number of research studies were conducted in the late 1980s and in the 1990s to assess nondietary pesticide exposures. These have established that residential contamination is an important component of total pesticide exposure in children.

Pesticide use was observed to be high among Missouri households in which a child had been diagnosed with cancer, as well as in a smaller number of households in which no cancer was reported. Nearly 98% of approximately 240 families surveyed used pesticides at least once per year; 67% 5 times or more per year. Eighty percent used pesticides during pregnancy, and 70% during the first 6 months of the child's life.

Though significant pesticide use was directed to controlling weeds (57%) and to gardens and orchards (33%), the primary area for pesticide use was inside the home itself (80%). Over 50% of surveyed families used flea collars and/or shampoos to control fleas or ticks on pets. A limited number used no-pest strips (9%) and Kwell (9%), a lice control product. The active ingredients in these products were dichlorvos and lindane, respectively, both suspected carcinogens.

Pesticide exposures of children in residences, not surprisingly, vary considerably. They depend on the nature and extent of pesticide use in the home, on lawn and gardens, and by occupants such as farm workers and commercial applicators. They also depend on the type of pesticides that are used, as they vary in volatility and persistence.

Like adults, children can be exposed to pesticides by inhalation, ingestion, and dermal contact. Toddlers are more likely to both ingest and dermally contact pesticide residues present in household dust and on surfaces because they play on floors and engage in hand-to-mouth behavior. As indicated in Chapter 2, such play and mouthing behavior (hands, toys, and other objects and surfaces) is considered to be the major pathway for childhood lead exposures and elevated blood lead levels.

Uptake of pesticides based on mouthing behavior is inferred from studies on house dust/outdoor soil ingestion and blood lead. Though infant children and toddlers spend more time indoors playing on dusty surfaces, ingestion rates of household dust have not been determined.

An attempt was made in USEPA's NOPES study (described previously), to assess pesticidal exposures to children by both inhalation and dust ingestion and to provide cancer risk assessments. The data suggest that dust ingestion constitutes a substantial portion of a child's pesticide exposure in some houses. It also suggests that many young children may be exposed to a greater range of pesticides through ingestion than inhalation since more pesticides are found in house dust than in air samples.

Many pesticides can be absorbed through the skin and move into the bloodstream. Though the potential for significant dermal exposure for infant children and toddlers exists, the magnitude of this exposure pathway has received relatively limited scientific evaluation.

Most investigations of residential pesticide exposure have focused on airborne concentrations based on fixed area sampling devices. Though some have demonstrated differences in air concentrations as a function of height (higher concentrations near the floor), such differences apparently do not persist after pesticide application. If they did, children playing at floor level would be expected to be exposed to higher air concentrations than adults.

Based on "personal dust cloud" phenomenon reported for office buildings, which indicate personal dust exposures 3 to 5 times greater than area samples, it would be desirable to measure breathing zone levels using personal monitors to assess levels of airborne pesticides to which children are actually exposed.

Significant exposures of children to pesticides have been reported in studies by the Centers for Disease Control in Arkansas and agricultural areas of Washington. In the former case, measurable urinary levels of PCP and a metabolite of *p*-dichlorobenzene were observed in nearly 100% of the 200 children evaluated. The source and nature of these exposures were unknown. In Washington agricultural areas, 47% of the children of applicators were found to have detectable levels of an organophosphate insecticide in their urine, with children age 3 to 4 years having significantly higher exposures than their older siblings.

7. Health effects

There are apparently no published studies that document airborne levels or potential adverse health effects associated with the use of antimicrobial compounds. However, there have been numerous cases of inadvertent ingestion poisoning of children (reported by Poison Control Centers).

Health concerns associated with indoor pesticide exposure include: acute symptoms due to high-level exposures which occur immediately after application; long-term risk of cancer from chronic exposures to substances such as chlordane, heptachlor, and PCP; and immunologic effects.

a. Acute symptoms. Mercury is a very toxic substance, and exposures associated with use of mercury biocides in latex paint have been a public health concern. Inhalation of high concentrations of mercury vapor has been reported to cause pain and chest tightness, difficult breathing, coughing, nausea, abdominal pain, vomiting, diarrhea, and headaches. Chronic exposure to low concentrations of mercury has been associated with vasomotor disturbances, muscle tremors, and personality changes.

Though the use of mercury biocides in latex paints was voluntarily discontinued in the U.S. in 1990, it had previously been widely used to paint interior walls of homes. Potentially significant mercury exposures of new home occupants likely occurred, since 25 to 30% of interior latex paints contained mercury; a high percentage of paints used PMA, apparently the most mercury-labile biocide used in paint. The health effects of such exposures are unknown and unknowable. A single case report suggests that clinical cases of mercury poisoning in children likely occurred and that numerous children and other individuals may have experienced subclinical symptoms.

Acute health effects have been reported for misapplication uses of home foggers, termiticides, wood preservatives, and other pesticide treatments. Major occupant health complaints have been associated with home fogging devices. They include headache, nausea, dizziness, and eye and skin irritation. It has been estimated that acute symptoms apparently due to the misapplication of pesticides indoors occurred in occupants of some 2.5 million homes in the U.S. during 1976–1977.

Many of the compounds that have been, and continue to be, used as insecticidal pesticides are chlorinated hydrocarbons or organophosphorous compounds. Though they have different chemical structures and toxicities relative to each other, both compound groups have similar modes of action. They are neurotoxic compounds that interfere with nerve signal transmission by inhibiting the enzyme acetylcholine esterase. Depending on the exposure dose and innate toxicity of substances to which humans have been exposed, a variety of symptoms that affect the central nervous system can be expected. For example, severe acute symptoms include muscular tremors.

In many reported acute pesticide exposure cases, symptoms may have been caused by exposure to active pesticidal ingredients, solvents used in pesticidal formulations, or both. Solvents used with pesticides are themselves neurotoxic. In diagnosing the cause of symptoms allegedly associated with such exposures, tests showing significantly depressed acetylcholine esterase activity indicate that the active pesticidal ingredient(s) was the more likely cause of some, if not all, reported symptoms.

b. Cancer. Chlorinated hydrocarbons such as chlordane, heptachlor, lindane, dieldrin, and PCP are known animal and suspected human carcinogens. As a consequence, their use has been restricted by USEPA. Subterranean termite treatments with chlordane and heptachlor were permissible under USEPA rules in the 1970s and 1980s because they were believed to pose a low probability of human exposure. However, data obtained in studies of military housing indicate that these low-volatility substances are in fact mobile, entering building spaces through the movement of soil gases. Although exposure concentrations are low, such compounds have relatively long half-lives (12 to 13 years in the ambient environment) so that exposures can continue to occur for decades. This would also be the case with exposures associated with PCP-treated logs and other wood materials.

In response to exposure concerns, a committee of the National Research Council conducted a risk assessment of pesticides used in termite control. They proposed guideline exposure levels of 5 $\mu g/m^3$ chlordane, 2 $\mu g/m^3$ heptachlor, 1 $\mu g/m^3$ aldrin/dieldrin, and 10 $\mu g/m^3$ chlorpyrifos, using occupational exposure limits and relative tumor incidence in animals. Based on this assessment, chlorpyrifos, the active ingredient used in most termiticidal treatments in the U.S. between the years 1980 and 2000, would have posed the lowest risk of cancer to humans in long-term exposures associated with residences.

Chronic exposure to pesticides is a major public health concern since many pesticides have been shown to be animal carcinogens or potential human carcinogens. Of 51 pesticides evaluated by the National Cancer Institute and National Toxicology Program (as of 1990), 24 have been demonstrated to be carcinogenic in chronic animal exposure studies. These include herbicides such as atrazine, amitrole, and trifluralin; organochlorine insecticides such as aldrin, chlordane/heptachlor, lindane, dieldrin, chlordecone, and DDT; organophosphorous insecticides such as dichlorvos and methyl parathion; fungicides such as captan, chlorothalonil, hexachlorobenzene, PCP, *o*-phenylphenol, and 2,4,6-trichlorophenol. Most pesticides identified as actual or potential animal carcinogens are still registered for use in the U.S.

Most information linking pesticides and human cancers has come from epidemiological studies of farmers/farm workers/farm families. Males typically have higher-than-expected rates of cancers of the lymphatic and hematopoietic systems, lip, stomach, prostate, brain, testes, soft tissues, and skin. Females are reported to have an excess of lymphatic, hematopoietic, lip, stomach, and ovarian cancers. There is strong evidence that selected organochlorine, organophosphorous, and arsenical insecticides, and triazine and phenoxyacetic acid herbicides may play a role in human cancer.

Organochlorine insecticides have been epidemiologically linked with leukemia (DDT, methoxychlor), non-Hodgkin's lymphoma, lymphoma (DDT, chlordane, toxophene), and lung cancer (chlordane). DDT exposures have also been linked with soft tissue sarcomas and pancreatic, breast, and liver cancers. Organophosphorous insecticides have been linked with leukemia (crotoxyphos, dichlorvos, Famphur) and non-Hodgkin's lymphoma (diazinon, dichlorvos, malathion). Phenoxyacetic acid herbicides such as 2,4-D have been linked with non-Hodgkin's lymphoma and prostate cancer.

As indicated previously in this section, potentially carcinogenic pesticides have been, or are being, used indoors; are present as a consequence of subterranean termiticidal applications; may be released from treated products (wood preservatives, pet collars, natural fibers and fabrics); or may be passively transported indoors on shoes and clothing. Because pesticides degrade slowly indoors, the potential exists for chronic exposures to low concentrations of a number of potentially carcinogenic pesticides. Cancer risks may, as a consequence, be increased in both children and adults. Because of the nature of exposure pathways and increased sensitivity to chemical exposures, children may be at special risk.

c. Childhood cancer. Children may be exposed to pesticides through their diet, parental pesticide use both inside and outside the home, and a "toxics brought home" phenomenon associated with parental occupational exposure. In a study conducted in the U.S. and Canada, acute nonlymphoblastic leukemia among children up to age 18 was observed to be strongly associated with occupational paternal pesticide exposure of >1000 days, with much higher risks for children under age 6. In a similar U.S. study among children under age 10, the use of a household pesticide ≥ once/week by either parent during the pregnancy of the affected child was observed to significantly increase the risk of acute lymphocytic leukemia. A higher risk was observed when either parent used a garden pesticide or herbicide ≥ once/month. In both studies, pesticide exposures inside the home appeared to be responsible for increased risk of developing leukemia. It has also been suggested that pesticide exposures can cause neuroblastoma, brain tumors, and hematological abnormalities in children.

Leukemia in very young children has been epidemiologically associated with the use of pest strips during the last 3 months of pregnancy, as well as other exposure periods. Pest strip exposures have also been associated with leukemia and brain cancer in older children. Dichlorvos, the pesticide used in pest strips at the time of these studies, is a known animal carcinogen and has been associated with leukemia in adult men. Lymphoma in children has been associated with use of insecticides.

Childhood brain cancer has been linked with pesticide exposures in some studies but not others. Recently, significant positive associations have been observed between childhood brain cancer and the use of herbicides to control weeds, as well as a variety of pesticides: bug bombs, termiticides, flea collars on pets, Kwell (a lice control agent), and garden/orchard pesticides (including carbaryl). Notably high statistical associations were observed with potential infant exposures (particularly from birth to 6 months of age) to pesticides used on pets.

d. Immunological effects.   Concern has been expressed that pyrethrins and synthetic pyrethroids can induce allergic reactions. Such responses have yet to be systematically investigated.

In case-control studies of very small populations, significant immunological effects at the cell level have been reported for apparent exposures to PCP, chlordane/heptachlor, and chlorpyrifos. Individuals exposed to PCP-treated log homes had activated T-cells, autoimmunity, functional immunosuppression, and B-cell disregulation. Activated T-cells, autoimmunity, and functional immunosuppression have also been reported for exposure to chlordane/heptachlor. Immunological abnormalities have also been reported in individuals apparently exposed to chlorpyrifos. Dose–response relationships for these chemicals have yet to be demonstrated to establish a definitive link between exposure and immunological changes.

The health significance of such changes is unknown. It may include an increased risk of infection, development of sensitivity to a variety of allergens, increased prevalence rates of chronic allergic rhinitis and asthma, development of autoimmune diseases such as lupus, and cancer. Since a number of studies have demonstrated a link between decreased immune function and increased risk of cancer, this may explain, in part, the reported increased risks of leukemia and lymphoma in children.

e. Other health concerns.   A new concern has recently been raised relative to chlorinated hydrocarbon pesticide exposures. These pesticides, as well as other chlorinated substances such as dioxins and PCBs, have been observed to be endocrine disruptors in animals; i.e., they mimic the action of the female hormone, estrogen. Relatively low exposure levels of estrogen-mimicking substances may cause reproductive changes in males and increase the risk of breast cancer in females. Although animal studies indicate that risks may be real, no definitive studies have been conducted to show endocrine disruptive effects in humans. The issue of environmental exposure to endocrine disruptors is relatively new, and additional studies are needed to determine whether low-level exposures to chlorinated hydrocarbons such as chlordane, heptachlor, PCP, and others have any adverse reproductive effects or increased cancer risks in humans.

Several recent studies have attempted to evaluate the potential relationship between exposure to pesticides (e.g., chlorpyrifos) and allergens (e.g., house dust mites). Although early studies indicate that such exposures may increase allergy symptoms, they should be judged to be preliminary, with additional studies needed to confirm relationships and determine their mechanism of action (if any relationship is found).

Readings

Brown, S.K. et al., Concentrations of volatile organic compounds in indoor air — a review, *Indoor Air*, 4:123, 1994.

Godish, T., Aldehydes, in *Indoor Air Quality Handbook*, Spengler, J.D., Samet, J.M., and McCarthy, J., Eds., McGraw-Hill Publishers, New York, 2000, chap. 32.

Godish, T., Formaldehyde — sources and levels, *Comments in Toxicology*, 2, 115, 1988.

Lewis, F., Pesticides, in *Indoor Air Quality Handbook*, Spengler, J.D., Samet, J.M., and McCarthy, J., Eds., McGraw-Hill Publishers, New York, 2000, chap. 35.

Marbury, M.D. and Kreiger, R.A., Formaldehyde, in *Indoor Air Pollution: A Health Perspective*, Johns Hopkins University Press, Baltimore, 1991, 223.

Maroni, M., Siefert, B., and Lindvall, T., *Indoor Air Quality. A Comprehensive Reference Book*, Elsevier, Amsterdam, 1995, chap. 2.

Molhave, L., The Sick Building Syndrome (SBS) caused by exposure to volatile organic compounds, in *The Practitioner's Approach to Indoor Air Quality, Proc. Indoor Air Qual. Internatl. Symposium*, Weekes, D.M. and Gammage, R.B., Eds., American Industrial Hygiene Association, Falls Church, VA, 1990, 1.

Molhave, L., Volatile organic compounds, indoor air quality and health, *Indoor Air*, 1, 387, 1991.

National Research Council (Committee on Aldehydes), *Formaldehyde and Other Aldehydes*, National Academy Press, Washington, D.C., 1991.

Spengler, J.D., Samet, J.M., and McCarthy, J.F., Eds., *Indoor Air Quality Handbook*, McGraw-Hill Publishers, New York, 2000, chaps. 20–25, 31–36.

Wallace, L.A., Volatile organic compounds, in *Indoor Air Pollution. A Health Perspective*, Samet, J.M. and Spengler, J.D., Eds., Johns Hopkins University Press, Baltimore, 1991, 253.

World Health Organization, Indoor air quality: organic pollutants, *EURO Reports and Studies 171*, WHO Regional Office for Europe, Copenhagen, 1989.

Questions

1. Describe major sources of formaldehyde exposure in buildings.
2. Characterize formaldehyde exposure potentials associated with UFFI houses, new mobile homes, older mobile homes, tobacco smoking.

3. The aldehydes, formaldehyde and acrolein, are potent mucous membrane irritants. Why?
4. In what environments are glutaraldehyde exposures a potential health concern?
5. What are the primary health concerns associated with indoor formaldehyde exposures?
6. Describe how temperature, humidity, and outdoor temperature affect indoor formaldehyde concentrations.
7. When multiple formaldehyde sources are present, air concentrations are not equal to the sum of each source's emission potential. Why?
8. What is the significance of ozone and aldehydes in indoor chemistry?
9. How may indoor chemical reactions result in building-related health complaints? Describe specific concerns.
10. What is the common chemical sense? How may it be related to SBS-type symptoms?
11. Use of the TVOC theory in diagnosing problem buildings should be limited. Why?
12. What is the significance of sink effects in exposures to VOCs and SVOCs?
13. What common indoor VOC/SVOC contaminants have a potential for causing cancer in humans and pets?
14. Pesticides used outdoors are commonly found indoors. Why is this the case?
15. What biocides/pesticides are present in your home?
16. Why would you expect air concentrations of SVOCs to be low relative to concentrations of VOCs commonly observed indoors?
17. How can you be exposed to airborne mercury in your home?
18. Describe three ways that occupants of residential environments are exposed to pesticides indoors.
19. Though pentachlorophenol was commonly used as an exterior wood preservative, it was present in the blood of a large percentage of Americans in the late 1970s. How is it possible that so many Americans were exposed?
20. Describe the potential health significance of chlorinated hydrocarbon pesticides and exposures that occur indoors.
21. What are the sources and potential significance of polyvalent alcohols and their derivatives in indoor air?
22. Describe potential pathways for pesticide exposures to young children in residences?
23. What health concerns are potentially associated with low-level exposures to plasticizer chemicals?
24. Describe factors that contribute to elevated pesticide residues in residential floor dust.
25. What factors account for elevated levels of cyclodiene pesticides in indoor air and floor dust?
26. Describe childhood cancer concerns that may be associated with residential pesticide use/exposure.
27. Describe the potential significance of the "toxics brought home" phenomenon.
28. What are the immunologic concerns associated with exposures to pesticides such as chlorpyrifos, chlordane/heptaclor, and other pesticide chemicals indoors?

chapter five

Biological contaminants — illness syndromes; bacteria; viruses; and exposures to insect, mite, and animal allergens

There is increasing evidence that a significant proportion of illness symptoms and disease associated with building environments is due to particulate-phase and, to a much lesser extent, gas-phase exposures to substances produced by a variety of organisms. These exposures include: airborne mold (fungi), bacteria and viruses, which may be viable (living) or nonviable; fragments of bacteria and mold which contain antigens, endotoxins, glucans, and/or mycotoxins; microbially produced volatile organic compounds; and immunologically active particles produced by insects, arachnids (mites, spiders), and pets (cats, dogs, etc.).

Contaminants of biological origin such as bacteria, viruses, and fungi can cause infectious disease by airborne transmission. Exposures to mold spores/fragments and allergens produced by insects, arachnids, and pet danders may cause immunological responses such as chronic allergic rhinitis and asthma. Exposures to high concentrations of small fungal spores and the spores of the higher bacteria may cause hypersensitivity pneumonitis. Exposures to fungal glucans, bacterial endotoxins, or microbial volatile organic compounds (MVOCs) may cause inflammatory responses in the respiratory system. Mycotoxin exposure may cause severe toxic effects.

Of special concern are those biological contaminants that cause immunological sensitization manifested as chronic allergic rhinitis, asthma, and hypersensitivity pneumonitis. Chronic allergic rhinitis is an indoor environment (IE) concern because of its very high and increasing prevalence rate in developed countries and the similarity of allergy symptoms to those caused by gas-phase irritants. Likewise, asthma is a major public health concern because of its relatively high and increasing prevalence rate and the seriousness of the disease. The nature of these ailments is described below. In addition, special attention is given in this chapter to illness and exposure concerns associated with bacteria and bacterial products, viruses, and allergen-producing arachnids, insects, and mammals. Mold is discussed in Chapter 6.

I. Illness syndromes

A. Chronic allergic rhinitis

Chronic allergic rhinitis or common allergy is a relatively nonserious illness caused by exposures to substances that have antigenic/allergenic properties. In sensitized individuals, inhalation exposures to specific allergens result in inflammatory symptoms of the nose, throat, sinuses, eyes, and upper airways characterized by congestion, runny nose, phlegm production, sneezing, and in many cases, coughing.

Inhalant allergies are commonly caused by immunologic sensitization to dust mite antigens, mold spores and hyphal fragments, pet danders, and pollen. They may also be caused by exposure to antigens produced by insects (e.g., cockroaches), birds, rats, and mice.

Symptoms are produced as a result of immunological sensitization and subsequent exposures that produce inflammatory responses by one of four different mechanisms. The most important of these is the production of IgE antibodies in response to exposures to an antigenic substance, which can cause a detectable immune response. Antigens that evoke an IgE response are described as allergens.

Antigens/allergens are relatively large molecules (typically proteins). However, some highly reactive compounds, such as formaldehyde (HCHO), toluene diisocyanate (TDI), and trimellitic anhydride (TMA) can complex with proteins to produce hapten antigens that induce allergenic responses.

Allergy mediated by IgE antibodies is very common. Approximately 40% of the population of the U.S. has been estimated to have IgE antibodies specific for one or more allergens, with 50 to 65% of these experiencing clinical symptoms. Approximately 10% of the U.S. population has experienced severe chronic allergic rhinitis; that is, symptoms severe enough to be diagnosed and treated by a physician. Accordingly, tens of millions of individuals in the U.S. experience seasonal or persistent allergy symptoms each year. The prevalence rate for chronic allergic rhinitis appears to have increased significantly over the past two decades.

Sensitivity to allergens varies with age. In children under the age of 5, sensitivity to inhalant allergens has been reported to be about 22%, rising to 45% between the ages of 5 and 13. However, peak sensitivity reportedly occurs among individuals in the age range of 20 to 45 years.

Whether an individual becomes sensitized depends on the nature of the antigen, the amount of antigen, route and frequency of the exposure, and one's genetic background. Some individuals have a high propensity for becoming sensitized to (often many) antigens. This genetically determined predisposition to allergy is called atopy. Most individuals with severe allergic symptoms are atopic.

B. Asthma

Asthma is a disease that affects the respiratory airways of the lung. It is characterized by severe episodic constriction of bronchial tubes, which results in chest tightness, shortness of breath, coughing, and wheezing. Symptoms, which vary in severity, may occur within hours of exposure to allergens or nonspecific irritating substances (or activities) or have a delayed onset (4 to 12 hours later).

The prevalence of asthma in the U.S. has been estimated to be about 4.3%, or about 10 to 12 million individuals a year experiencing mild to severe asthmatic symptoms. The prevalence rate has increased 50% since 1980, with the annual death rate doubling to approximately 5000 per year (Figure 5.1). The increase in the prevalence of asthma has been disproportionately higher in females, particularly for black females (prevalence rate 6%). The prevalence rate for white females is about 4.7%.

Asthma prevalence in children varies both regionally and across racial groups. Data obtained from the National Health and Nutritional Examination Survey II (NHANES) conducted for the period 1976–1980 showed that the cumulative percentage of asthma symptoms (sometime in their lives) was approximately 8.3% in midwestern children and 11.8% in southern children in the age range of 3 to 11 years. Highest prevalence rates (13.4%)

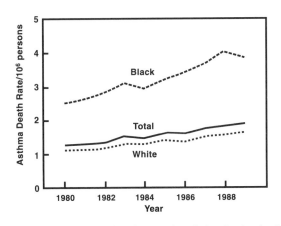

Figure 5.1 Changes in incidence of asthma-related deaths in the U.S. since 1980. (From CDC, Asthma in the U.S., 1980–1990, *Morbidity Mortality Weekly Report*, 39, 493, 1992.)

were reported for black children. Since this survey was conducted, significant increases in prevalence rates have been reported, with most of the increase occurring in inner-city black and Hispanic populations. Unlike adults, where asthma is more prevalent among females, male children have higher prevalence rates than female children, with the difference being twofold at certain ages.

Asthma is the major medical cause of absenteeism in school-age children. It is also the leading cause of childhood hospital admission and long-term medication use. Health-care costs for treatment of asthmatic symptoms among all age groups in 1990 were estimated to be 6.2 billion dollars or 1% of all health-care costs.

There is strong evidence that asthma is associated with sensitization to common inhalant allergens. The primary allergens that appear to cause sensitization are produced by dust mites in humid/wet regions; by the German cockroach in many U.S. cities; and pet danders in northern and dry regions. Sensitization appears to be the major risk factor for nonspecific bronchial hyperreactivity and symptomatic asthma.

Though sensitization has been shown to be an important risk factor, no definitive link has been established between allergen exposure and symptoms or severity of asthma. The absence of an apparent dose–response relationship between symptom severity and current exposure may be due to a variety of factors. These include individual differences in response, inadequate exposure measurements, and concurrent exposures to a variety of allergens.

The link between allergen exposure and asthmatic symptoms is an indirect one. Significant decreases in prevalence and severity of asthmatic symptoms have been observed in hospital patients where allergen exposure is low, and in a variety of intervention studies.

Asthma appears to be rare among aboriginal rural populations. Its prevalence increases dramatically with urbanization and life in more modern housing. Increases in the prevalence of asthma may be due to changes in lifestyle and housing, with associated increases in allergen exposures.

C. Hypersensitivity pneumonitis

Hypersensitivity pneumonitis, or extrinsic allergic alveolitis, is a group of immunologically mediated lung diseases associated with repeated exposures to a variety of biogenic aerosols that affect tissues in the periphery of the lung.

Once sensitization has occurred, acute symptoms develop in response to high concentration exposures. Respiratory and systemic symptoms occur within 4 to 6 hours after exposure. These include fever, chills, a nonproductive cough, shortness of breath, myalgia (i.e., ache all over), and malaise. Symptoms may persist for 18 to 24 hours with spontaneous recovery. In office and industrial environments, symptoms diminish during the work week and begin anew on reexposure after weekends and vacations.

When exposure is chronic, hypersensitivity pneumonitis is characterized by irreversible lung tissue damage, with progressive shortness of breath, cough, malaise, weakness, anorexia, and weight loss. Hypersensitivity pneumonitis occurs mostly in adults and has been linked to a number of occupations. Malt worker's lung, cheese washer's lung, farmer's lung, and bagassosis are forms of hypersensitivity pneumonitis.

Outbreaks of hypersensitivity pneumonitis have been reported in a variety of building environments. These have been associated with contaminated components of heating, ventilating, and air-conditioning systems (including condensate drip pans and ductwork) and a variety of moldy materials.

Hypersensitivity pneumonitis appears to be caused by exposures to very high levels of bacterial (thermophilic actinomycetes) spores or relatively small (<3 µm) mold spores that are deposited in lung alveolar tissue, causing inflammatory responses and tissue damage.

II. Bacteria and viruses

Bacteria and viruses, because of their small size, readily become airborne and remain suspended in air for hours. The presence of bacteria has been confirmed by air sampling; the presence of viruses is inferred from the spread of respiratory infections. Airborne bacteria and viruses in indoor spaces are of concern because they may transmit infectious diseases such as tuberculosis (TB), meningitis, influenza, and colds. Bacterial endotoxins may also pose significant exposure and health concerns.

A. Bacteria

Bacteria are single-celled organisms that are characterized by their lack of a true nucleus and their relatively small size. They reproduce by simple fission. Bacteria vary in size and shape. In bacterial species that have historically characterized this group of organisms, cells assume three distinct morphological shapes: spherical (coccus), rod-shaped (bacillus), and spiral/curved (spirillum, spirochete, vibrio). Species in the actinomycetes, sometimes called the higher bacteria, form branching, very slender filaments and, in some families, produce fungal-like mycelia. The "hyphae" of actinomycete species break into spherical or rod-shaped segments that function as asexual spores.

Individual cells in the "lower bacteria" vary in size; coccoid bacterial cells range from 0.4 to 2 µm; the smallest bacilli have a length of 0.5 µm and a diameter of 0.2 µm; pathogenic bacilli have diameters no greater than 1 to 3 µm; nonpathogenic bacilli have diameters and lengths of up to 4 and 20 µm, respectively; spirilla typically have diameters of <1 µm with lengths from 1 to 14 µm.

Microscopic bacteria form colonies that can be easily seen with the naked eye on culture media in relatively short periods of time (e.g., 24 hrs). The "lower bacteria" look "slimy" or glistening; the actinomycetes look somewhat like fungi (Figure 5.2).

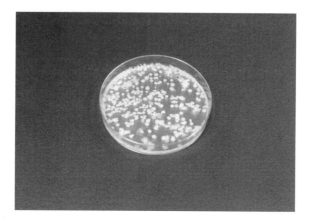

Figure 5.2   Colonies of higher bacteria (actinomycetes).

Most bacteria are heterotrophic; i.e., they derive their nutrition by metabolizing simple–complex organic molecules. Bacteria can grow on many substrates, and as a consequence are widely distributed in nature. In decomposing dead organic matter or destroying living tissue, they serve as organisms of deterioration or decay; they are commensal when they live on the by-products of organisms without causing disease; they are parasitic when they infect an organism and cause disease.

Bacteria vary in their environmental requirements and tolerances. These include nutritional substrates, availability of oxygen, moisture levels, pH, and temperature. Water is required for the growth of individual cells and for rapid reproduction. Temperature optima vary for different species/types. Mesophilic bacteria grow well in the temperature range of 20 to 40°C; thermophiles at 40 to 80°C. Bacteria do not tolerate ultraviolet light and may be killed in a few hours when exposed to direct sunlight.

1.   Bacterial toxins

Bacteria produce a variety of metabolites during growth. In the actinomycete group, these include antibiotics which limit the growth of other bacteria. Streptomycin produced by the members of the genus *Streptomyces* is a classic example.

Bacteria may produce both exotoxins and endotoxins. Exotoxins are produced within bacterial cells and excreted onto or into substrates. The best known exotoxin is botulism toxin, produced by *Clostridium botulinum*.

Endotoxins are produced in the outer membrane of Gram-negative bacteria. The cell wall is covered by a lipopolysaccharide layer that consists of a complex of lipids, carbohydrates, and proteins. Endotoxins are large molecules of Lipid-A and a core polysaccharide (2-keto-3-deoxy-D-mannooctalosomic acid). They are heat stable and are only released after bacterial walls have been damaged or destroyed. *Escherichia coli* (*E. coli*, found in human intestines) and the genera *Pseudomonas* and *Salmonella*, are common bacteria that produce endotoxins. Endotoxins vary with species, with a greater variation in the structure of the polysaccharide and less with the lipid.

Endotoxin levels in air and other media are determined by bioassay techniques using *Limulus* (horseshoe crab) amebocyte lysate (LAL), which clots or increases in turbidity on exposure to endotoxin. Endotoxin exposures have been reported to be high in airborne dusts associated with cotton processing and hog and poultry confinement operations. High concentrations have also been reported for metal-cutting fluid mists, wastewater treatment mists, and mists associated with the processing of fiberglass insulation. Highest concentrations in these occupational environments range from 100 to 16,000 endotoxin units per cubic meter (EU/m^3). Outdoor endotoxin levels are reported to be <2 EU/m^3. Endotoxin levels ranging from 100 to 408 EU/m^3 have been reported in problem office buildings in the Netherlands. Though the source is unknown, it is suspected that elevated endotoxin levels are associated with cool-mist humidifiers used to climate-control mechanically ventilated buildings in European countries. Though modular and central humidifiers are used by millions of homeowners and residential tenants, cool-mist humidifiers are not used in mechanically ventilated buildings in North America. Exposure to high endotoxin levels is reported to be the probable cause of humidifier fever, an illness syndrome somewhat similar to hypersensitivity pneumonitis (but without progressive lung damage). Outbreaks of humidifier fever are commonly reported in northern European countries.

2. Microbial VOCs

A variety of microbial volatile organic compounds (MVOCs) are produced during bacterial growth. They are associated with the sour smell of wash waters; the septic odor of sewage; the smell of human wastes and other dungs; the putrid odors of animal and plant decay; the earthy odor of soil, human body odor; etc. Volatile emissions from bacterial growth may include inorganic gases such as methane, hydrogen sulfide, and ammonia from anaerobic decomposition; amines such as cadaverine associated with the decay of animal flesh; and a variety of alcohols and ketones. Geosmin, an MVOC associated with some fungi, is the major odoriferous MVOC produced by the "higher bacteria," the actinomycetes. It is responsible for the characteristic odor of soil.

3. Indoor exposures

Relatively few studies have been conducted to assess airborne bacterial levels in indoor spaces and to identify major genera and species present. Most assessments of airborne bacterial concentrations have been conducted using culturable/viable sampling procedures. Typically, concentrations vary from several hundred to several thousand colony-forming units per cubic meter (CFU/m^3). Under building closure conditions, humans themselves appear to be the major source of airborne bacteria. Gram-positive cocci such as

Staphylococcus, Streptococcus, and *Micrococcus,* shed from human skin and mucous membranes of the upper respiratory system, comprise 85 to 90% of culturable/viable bacteria collected during air sampling in indoor spaces. The reported viability of airborne bacteria is very low (<1%), so total airborne concentrations may be two orders of magnitude higher than culturable/viable concentrations.

Since bacteria require liquid water to grow and multiply, sources of indoor airborne bacteria are relatively limited. Major sources include humans and materials/equipment in which liquid water is present. This includes, in the latter case, air-conditioning system cooling coil condensate pans, central and free-standing cool-mist humidifiers, hot water heaters (thermophilic bacteria), and outdoor sources such as cooling towers and evaporative condensers. Cooling towers, evaporative condensers, and hot water heaters are known sources of *Legionella pneumophilia,* the cause of Legionnaires' disease infections. In residences, it is likely that moist/wet soil under crawlspaces may be a source of actinomycete species and the odor of geosmin in indoor spaces.

4. Health concerns

The contamination of indoor air by viable bacterial cells or substances produced by bacteria poses a variety of health concerns. These include the spread of infectious disease by bacterial aerosols, outbreaks of hypersensitivity pneumonitis associated with thermophilic actinomycetes, illness symptoms and humidifier fever associated with bacterial endotoxins, infections in surgical patients, and potential health effects associated with MVOCs.

a. Infectious disease. A number of major diseases are spread by means of bacterial aerosols. Selected diseases and their causal organisms are summarized in Table 5.1. Of special recent concern have been TB and Legionnaires' disease (in both the greater public health and IE contexts). Tuberculosis and Legionnaires' disease are discussed in detail below.

Of note are spinal meningitis outbreaks in high-population-density buildings, such as army barracks, and the specter of exposure to anthrax

Table 5.1   Major Infectious Diseases Associated with
Bacterial Aerosols

Disease	Causal organism
Tuberculosis	*Mycobacterium tuberculosis*
Pneumonia	*Mycoplasma pneumoniae*
Diphtheria	*Corynebacterium diphtheriae*
Anthrax	*Bacillus anthracis*
Legionnaires' disease	*Legionella pneumophila*
Meningococcal meningitis	*Neisseria meningitides*
Respiratory infections	*Pseudomonas aeruginosa*
Wound infections	*Staphylococcus aureus*

spores as a weapon of war or act of terrorism. Although rare, meningitis is a relatively deadly, and as consequence, "scary" disease. Anthrax, a disease primarily associated with animals, has been developed into a highly dangerous germ warfare agent. Release of even a small quantity of anthrax spores would create an episode of deaths and injury of nightmarish proportions.

b. Tuberculosis.   Until the beginning of the 20th century, TB was a major cause of death in the U.S., as well as many other countries. Due to advances in diagnosis, prevention, and treatment, the prevalence of TB declined dramatically. It is, however, still an important disease in the U.S., with 20,000 cases and a few thousand deaths reported each year. While the incidence rate has now stabilized, it had been increasing in the U.S. since 1970 due to the influx of immigrants from countries where TB is still common; the development of resistance by the causal organism to a variety of antibiotics; and increases in susceptible populations. (immunocompromised individuals such as AIDS patients and individuals living under poor sanitary conditions, such as the homeless and alcoholics).

Tuberculosis is a chronic infection of the lungs but may also affect the intestines, kidneys, lymph nodes, bones, and joints. The causal organism, *Mycobacterium tuberculosis,* is a slender, rod-shaped, nonmotile, nonspore-forming bacillus. It is relatively resistant to the effects of drying and may remain alive in dust and dried sputum in dark places for weeks or months. Since it has no natural existence outside an infected individual, humans are the most important reservoir of infection, with exposure to aerosols being the most common mode of transmission. Studies have shown that exposure to one aerosol droplet may be sufficient to cause infection.

Outbreaks of TB have occurred in hospitals, nursing homes, homeless shelters, correctional facilities, and residential AIDS care centers. Individuals in such environments are at a much higher risk of exposure and infection than in other environments where the infection rate is much lower. Because the risk of TB infection in patients and staff in health-care facilities can be relatively high, the Occupational Safety and Health Administration (OSHA) has proposed an occupational exposure standard for TB which is scheduled to go into effect in the year 2000.

c. Legionnaires' disease.   Legionnaires' disease was initially identified after a major investigation of an outbreak of a pneumonia-like illness (by the Centers for Disease Control) that affected approximately 220 individuals at a state American Legion convention at the Hotel Bellevue in Philadelphia, PA, in 1976. The disease was characterized by symptoms of fever, cough, shortness of breath, chest pain, headache, myalgia, diarrhea, and confusion. It is a progressive pneumonia with a fatality rate of approximately 15%.

Nearly 10% of reported cases of Legionnaires' disease are associated with outbreaks affecting several or more individuals, but in most cases it occurs sporadically, affecting single individuals. Approximately 25% of reported cases occur in hospital patients, with about 75% of cases being community-acquired.

The causal agents of Legionnaires' disease are species of bacteria commonly found in freshwater environments (circa 20 to 25% of samples are positive). There are approximately 17 species of the genus *Legionella* that cause human disease, with *L. pneumophila* being responsible for approximately 80% of the 1000 or so cases reported annually in the U.S.

Members of the *Legionella* genus are Gram-negative, aerobic, rod-shaped bacteria that live as facultative intracellular parasites in other aquatic microscopic organisms. Because of its relative tolerance to chlorine, *L. pneumophila* is commonly found in potable water systems.

Legionella species grow well at elevated temperatures (circa 35°C). As a consequence, abundant populations are common in the waters of cooling towers and evaporative condensers, hot water heaters, whirlpools, spas, and hot tubs.

Legionnaires' disease appears to be an opportunistic infection since host risk factors play an important role in development of the disease. Risk factors include tobacco smoking, alcohol consumption, middle age, male gender, preexisting lung disease, diabetes, AIDS, immunosuppressive therapy, and travel.

The primary route of infection is inhalation exposure to aerosols produced by *Legionella*-contaminated water. Though there is a potential for exposure to *Legionella* in residential and motel/hotel room showers, such exposures have yet to be documented and reported. They have been documented in hospitals, where control of *Legionella* is a major infection control concern.

d. *Pontiac fever.* Pontiac fever is a nonpneumonic, self-limiting (2- to 3-day duration), influenza-type illness apparently caused by exposure to certain strains of *L. pneumophila*. The attack rate is approximately 95% compared to 5% for Legionnaires' disease. Pontiac fever, unlike Legionnaires' disease, is not fatal.

The disease was first identified a decade after a major outbreak occurred among health department employees in Pontiac, MI. Analysis of preserved blood samples indicated that both Pontiac fever and Legionnaires' disease shared the same etiological agent. Symptoms consist of fever, chills, headache, and myalgia. After initial illness, affected individuals do not become symptomatic again. In the first documented case, the source of exposure was believed to be an evaporative condenser whose aerosol became entrained in the building's air handling system. Other outbreaks have been reported in an office building, an automobile assembly plant, and among maintenance workers inside a steam turbine condenser. Based on the relatively few reports in the scientific literature, Pontiac fever is either rare or rarely diagnosed.

Pontiac fever appears to be associated with large numbers of dead bacterial cells or cells with reduced virulence that cannot produce sustained growth in the lungs.

e. Hypersensitivity diseases. Because of the presence of antigens on their cell walls and membranes, bacteria have the potential to cause allergic reactions similar to those observed with mold, insect excreta, and animal danders. Though a few studies have shown that exposure to various Gram-negative and Gram-positive bacteria can cause allergic responses (such as the release of histamine), bacteria have not been clinically identified as a cause of chronic allergic rhinitis or asthma.

The higher bacteria, especially thermophilic actinomycetes, have been definitively identified as a major causal organism in the development of hypersensitivity pneumonitis. Exposures to thermophilic actinomycete species *Faenia rectivirgula, Saccharomonosporo viridis,* and *Thermoactinomyces vulgaris* are known to cause farmer's lung, with *T. vulgaris, T. sacchari,* and *T. thalophilus* responsible for bagassosis, a hypersensitivity disease of sugar cane workers. Hypersensitivity pneumonitis has also been associated with composting operations and air conditioning systems. In residences, cases of hypersensitivity pneumonitis have been associated with contamination of a sonic humidifier by the bacterium, *Klebsiella oxytoca* and a *Bacillus subtillus* infestation of bathroom wood materials disturbed during remodeling.

The relationship between thermophilic actinomycete exposure and hypersensitivity pneumonitis appears to be due to very high exposures to small-diameter spores. Actinomycete spores have aerodynamic diameters in the range of 1 to 1.5 μm, well in the respirable range wherein spores have a high probability of deposition in the lower respiratory airways and lung tissue (where hypersensitivity pneumonitis pathology and immediate symptoms occur). This is consistent with mold species as well. Molds which produce small spores, such as *Aspergillus fumigatus* and *Penicillium* spp. have also been identified as causes of hypersensitivity pneumonitis. The lower bacteria (which consist of small aerodynamic diameter cells) have also been identified as causing hypersensitivity pneumonitis.

The aerodynamic size phenomenon may be a significant one in terms of the potential role of bacteria in causing hypersensitivity diseases such as chronic allergic rhinitis and asthma. In both cases, disease/symptom induction is typically associated with allergens with larger aerodynamic diameters (>2.5 μm). Such allergens tend to be deposited in the nasopharyngeal and upper respiratory airways where allergy and asthma symptoms manifest themselves. Because of their small size, bacteria may not pose a significant immunological sensitization and exposure risk for allergy and asthma.

f. Endotoxins and humidifier fever. As indicated previously, endotoxins are released into the environment when Gram-negative bacteria undergo lysis or disruption of their outer membranes. Endotoxins are potent cytotoxic compounds that can kill pulmonary macrophages and affect the immune responses of neutrophils and basophils. In addition, they stimulate cell division and cause membrane dysfunction. Clinical manifestations of inhalation exposure include fever, increased respiratory rate, chest tightness, airway hyperreactivity, and bronchitis.

A consensus has recently developed that endotoxins produced by Gram-negative bacteria such as *Pseudomonas* spp., *Flavobacterium* spp., *Klebsiella* spp., and *Enterobacter aerogenes* and other *Enterobacter* species cause humidifier fever, a respiratory disease reported in European office buildings using cool-mist humidification systems. Unlike hypersensitivity pneumonitis, chronic exposures to endotoxin do not appear to cause chronic lung damage, although some animal and epidemiological studies suggest that very high-level exposures can cause lung changes similar to emphysema.

Endotoxin exposures have been reported to be the causative agent in what has been described as toxic pneumonitis, inhalation fever, and organic dust syndrome (associated with dusty work). Symptoms, which include fever, chills, myalgia, and other influenza-like symptoms, are short-lived (disappear in a few days). The number of cases as a function of populations exposed may be close to 100%. A guideline value of 200 EU/m^3 for endotoxin exposure has been recommended.

Endotoxin exposures appear to be the cause of reactive airway disease syndrome (RADS), a form of nonallergenic asthma. Reported symptoms include respiratory airway irritation; stuffy, swollen nose; dry cough; increased airway responsiveness; and, in advanced cases, pulmonary impairment. Reactive airway disease syndrome is often accompanied by symptoms of headache, fatigue, and joint pain. A guideline value of 10 EU/m^3 has been proposed to reduce the incidence of RADS symptoms among those potentially exposed.

Endotoxin exposure has been reported to potentiate allergic and asthmatic symptoms associated with exposure to dust mite allergens in dust mite-sensitive individuals. Dust endotoxin levels >1.1 EU/m^3 have been associated with severe shortness of breath, increased medication use, and pulmonary function changes.

Endotoxin may also contribute to sick building symptoms. Studies conducted in the Netherlands report significantly higher prevalence rates of mucous membrane symptoms among occupants of buildings with dust endotoxin levels >100 EU/m^3. Consistent with these are studies of Danish municipal buildings wherein symptoms of fatigue, headache, heavy headedness, hoarseness, and sore throat were closely associated with airborne concentrations of both Gram-negative bacteria and endotoxin.

B. Viruses

Viruses, the smallest living things, are best known for their ability to cause human disease. They consist of single or double strands of DNA or RNA covered by a protein coat. Since viruses have no cytoplasm, they must infect host cells in order to replicate. As such, they are obligate parasites.

Viruses vary in size from about 20 nm, with a few genes, to 200 to 300 nm, with hundreds of genes. They may appear as highly complex structures with unique geometries or crystalline-like character.

A number of viral diseases are transmitted in aerosols derived from infected individuals. Selected viral diseases and associated causal viruses

Table 5.2 Major Infectious Diseases
Associated with Viral Aerosols

Disease	Virus type
Influenza	Orthomyxovirus
Cold	Coronavirus
Measles	Paramyxovirus
Rubella	Togavirus
Chicken pox	Herpes virus
Respiratory infection	Adenovirus

transmitted through air are summarized in Table 5.2. The airborne nature of the transmission of flu, cold, and respiratory infection viruses explains the high attack rates of these diseases, particularly in high-population-density environments such as schools and office buildings.

Viruses mutate readily. Therefore, new virulent forms of cold and flu viruses appear, and permanent immunity does not occur as a consequence of a single infection. Primary risk factors are susceptible hosts; the presence of an infected individual with an active case of the disease; the proximity of potential hosts and an infectious aerosol source; and, in many buildings, inadequate ventilation.

III. Settled organic dust

Settled floor and other surface dust has been implicated as a potential causal factor for sick building-type symptoms (SBS) by European investigators. The component of surface dusts that appears to be closely associated with SBS is described as macromolecular organic dust (MOD). This dust appears to be immunogenic, consisting of proteins, DNA, and other large molecules of biological origin. Modified human serum albumin appears to comprise the major portion of MOD.

Settled dust is an apparent source of volatile organic and semivolatile organic compounds (VOCs and SVOCs), including aldehydes (C_2–C_{11}) and carboxylic acids. Microbial decomposition of lipids or *de nova* synthesis has been suggested as the origin of microbial VOC (MVOC) emissions from settled dust. Significant associations between surface dust MVOC emissions and symptoms of mucous membrane irritation and difficulty concentrating have been reported (Table 5.3). A strong correlation between compounds emitted from MOD and mucous membrane symptoms among occupants of office buildings has also been reported.

IV. Mites

Exposure to house dust mite allergens has been reported to be the major cause of chronic allergic rhinitis, asthma, and atopic dermatitis in North America and Europe. The prevalence of mite sensitization among asthmatic

Table 5.3 Least Squares Statistical Relationships Between VOC Emissions from Floor Dust and Mucous Membrane Symptoms and Difficulty Concentrating Among Occupants of Office Buildings

| Average mucous membrane irritation | | Difficulty concentrating | |
VOC	Modeling power	VOC	Modeling power
2-Methylpropanal	0.640	Pentanoic acid	0.562
Hexanoic acid	0.586	Hexanoic acid	0.558
2-Alkanone M^+ = 142	0.586	Hexanal	0.526
3-Methylbutanal	0.498	Heptanoic acid	0.422
Octane	0.412	2-Methylbutanal	0.418
Pentanoic acid	0.375	Butyric Acid	0.380
Heptanoic acid	0.362	Benzaldehyde	0.366
2-Undecanone	0.332		
5-Methyl-3-methylene-5-Hexene-2-on M^+ = 124	0.305		

Source: From Wilkins, C.K. et al., *Indoor Air,* 3, 283, 1993. With permission.

and nonasthmatic emergency room patients in two cities in the U.S. can be seen in Table 5.4.

House dust mites are arachnids; they are related to spiders, ticks, and scorpions. Major species identified in floor dust samples include *Dermatophagoides farinae, D. pteronyssinus, D. microceras, Euroglyphus maynei,* and *Blomia tropicales. D. farinae, D. pteronyssinus,* and *E. maynei* are the most common house dust mites found worldwide.

Dust mites phagocitize (eat) decomposing (defatted) human skin scales. As a consequence, they are commensal with humans, depending on us for their food supply. As a result, they are abundant in high-use areas of building spaces, particularly residences. In temperate regions of the world, high mite populations occur in carpets, soft furniture, and mattresses. *D. farinae* and *D. pteronyssinus* are the most common mite species found in floor dust samples in the U.S. Most residences are cohabited with both species, with one species usually dominating.

Table 5.4 Prevalence (%) of IgE Antibodies to Common Indoor Allergens Among Emergency Room Patients

Location	Patient type	House dust mite	Cockroach	Cat
Atlanta, GA	Asthmatic (N=81)	60	26	7.4
	Nonasthmatic (N=63)	21	9.5	4.7
Wilmington, DE	Asthmatic (N=114)	32	22	32
	Nonasthmatic (N=114)	15	6	5

Source: Data extracted from Chapman, M.D., Cockroach allergens: a common cause of asthma in North American cities, *Mosby-Yearbook,* 1, 1993.

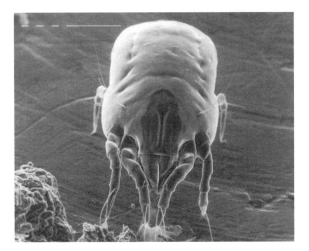

Figure 5.3   Dust mite. (Courtesy of Arlian, L., Wright State University.)

Dust mites are relatively small and cannot be seen with the unaided eye. Adult females of *D. farinae* are approximately 0.36 mm long and 0.43 mm wide; females of *D. pteronyssinus* are somewhat smaller. A dust mite magnified under a scanning electron microscope can be seen in Figure 5.3.

Dust mites, like their insect cousins, have multiple developmental stages. These include egg, larva, several nymph stages, and ultimately the adult. Each life cycle stage is characterized by an active feeding period, with quiescent periods between stages. A quiescent period of up to several months for a dessication-resistant nymphal stage allows individuals to survive dry conditions and contributes to the development of abundant mite populations when building climatic conditions are more optimal. The lifetime of *D. farinae* females averages about 75 days; for *D. pteronyssinus*, 31 days. Dust mites have significant microclimate requirements. Since they live in microenvironments where no liquid water is present, they must extract water (by means of specialized glands) from unsaturated air to compensate for body losses. The maintenance of body fluid requirements and survival depends on the relative humidity (not absolute humidity) of their surroundings. The critical relative humidity for nonfeeding mites is approximately 70%. At relative humidities of <55 to 60% they may dehydrate and die.

Because of high relative humidity requirements, mite populations are small in desert or semiarid regions as well as at high altitudes or high latitudes. They are, of course, very abundant in humid regions, and show seasonal variability which coincides with changes in relative humidity. High dust mite populations occur during humid seasons, as can be seen in Figure 5.4. Note that room/household relative humidities are lower than the critical 70% value. This reflects differences between inhabited microenvironments and the surrounding building environment. Higher relative humidities may occur in bedding and in carpeting near cooler floor surfaces.

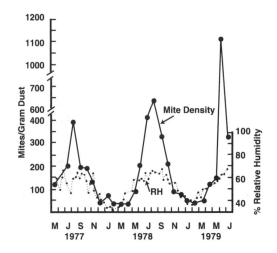

Figure 5.4 Seasonal changes in dust mite populations. (From Arlian, L. et al., *J. Allergy Clin. Immunol.*, 69, 527, 1982. With permission.)

Fecal pellets and mite body parts are the major sources of mite allergens. Four allergen groups have been identified. Groups I, III, and IV allergens are digestive enzymes; Group II allergens are associated with body parts.

Group I allergens such as Der p I and Der f I have received considerable research attention. They are commonly measured in floor dust samples to determine the dust sensitization potential for exposed populations. Combined concentrations of Der f I and Der p I of <2 µg/g dust are considered to be a low sensitization risk; 2 to 10 µg/g, moderate risk; and >10 µg/g, high risk for both sensitization and acute asthmatic attacks (Table 5.5). Relative concentration categories of mite allergen (as well as cat and dog allergens) in house dust is summarized in Table 5.6.

Mite numbers in floor dust have been used as indicators of sensitization risk. The threshold for immunological reactions has been reported to be approximately 100 mites/g. Since mite populations vary seasonally, low mite populations may not be indicative of allergen levels which persist for months after populations begin to decline.

Table 5.5 Exposure Guideline Values for Common Indoor Allergens

	Sensitization risk			Risk of asthmatic attack
Allergens	low	moderate	high	high
Dust mite	2 µg/g	2–10 µg/g	>10 µg/g	>10 µg/g
Cockroach	<2 U/g	>2 U/g	>2 U/g	>2 µg/g
Cat	<1 µg/g	1–8 µg/g	>8 µg/g	>8 µg/g
Dog	?	?	?	?

? = unknown.

Table 5.6 Relative Concentration Categories of Mite, Cat, and Dog Allergens in House Dust (µg/g)

Category	House dust mite		Cat	Dog
	Der p1	Der f 1	Fel d 1	Can f 1
Very low	<0.5	<0.5	<0.1	<0.3
Low	0.5–5	0.5–5	0.1–1.0	0.3–10.0
Intermediate	>5–15	>5–15	>1–10	>10–100
High	>15–20	>15–20	>10–100	>100–1000
Very high	>20	>20	>100	>1000

Source: From ECA (European Collaborative Action "Indoor Air Quality and its Impact on Man" COST Project 613), *Biological Particles in Indoor Environments*, Report No. 12 EUR 14988EN, Office for Publications of the European Communities, Luxembourg, 1993.

Highest mite populations occur in residential buildings; significantly lower concentrations are reported in institutional buildings such as schools, nursing homes, and office buildings. In office/commercial buildings, mite populations are rarely >20 to 40 mites/g dust, whereas mite populations >100 mites/g are common in many residences. Allergen levels >2 µg/g are very uncommon in institutional and commercial buildings. Dust mites are approximately 4 times more prevalent in residences with wall-to-wall carpeting compared to those with bare floors or area rugs. Carpeting appears to provide a favorable environment for the collection and decomposition of human skin scales. It also provides an environment with relative humidity levels above minimum requirements for dust mite survival.

Mite fecal pellets are large (10 to 35 µm). As a consequence, disturbance of floor and other surface dusts is necessary to disperse them in air. Because of their large size and mass, they settle out rapidly and do not form true aerosols in the fashion of cat dander and mold spores. Consequently, exposure is episodic and is typically associated with bed disturbances during sleep, bed-making, and house-cleaning activities. Only the smallest of particles in the fecal pellet range have the potential to enter and deposit in the upper respiratory system.

Though mite allergens are the most common cause of immunological sensitization resulting in symptoms of chronic allergic rhinitis and asthma, other allergen-producing mites may be present in building environments. Most notable of these are stored product mites such as *Tryrophagus putrescentiae* and *Glycophagus domesticus*.

V. Insects

A number of insects have been identified as sources of inhalant allergens that may cause chronic allergic rhinitis and/or asthma. These include cockroaches, crickets, beetles, moths, locusts, midges, and flies. The best-studied

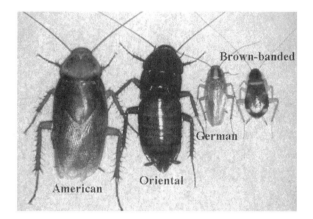

Figure 5.5   Major cockroach species. (Courtesy of Ogg, C., University of Nebraska.)

association between insect allergens and respiratory disease has been that of cockroaches.

Cockroach species of concern include the German cockroach (*Blattella germanica*), American cockroach (*Periplaneta americana*), and Asian cockroach (*Blattella orientalis*) (Figure 5.5). Each of these may be found in indoor environments. The German cockroach appears to be the most important, particularly in the northern U.S. where it is the most prevalent species, especially in large, densely populated cities. They are commonly found in houses, restaurants, and schools.

As adults, German cockroaches attain a length of approximately 17 mm. Like other cockroaches, the body is oval-shaped and flattened, with spiny legs and long, filamentous antennae. German cockroaches consume a variety of foods and may infest buildings by being brought in on materials infested with one or more life stages. They are most commonly found in kitchens and bathrooms. Populations of thousands to tens of thousands of individuals may develop within a single building. Being primarily nocturnal creatures, they feed at night, hiding in cracks and other dark spaces during the day.

Female cockroaches, on average, live about 200 days, with the male life span somewhat shorter. Females produce 4 to 6 egg cases in their lifetimes, with each case containing 30 to 40 eggs. One female cockroach may produce several hundred young within a period of approximately 6 months. As a consequence, only a few initial individuals are needed to cause a significant infestation problem.

Cockroach allergens are found in fecal material and saliva, as well as other body parts. Two major German cockroach allergens have been identified, Bla g I and Bla g II.

Cockroach allergens are determined from floor dust samples using monoclonal antibody assays. Concentrations are reported in units per gram (U/g). Preliminary evaluations by medical scientists indicate that a floor dust level of 2 U/g Bla g II is a risk factor for sensitization (Table 5.5). Some scientists, however, believe that any measurable level of Bla g II in floor dust is of medical concern.

It is rare to find floor dust levels in suburban areas >2 U/g; consequently, there are few cockroach allergen-sensitized children in such environments. In low-income urban areas, floor dust from 25 to 75% of houses sampled contained ≥ 2 U/g Bla g II; consistently, a high percentage of children in these homes have been sensitized. Such sensitization appears to be a major risk factor for symptomatic asthma. In low-income areas, 40 to 60% of asthma patients are sensitive to cockroach allergens. The prevalence of sensitization to cockroach allergens among asthmatic and nonasthmatic emergency room patients in Atlanta, GA, and Wilmington, DE, can be seen in Table 5.4. Highest cockroach allergen levels are found in kitchens, with concentrations as high as 1000 U/g in heavily infested houses. Highest concentrations occur in houses where cockroaches have been observed; however, 20% of houses without any visible cockroach infestation have detectable Bla g II in dust samples.

Little is known of the aerodynamic behavior of particles containing cockroach allergens. Scientific speculation suggests that such particles are likely to be large (>10 μm). As a result, they may not remain airborne for long (<1 hour). Evidence for a link between the presence of cockroach allergens and asthmatic/allergic symptoms is indirect.

The American cockroach, *Periplaneta americanus* (Figure 5.5), is the larger of the building (mostly residences)-infesting cockroaches. It is approximately 38 mm long, with reddish brown wings. Because it is a common inhabitant of sewer systems, basements in buildings may become infested following heavy rains. It is commonly found in restaurants, supermarkets, bakeries, etc. Unlike the German cockroach, it is also found outdoors, particularly in states with moderate winter temperatures. Adult females typically live an average of 440 days.

Limited information is available on the allergenicity of the American cockroach. Antibody tests indicate that *P. americanus* allergens cross-react with *B. germanica* antibodies, thus making it more difficult to distinguish between the two types of sensitization.

VI. Animal allergens

Many mammalian and avian species produce allergens that cause immunological sensitization and symptoms of chronic allergic rhinitis and asthma. These include the danders of cats and dogs; rodents such as rats, mice, and guinea pigs; and birds (feathers) kept indoors as pets.

There are approximately 100 million domestic animals in, or in close proximity to, residences in the U.S. Indeed, 35 to 50% of residences in the U.S. keep mammalian pets, most of which are housed indoors. Not surprisingly, sensitization to cat and dog allergens is common. Sensitization to cat dander among asthmatics varies from 9 to 41%. The highest level of sensitization is among suburban dwellers, and the lowest is in low-income urban areas (where cockroach sensitization is high; Table 5.4). It has been estimated that 6 to 10 million individuals in the U.S. are allergic to cat allergen.

A. Cat allergens

Fel d I, the major cat allergen, is an acidic glycoprotein found on hair and skin scales, and in saliva, sebaceous gland excretions, and voided urine. The antigen is dispersed by licking and grooming activity. Allergen production is voluminous, with a single cat in an indoor space producing 3 to 7 µg Fel d I/day. The size range of particles that contain cat allergen varies from <2.5 to 10 µm, with a very large percentage <2.5 µm. Because of this small size, cat allergen particles become airborne (even in undisturbed situations) and remain airborne for extended periods of time (hours). As such, humans have a high probability of being exposed to cat antigens that produce sensitization and allergic or asthmatic symptoms. Higher airborne cat allergen levels are associated with greater numbers of cats present and low building air exchange rates. Relative concentration categories of cat allergen in house dust are summarized in Table 5.6.

Cat allergen is ubiquitously found in building dust samples collected in residences, schools, medical facilities, retail establishments, and a variety of public places. This ubiquitous presence of cat allergen results from its passive transport on the clothing of individuals in cat-owning households (as well as visitors). Concentrations of cat allergen measured in the floor dust of elementary schools can be seen in Table 5.7. Of these, >20% of floor dust samples had cat allergen concentrations that are reported to cause a moderate to high risk of sensitization on exposure.

Concentrations of Fel d I in building dust of <1 µg/g are considered to pose a low sensitization risk; concentrations of 1 to 8 µg/g pose a major sensitization risk; and concentrations >8 µg/g pose a major risk of causing acute asthmatic symptoms (Table 5.6).

Concentrations of Fel d I in houses with cats commonly exceed 10 µg/g floor dust; in cat-free buildings, concentrations are typically <10 µg/g but often near or above the 1 µg/g threshold for sensitization. Highest concentrations are found on living/family room floor surfaces in residences.

Table 5.7 Prevalence (%) of Cat and Dog Allergens in Floor Dust Samples in Midwestern U.S. Elementary School Classrooms

Allergen	Allergen concentration (µg/g dust)			
	<1	1–8	>8	Range
Cat (Fel d 1)	77.1	20.0	1.4	ND–57.2
Dog (Can f 1)	2.9	88.5	8.6	0.5–57.3

ND = not detectable.

Source: From Godish, D., unpublished data, 1998.

B. Dog allergens

The major dog allergens, Can f I and albumin, are primarily found on the animal's hair and in saliva, with much lower concentrations in the urine and feces. The source of the antigenic protein was unknown at the time of this writing.

Approximately 2.3% of randomly sampled individuals in the U.S. population show skin test sensitivity to dog allergens. This is comparable to sensitization rates reported for cats. The absence of well-characterized and standardized extracts used in allergy testing has limited efforts to investigate the prevalence, and evaluate the significance, of dog allergen sensitivity.

Concentrations of Can f I in floor dust samples have reportedly ranged as high as 10,000 µg/g in residences with dogs to <0.3 µg/g in residences without dogs. Relative concentration categories of dog allergen concentrations in house dust are summarized in Table 5.6. As with cats, dog allergens are ubiquitously present in floor dust samples in residences without dogs, schools, and a variety of public places. The mechanism of such contamination is passive transport on the clothing of residents of dog-owning households. Levels of dog allergen in floor dust of elementary schools can be seen in Table 5.7. (Note that dog allergen levels are significantly higher than those of cat allergen.)

Though dog allergen is ubiquitously present in building spaces and has an apparent sensitization prevalence similar to cat allergen, no consensus guidelines have been established on floor dust levels required to cause sensitization. Using other common indoor allergens as a guide, it is likely (by analogy) that several µg dog allergen/g floor dust may be sufficient to cause sensitization.

C. Rodent allergens

Rodent allergens may pose exposure risks to individuals living or working in building environments with significant rodent infestation problems. These include environments where rodents are kept as pets (e.g., residences and schools) and research laboratories. Rat allergens Rat n IA and Rat n IB, and mouse allergens, Mus m I and Mus m II, have been reported from hair and urinary excretions. Mouse allergens have been identified in air samples collected in urban households.

VII. Passive allergen transport

The passive transport of cat, dog, and possibly cockroach antigens from source buildings to buildings such as schools and other public places represents a potential exposure problem of unknown public health significance. School buildings with high population densities of very young to older children as well as teachers should be of particular concern.

In a survey of school teachers in Indiana, a large percentage of teachers (circa 30%) reported experiencing allergy-type symptoms during the school day, with symptoms showing a strong positive correlation with teacher perceptions of the extent of surface dustiness and overall classroom cleanliness. Teachers and children not significantly exposed at home would be at risk to elevated classroom allergens. This exposure risk may be increasing as schools in the U.S. install carpeting as older schools are renovated and new schools are built.

Readings

Arlian, L.G., Biology and ecology of house dust mites, *Dermatophagoides* spp. and *Euroglyphus* spp., *Immunol. Allergy Clin. N. Am.*, 9, 339, 1989.

Burge, H.A., Health effects of biological contaminants, in *Indoor Air and Human Health*, Gammage, R.B. and Berven, B.A., Eds., CRC Press/Lewis Publishers, Boca Raton, 1996, 171.

Burge, H.A., Ed., *Bioaerosols. Indoor Air Research Series*, Center for Indoor Air Research. Lewis Publishers, Boca Raton, 1995

Cox, C.S. and Wathes, C.M., Eds., *Bioaerosols Handbook*, CRC Press/Lewis Publishers, Boca Raton, 1995.

Etkin, D.S., Biocontaminants in indoor environments, *Indoor Air Quality Update*, Cutter Information Corp., Arlington, MA., 1994.

Kubica, G.P., Exposure risk and prevention of aerial transmission of tuberculosis in health care settings, in *Indoor Air and Human Health*, Gammage, R.B. and Berven, B.A., Eds., CRC Press/Lewis Publishers, Boca Raton, 1996, 141.

Maroni, M., Siefert, B., and Lindvall, T., Biological agents, in *Indoor Air Quality. A Comprehensive Reference Book*, Elsevier, Amsterdam, 1995, chap. 5.

Milton, D.K., Bacterial endotoxins: A review of health effects and potential impact in the indoor environment, in *Indoor Air and Human Health*, Gammage, R.B. and Berven, B.A., Eds., CRC Press/Lewis Publishers, Boca Raton, 1996, 179.

Morey, P.R., Feeley, J.C., and Otten, J., Eds., Biological contaminants in indoor environments. STP 102. American Society of Testing Materials, Philadelphia, 1990.

Pape, A.M, Patterson, R., and Burge, H., Eds., *Indoor Allergens: Assessing and Controlling Adverse Health Effects*, National Academy Press, Washington, D.C., 1999.

Platt-Mills, T.A.E., Estimation of allergen concentration in indoor environments: prediction of health-related effects, in *Indoor Air and Human Health*, Gammage, R.B. and Berven, B.A., Eds., CRC Press/Lewis Publishers, Boca Raton, 1996, 197.

Spengler, J.D., Samet, J.M., and McCarthy, J.F., Eds., *Indoor Air Quality Handbook*, McGraw-Hill Publishers, New York, 2000, chaps. 42, 43, 47, 48.

Questions

1. How are antigens and allergens related?
2. What relationship if any is there between chronic allergic rhinitis and asthma?
3. What is the biological mechanism involved in the induction of chronic allergic rhinitis and allergy symptoms?
4. Most allergens are proteins. However, one can become sensitized to substances such as formaldehyde, toluene diisocyanate, and trimellitic anhydride. How?

5. How significant and serious a health problem is chronic allergic rhinitis?
6. Describe prevalence characteristics of asthma in the U.S.
7. Describe characteristic illness symptoms and patterns of hypersensitivity pneumonitis and potential causes.
8. What is the role of particle size in causing hypersensitivity pneumonitis in an exposed individual?
9. Why don't bacterial exposures result in symptoms of chronic allergic rhinitis?
10. Bacterial exposures in indoor air pose what health concerns?
11. How do bacterial exotoxins and endotoxins differ in their formation and health effects?
12. What is humidifier fever? What is its likely cause?
13. How are the actinomycetes uniquely different from lower forms of bacteria?
14. What are MVOCs and what is their significance?
15. Why is TB a significant concern in hospitals, homeless shelters, and long-term residential care facilities?
16. Describe illness symptoms associated with Legionnaires' disease, exposure pathways, and human risk factors.
17. Describe the disease syndrome, risk factors, and cause of Pontiac fever.
18. How do hypersensitivity pneumonitis and humidifier fever differ?
19. Describe illness syndromes associated with endotoxin exposures.
20. What role (if any) do viruses play in indoor health concerns?
21. Describe settled organic floor dust relative to its composition, immunological nature, and potential to cause adverse health effects.
22. Describe what dust mites are, their ecological requirements, and their role in causing human disease.
23. Allergens in floor dust of inner city and suburban residences differ. Describe these differences and their relationship to childhood asthma.
24. What is the public health significance of passive transport and subsequent deposition of allergens?

chapter six

Biological contaminants — mold

Biological contaminants were described in Chapter 5 in the context of illness syndromes and disease risks associated with exposure to airborne organisms and products of biological origin. Discussion of airborne bacteria and viruses was in the context of their role in causing infectious disease and potentially contributing to other problems; mites, insects and animal allergens, in the context of causing chronic allergic rhinitis and asthma. Because mold is such a significant indoor environment concern, this chapter is devoted to mold-related health concerns and risk factors for mold infestation.

I. Biology of mold

The terms mold and mildew are commonly used to describe the visible manifestations of the growth of a large number of organisms that are scientifically classified as fungi. Terms such as *yeast* and *mushrooms* are used to describe, respectively, single-celled fungi (widely used for baking and brewing) and the large reproductive structures of a major class of fungi that are used for food or are known for their high toxicity.

Fungi form true nuclei, which distinguishes them from lower organisms such as bacteria. They differ from plants in that they do not produce chlorophyll and thus cannot manufacture their own food; from animals in that (except for reproductive cells in some species) they are not motile.

Structurally, fungi exist as masses of threadlike filaments or hyphae. The collective mass of hyphal filaments is described as mycelium, the vegetative part of the organism that infests a substrate and extracts food for the organism's growth. Though hyphal filaments are microscopic, the mycelium is typically visible to the naked eye. Masses of mycelia can be distinguished as fungal colonies (Figure 6.1). In many species, the hyphae are colorless; in other species they contain pigments.

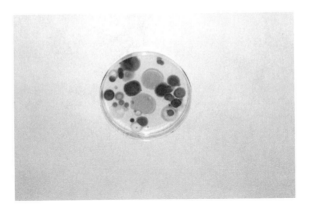

Figure 6.1 Fungal colonies.

A. Reproduction

Specialized reproductive structures develop in the life history of all fungal species. These structures may be produced as a consequence of sexual or asexual processes. Most species of fungi undergo both sexual and asexual reproduction at some time in their life history. Sexual processes occur in specialized cells or structures, with the production of spores that will ultimately be dispersed. Sexual spores are produced in specialized "fruiting" structures. The most noticeable of these are mushrooms and bracket fungi on trees or tree debris.

Asexual spores, as well as the fruiting structures (if any) on or within which they are borne, vary in size and shape as well. Asexual spores are produced in large masses, particularly when a mold colony is maturing and experiencing environmental stress, e.g., depletion of substrate, lower substrate moisture content or air relative humidity, or competition with other colonies or species. In many cases, asexual spores produce pigments which may, along with pigment in the mycelium, characterize colonies in culture or on natural substrates. Both spore-forming structures and asexual spores for species in the genera *Penicillium* and *Aspergillus* can be seen in Figures 6.2 and 6.3.

Fungi may also produce spores or other structures that are designed to survive harsh environmental conditions. These include thick-walled, dormant clamydospores and hardened mycelial masses called sclerotia.

A number of species of fungi grow as single cells during all or part of their life history. Most notable of these are yeasts, which reproduce asexually by budding and sexually by producing sac-like ascospores. In genera such as *Candida*, under certain environmental conditions the organism develops yeast-like growth, while under others, it develops a typical mycelium.

B. Dispersal

Most fungal species produce sexual and/or asexual spores which serve to both reproduce the organism and disperse it to new substrates. As a conse-

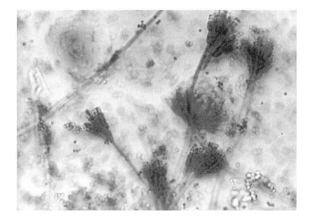

Figure 6.2 Asexual spores and spore bearing structures of *Penicillium*. (Courtesy of University of Minnesota.)

quence of evolutionary processes, fungal spores have size and aerodynamic properties that enhance airborne dispersal. Spores vary in size (2 to 100 μm), with the smallest being single cells and the largest being multicellular. Spores of *Alternaria* and *Cladosporium,* which illustrate different spore shapes and sizes, can be seen in Figures 6.4 and 6.5.

Spore dispersal may occur by passive or active release mechanisms, with subsequent entrainment and movement by horizontal or convective air currents. Once airborne, spores may be carried varying distances, depending on their aerodynamic properties as well as existing atmospheric conditions. In still air (such as may occur in houses) they settle out relatively rapidly. Based on Stokes Law, the largest and heaviest spores would settle out more quickly, with smaller spores being suspended for longer periods of time.

Figure 6.3 Asexual spores and spore-bearing structures of *Aspergillus*. (Courtesy of University of Minnesota.)

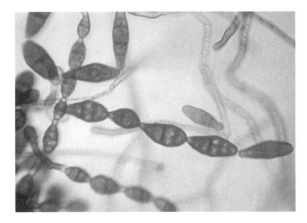

Figure 6.4 Asexual spores of *Alternaria*. (Courtesy of University of Toronto.)

However, spores with surface ornamentation and nonspherical shape tend to experience significant drag and thus settle out more slowly than Stokes Law calculations would predict. Settling times for particles of different aerodynamic diameters are given in Table 6.1.

When spores are released, they may become airborne as single structures or in strands of multiple attached spores. The latter is commonly the case with asexual spores of *Penicillium, Aspergillus,* and *Cladosporium*. Such clusters may have aerodynamic behavior characteristics that differ from single spores. In theory, they should settle out more rapidly.

As is the case with most species of living things, fungi produce an excess of reproductive propagules to ensure survival. Not surprisingly, many do not survive the often harsh environmental conditions that exist in the time period between their release from mycelia and their deposition on substrates. Ultraviolet light and low atmospheric moisture pose significant risks to their

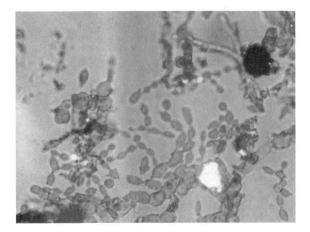

Figure 6.5 Asexual spores of *Cladosporium*. (Courtesy of University of Minnesota.)

Table 6.1 Settling Velocities for Particles of
Different Aerodynamic Diameters

Particle diameter (µm)	Settling rate (cm/s)
1	0.004
5	0.08
10	0.30
20	1.20

viability. Loss of viability during dispersion explains, in part, the often large differences in mold concentrations observed in concurrent airborne mold sampling using culturable/viable and total mold spore sampling methods. In the latter case, concentrations of total airborne mold vary from several times to one or more orders of magnitude higher than culturable/viable mold. Spore viability ratios (concentration of live spores divided by total number of spores) for species such as *Aspergillus, Penicillium,* and *Cladosporium* are in the range of about 0.10 to 0.15 (10 to 15%) and are much lower in species such as *Epicoccum.*

Spores must fall on suitable substrates for germination and subsequent infestation to take place. If environmental conditions are not suitable, spores will, in time, lose their viability. Spores can become resuspended so that they may have more than one chance to land on a substrate with suitable environmental conditions for growth. Indoors, resuspension occurs as a consequence of occupant activities, e.g., moving or dusting objects, children and pets playing, and air currents generated by forced air mechanical systems such as fans and heating/cooling systems, etc.

Different species respond to a variety of environmental conditions that affect their dispersal. As a consequence, the composition of airborne mold samples reflects not only the presence of individual species but also dispersal mechanisms and periodicity factors. The proximity of sources significantly affects sample concentrations. Samples collected near wood lots or in houses with significant structural deterioration tend to have high basidiospore (produced by members of the *Basidiomycetes*) concentrations. Samples collected near crop harvesting operations may have enormously high concentrations of *Alternaria, Epicoccum,* and *Cladosporium.* In houses, samples collected near sources of *Penicillium* or *Aspergillus* will have high concentrations of these two fungal types. In other cases, airborne mold spore concentrations may be low despite the fact that significant infestation is present. This is the case with *Stachybotrys chartarum,* a toxigenic species with large, initially sticky spores, which may cling together and settle out rapidly (Figure 6.6).

C. Nutrition

Fungi obtain food by parasitizing other organisms (mainly plants), or from decomposing organic matter, or, in specialized cases (lichens), mutualistic symbiosis with algae. Most fungi are saprobes, obtaining their nutrient

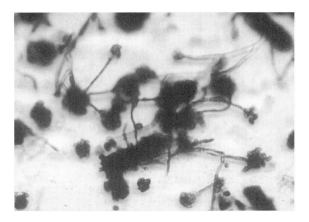

Figure 6.6 *Stachybotrys* spores/spore-bearing structures. (Courtesy of University of Minnesota.)

needs from nonliving organic matter. Most saprobes are saprophytes, that is, they decompose dead plant matter. Some species are both parasitic and saprobic, living on dead organic matter but invading living tissues when a host is present.

Most parasitic fungi are plant pathogens. A number of these are notable because they pose major threats to agricultural crops, which require continued selective breeding programs or intensive use of fungicides for their protection. A small number of species can parasitize humans, causing infections of the skin (e.g., athlete's foot) or more serious diseases such as histoplasmosis and life-threatening pulmonary infections (aspergillosis).

The major ecological role of fungi is decomposition, a role shared with bacteria and insect larvae. During decomposition, fungal hyphae infest substrates, producing extracellular enzymes that digest complex organic molecules into smaller and simpler molecules that are absorbed and metabolized. Decomposition is a successional phenomenon, with different species dominating as the decomposing substrate changes. As a consequence, substrate samples often show the presence of a number of species.

Fungal species have varied growth requirements. Therefore, individual species may be prevalent on some substrates and in certain environments but not others. These growth requirements affect species prevalence in both outdoor and indoor environments.

Many saprobic fungi can use a variety of substrates. Substrate colonization in individual situations is dependent on their ability to compete with other species, as well as the presence of suitable temperatures and availability of water.

D. Environmental requirements

Water is a major limiting factor in the growth of fungi. Humidity >70% is required for spores to germinate, and optimal substrate moisture is necessary for initial infestation and subsequent growth. Though water is required for growth, fungal species have a broad tolerance range for its availability. The term *water activity* (a_w) is used to describe the moisture content of substrates. Water activity is the relative humidity of the substrate expressed as a decimal fraction (e.g., 95% = 0.95 a_w). The range of water activities that individual species grow under varies from approximately 0.55 to 1.0 a_w. Species that require high water activities (>0.95 a_w) are described as *hydrophilic*; those that can tolerate lower water activities are *xerophilic*.

Fungi grow over a range of temperature conditions (<40 to 140°F, <5 to 60°C). Mesophilic fungi, the largest group, have an optimum temperature range of 68 to 86°F (20 to 30°C). Thermophilic fungi that are human pathogens grow well at temperatures of 95 to 104°F (35 to 40°C); true thermophilic fungi grow at temperatures of 113 to 140°F (45 to 60°C), e.g., compost heaps. Cryophilic fungi can grow at relatively low temperatures (<40°F, <5°C), e.g., on materials in cold storage and in regions having cold climates.

E. Classification

Up until the last several decades, fungi were classified as members of the Plant Kingdom because some contain cellulose in their cell walls. In today's classification, the approximately 70,000 identified species are placed in either the Kingdom Protista or Kingdom Fungi. The slime molds, which have affinities to both fungi and animals (division Myxomycota), and the water molds (division Oomycota), which produce motile spores, are classified in the Kingdom Protista. The Kingdom Fungi includes five major fungal types or divisions. These are the Chytridiomycota, Zygomycota, Ascomycota, Basidiomycota, and Deuteromycetes. The Chytridiomycota are a primitive group of fungi which do not produce airborne spores and thus pose little risk of human exposure and health effects.

Fungal species in the Zygomycota produce thick-walled sexual spores (zygospores) and asexual spores called sporangiospores. The mycelium is characterized by the absence of cross walls (i.e., it is nonseptate). Zygomycota species are saprobic and can grow on a variety of substrates, particularly in areas with high relative humidity. Two major genera in this group (*Rhizopus* and *Mucor*) are commonly isolated from indoor environment samples. In culture, they rapidly overgrow colonies of other genera. Genera in this group colonize soil and house dust. *Rhizopus stolonifera* colonizes bread, and thus is referred to as bread mold.

The Ascomycota represent a large and widely distributed group of organisms characterized by the production of sexual spores in a sac-like structure called an *ascus*. Members of this group are, for historical reasons, called ascomycetes. They range from the single-celled yeasts to species that produce mushroom-like fruiting bodies (morels). Many members of the Ascomycota reproduce asexually by producing conidia (spores) in large masses on specially differentiated hyphal structures. The asexual spores produced by Ascomycota are common indoor air contaminants (e.g., those from *Penicillium* and *Aspergillus*). Many of the fungal allergens identified by medical scientists have been associated with species of Ascomycota.

The division Basidiomycota, historically known as basidiomycetes, includes a large variety of fungal types and species. They are characterized by their sexual reproduction processes, which produce basidiospores on structures called basidia. Basidiospores are the primary means of dispersal, though asexual spores may also be formed. Basidiomycetes include yeasts like *Sporobolomyces*, which colonize indoor substrates; rusts and smuts, which are plant pathogens; many species of mushrooms, bracket fungi, and puff-balls; and a variety of other wood-decaying fungi which grow on structural timbers in human dwellings. The basidiomycetes rarely form fruiting bodies in culture (sexual or asexual) and, as a consequence, are difficult to identify. In addition, many do not grow well in culture.

The Deuteromycetes, also known as fungi imperfecti or asexual fungi, include a group of organisms classified by their asexual reproductive structures and apparent absence of sexual structures. This is an artificial classification, since the sexual stage is not commonly observed under culture conditions. Many deuteromycetes have been determined to be ascomycetes, while only a few are basidiomycetes. Classifications in this group have been problematic, since the asexual stage of Deuteromycete genera such as *Aspergillus* may be associated with different genera in the Ascomycota. Books used to identify species such as *Aspergillus* and *Penicillium* typically use the Deuteromycete name rather than the more scientifically correct name, which is based on the sexual stage.

II. Biologically significant fungal compounds

As fungi grow, they produce a variety of secondary metabolites. These are often species-dependent, but a number of compounds appear to be generic to fungi. These include pigments which may color spores or other structures or diffuse into substrates, a variety of volatile organic compounds (MVOCs) with their distinctive musty or earthy odors, alkaloids, antibiotics, and mycotoxins. Many secondary metabolites are produced by older parts of fungal colonies, particularly cells that are being restricted in their growth.

Some metabolites may have important direct functions. It is notable that genera found outdoors, such as *Cladosporium, Epicoccum, Alternaria,* and *Pithomyces,* are dark in color, producing black spores. Many mushrooms produce dark spores as well. The major benefit of dark spores and mycelium would be to protect the organism from ultraviolet light. Other metabolites may serve to reduce competition with different organisms (antibiotics and bacteria), other species, or members of the same species (alkaloids, mycotoxins).

A. MVOCs

During their growth, all fungal colonies release volatile organic compounds (VOCs or, more specifically, MVOCs to denote their microbial origin) that are characteristically microbial or fungal in the sense that they are responsible for the musty, mildewy, earthy, or mushroomy odors associated with mold infestations. Commonly reported fungal VOCs include: 3-methylfuran, heptanone, 1-octen-3-ol, octan-3-ol, 2-octen-1-ol, octen-3-one, octan-3-ol, 2-methyl-1-butanol, 2-hexanone, geosmin (1,10-dimethyl-*trans*-9-decalol), acetone, 2-butanol, dimethyl trisulfide, methanol, 1-propanol, 4-decanol, 2-methylisoborneol, trimethylhexane, 3,3-dimethyl-2-oxetanone, 3,3-dimethyl-1-octene, ethyl 2,4-dimethylpentone, and 2-methoxy-3-isopropylpyrizine. As can be seen, most fungal MVOCs are alcohols and ketones.

Microbial VOCs produced by mold colonies vary with individual species as well as substrates colonized. The most frequently reported MVOC in building investigations is 2-octen-1-ol. This compound, as well as 1-octen-3-ol, 2-hexanone, and light alcohols, tends to be associated with water-damaged materials. Fungal VOCs, reported as total MVOCs, range in concentration from 50 to 126 $\mu g/m^3$ in problem buildings, while outdoor levels average 8.6 $\mu g/m^3$. Their odor is readily detectable despite relatively low concentrations (even in mold-infested buildings), indicating a relatively low odor threshold.

B. Fungal toxins

Many fungal species produce toxins which, on exposure, adversely affect the physiological functioning of other organisms. Toxins such as antibiotics inhibit the growth of bacteria and provide a competitive advantage to the organism producing the antibiotic. Other toxins may inhibit the growth of fungal species in a somewhat similar fashion.

Fungal toxins can adversely affect humans and other animals. These include the highly poisonous toxins formed in the fruiting structure of a variety of mushroom species and mycotoxins produced by the mycelia of many common fungi.

1. Mycotoxins

Mycotoxins are generally produced when fungal mycelia are subject to nutrient limitation. Well-known mycotoxins and species producing them are summarized in Table 6.2. Fungal genera found indoors, such as *Penicillium* and *Aspergillus*, commonly produce mycotoxins. Best known of these are ochratoxin A and aflatoxin B, both relatively large molecules. Ochratoxin A is a colorless crystalline compound with a dihydroisocoumarin moiety linked to the amino acid phenylalanine and an atom of chlorine attached to the isocoumarin ring. The aflatoxins are a family of substituted coumarins containing a dihydrofuran moiety.

Of particular note are mycotoxins produced by *Stachybotrys chartarum*, a species of fungi that grows well on substrates containing cellulose (such as hay, straw and, in buildings, ceiling tile and gypsum board). *S. chartarum* has a characteristically dark mycelium and large, initially sticky asexual spores. Mycotoxins produced by *S. chartarum* include highly toxic members

Table 6.2 Mycotoxins Produced by Common Fungal Species

Genus/species	Mycotoxin
Penicillium sp.	Patulin
P. verrucosum	Ochratoxin A
	Citrinin
	Citroviriden
	Emodein
	Egliotoxin
	Verruculogen
	Secalonic acid D
Aspergillus	
A. flavus	Aflatoxin (B$_1$)
A. parasiticus	Aflatoxin (B$_1$)
A. versicolor	Sterigmatocysin
A. ochraceus	Ochratoxin A
A. clavatus	Patulin
Stachybotrys chartarum	T-2
	Nivalenol
	Deoxnivalenol
	Diacetooxyscirpenol
	Saratoxin H
	Spirolactone
Fusarium sp.	Zearalenone
	Fumonisins
Paecelomyces variatii	Patulin

of the trichothecene family such as saratoxin H and G and verucarin A and B, as well as others such as trichovarin A and B.

The trichothecenes are macrocyclic compounds with both olefinic and epoxy groups. In addition to *Stachybotrys,* the trichothecenes are produced by species in the genera *Trichothecium, Fusarium, Myrothecium, Trichoderma,* and *Cephalosporium.* Trichothecenes associated with species of the genus *Fusarium* (which occurs widely in nature as both a saprobe and parasite on plants) have received considerable scientific attention. Mycotoxins produced by *Fusarium* species include the highly toxic compounds T-2 and DON (deoxynivalenol). *Fusarium* spores are commonly collected in outdoor samples and less commonly indoors.

Though mycotoxins are produced by mycelia, high concentrations are often found in spores. Concentrations of aflatoxin as high as 200 ppm w/w have been reported in the asexual spores of *A. flavus* and *A. parasiticus.* Very high mycotoxin levels are also found in the spores of *Stachybotrys.* As such, spores have the potential to cause significant risk of mycotoxin exposure. Limited studies have confirmed the presence of mycotoxins in airborne samples in problem environments.

2. Fungal glucans

Glucans are polyglucose polymers present in the cell walls of mold hyphae and spores and certain bacteria. They are very soluble in water and retain their biological activity when mold spores are no longer viable.

Because of their macromolecular size and diversity, glucans are not easily measured in environmental samples except by bioassay procedures. Though they are likely to be ubiquitous in suspended aerosols both in indoor and outdoor air, little information is available on the range of airborne glucan concentrations indoors.

III. Exposure assessments

Numerous studies have been conducted to assess airborne mold levels in both outdoor and indoor environments. These studies have attempted to characterize the presence and prevalence of genera (and in some cases, species) present in airborne samples as well as total mold spore concentrations.

Most studies have been based on the use of culturable/viable sampling methods, wherein airborne mold spores/particles are impacted onto specially-formulated nutrient agar media. Such sampling is used to identify and describe the dominant types that grow on the culture media used and provide a relative measurement of their abundance. It has the advantage of providing investigators with the opportunity to identify mold colonies to the species level. However, culturable/viable sampling methods have significant limitations. Media used differ in their ability to support the growth of different species and genera. In addition, an often high percentage of airborne mold spores is not viable and cannot grow on any culture media. As a consequence, levels determined from airborne sampling are only a measure of the abundance of viable spores present and their culturability on the sampling media used.

Compared to culturable/viable sampling, total mold sampling methods which collect mold spores and other particles on greased microscope slides have been little used to assess airborne mold concentrations. However, because these sample values potentially represent most mold spores and hyphal fragments, including both viable and nonviable mold structures, they are a better indicator of airborne mold concentrations. Since allergenicity is independent of viability, total mold spore sampling is also a better indicator of potential health risks. It is not possible at this time to identify spores to species, nor in many cases to genus either, using total mold spore sampling.

Despite limitations associated with airborne mold sampling, studies conducted to date can be used (to a limited degree) to describe dominant mold types present in outdoor and indoor air and their relative abundance.

A. Outdoor prevalence

Significant differences in mold types present and total colony counts have been reported in comparison studies of outdoor and indoor airborne mold spores and structures. Such differences are typically the case when studies are conducted in the winter and under closed building conditions. There is only limited intrusion of outdoor airborne mold into indoor spaces when a building is under closure conditions. Mold taxa commonly reported in outdoor air samples include the phylloplane fungi (grown on the surface of leaves): *Cladosporium, Alternaria, Epicoccum,* and *Aureobasidium; Dresclera, Fusarium,* and a number of species of basidiomycetes; and plant pathogens such as *Botrytis, Helminthosporium,* and *Ustilago.* Less commonly the mold taxa *Aspergillus* and *Penicillium* are reported. Dozens of mold types are present in samples collected outdoors; however, individual samples tend to be dominated by a relatively few taxa, with *Cladosporium* and basidiomycetes being most abundant.

Outdoor mold concentrations as determined by culturable/viable sampling methods vary from nondetectable (under cold, snow-covered conditions) to tens of thousands of colony-forming units/cubic meter (CFU/m^3). High concentrations are reported during the autumn when decomposing plant material is abundant and grain harvesting is taking place. Under moderate weather conditions, outdoor culturable/viable mold concentrations are typically in the range of hundreds to thousands of $CFUs/m^3$. With the exception of the winter season, outdoor mold levels are almost always higher than those measured indoors under closure conditions.

B. Indoor prevalence

Indoor mold taxa and their concentrations are also quite variable. This is due to factors such as the intrusion of outdoor mold spora, the presence or absence of sources of mold infestation, the nature of mold infestation and its location, and building type. In the latter case, residential buildings typically have both higher concentrations and a greater range of taxa present than mechanically ventilated buildings. While few differences in both mold taxa present and airborne concentrations are seen between indoor and outdoor samples when significant intrusion of outdoor mold spora occurs, significant differences between indoor and outdoor samples may be observed under closure conditions.

Mold taxa commonly reported and found in the highest concentrations indoors are, in declining order: *Cladosporium, Penicillium,* and *Aspergillus.* Less commonly reported and in fewer numbers are the taxa: *Alternaria, Fusarium, Epicoccum,* yeasts, basidiomycetes, *Mucor, Rhizopus, Aureobasidium, Chaetomium, Acremonium, Monilia, Pithomyces, Paecilomyces, Trichoderma, Scopulariopsus,* and rarely, *Stachybotrys.* Abundance, however, varies with the nature and degree of infestation.

Highest airborne mold concentrations are reported in buildings where significant mold infestation has occurred, most notably houses that have experienced severe flood damage. Mold concentrations in flood-damaged houses may be in the tens of thousands of $CFUs/m^3$. In buildings experi-

Table 6.3 Culturable/Viable and Total Mold Spore Concentration Ranges Observed in Buildings

Building type	Conditions	Concentration ranges	
		Culturable/viable (CFU/m³)	Total mold spores (S/m³)
Residence	New structure	<300	1000–3000
	Not mold infested (avg.)	>300<1000	>3000<10,000
	Moderately infested	>1000<3000	>10,000<30,000
	Heavily infested	>3000	>30,000
Mechanically ventilated nonresidential	Not mold infested (avg.)	<300	1000–3000
	Moderate, localized infestations	>1000<3000	>3000<10,000
	Heavily infested	>3000	>30,000

encing less significant water damage and mold infestation, total airborne culturable/viable mold concentrations >1000 CFU/m³ are common. Concentrations <1000 CFU/m³ are typical in buildings in which there is no apparent mold infestation; when these are new houses and mechanically ventilated buildings, levels are usually <300 CFU/m³ (Table 6.3).

Measurement of airborne mold concentrations as determined by total mold spore/particle sampling methods in either indoor or outdoor environments has been limited. The method is amenable to identifying and quantifying major mold genera, but few studies have reported quantitative exposure by mold type. Most information is available as total mold spore counts. In general, total mold spore levels in residential structures vary from 5000 to 15,000 spores/particles per cubic meter (S/m³); in mechanically ventilated buildings without any visible mold infestation, airborne mold varies from 1000 to 3000 S/m³ (Table 6.3). In moderately infested residences, concentrations of 15,000 to 30,000 S/m³ can be expected. Heavily contaminated school buildings (25,000 to 200,000 S/m³) and flooded residences (200,000 to 1,000,000 S/m³) have been among the highest reports of airborne mold. When very high concentrations are observed, they are typically dominated by *Aspergillus* or *Penicillium* or both.

IV. Health concerns

Exposure to airborne mold and mold-produced environmental contaminants may pose significant health risks to humans. These include well-known mycotic diseases as well as other less well-defined health effects.

There are three basic categories of mycotic disease: infections, allergenic/immunological illness, and nonallergic illness.

A. Infections

A number of fungal species can cause infectious disease. However, compared to bacteria and viruses, fungal infections have played a relatively minor historical role in causing human suffering and death. In the past two decades there has been a significant decline in skin disease, and a significant increase in systemic disease, caused by fungal infections. This is due, in the latter case, to significant increases in the population of individuals who are susceptible to systemic fungal infections. This susceptible population includes individuals with immune system deficiencies who would have normally died early in life; patients on immunosuppressive drugs; and other at-risk patients with cancer, on long-term antibiotic therapy, or with certain infectious diseases. Acquired immunodeficiency syndrome (AIDS) is the most prevalent of such infectious diseases. The epidemic increase in AIDS infections has resulted in a corresponding increase in life-threatening fungal infections, most notably by species of the genera *Candida* and *Cryptococcus*.

Fungi that cause systemic disease are saprobic organisms that become infectious when disruption of a human host's physiology, microflora, or immune system occurs. Immune-compromised individuals who have AIDS or are being treated with immunosuppressive drugs are of special concern because they are potentially exposed to both true pathogens and opportunistic fungi (Table 6.4) in both indoor and outdoor air. Exposure to opportunistic fungi is of major concern in hospital environments where special air filtration systems are often installed to protect patients. Most notable are efforts to control exposure to *Aspergillus fumigatus*, which is responsible for most aspergillosis infections. Other *Aspergillus* species of concern include *A. flavus*, *A. niger*, and *A. terreus*. Outbreaks of nosocomial (hospital-acquired) aspergillosis have been associated with new construction as well as renovation activities.

Life-threatening fungal infections in immune-compromised and other patients are often associated with *Candida* spp. *Candida* grows in a yeast-like fashion and is a normal part of the microflora of humans. As such, filtration

Table 6.4 Pathogenic and Opportunistic Fungal Species
that Cause Systemic Infections

Pathogenic	Opportunistic
Cryptococcus neoformans	*Candida* sp.
Histoplasma capsulatum	*Aspergillus* sp.
Coccidiodes inmitis	*Trichosporon* sp.
Blastomyces dermatidis	*Fusarium* sp.
Paracoccidiodes braziliences	*Pseudoallescheria boydii*
	Mucor sp.
	Rhizopus sp.
	Absidia sp.
	Rhizomucor sp.

systems installed in hospital wards for *A. fumigatus* and other organisms provide no protection against *Candida.*

B. Allergenic and immunologic illness

Allergenic and immunological illnesses caused by exposure to fungal allergens include allergic rhinitis, allergic sinusitis, allergic bronchiopulmonary mycosis, asthma, and hypersensitivity pneumonitis.

1. Allergic rhinitis

Over 60 species of fungi are known to cause chronic allergic symptoms in humans. These include genera in each of the major fungal divisions (Table 6.5). Fungal allergens are glycoproteins with molecular weights ranging from 10 to 80 daltons. They are located in cell walls, membranes, and cytoplasm. Some are extracellular enzymes. Single fungal species may produce dozens of allergens. Allergens produced depend on individual strains of the fungal species and their age, as well as the substrate on which the organism grows. The prevalence of mold allergy in humans has not been well-defined. Though it has been estimated that up to 30% of individuals with chronic allergic rhinitis are immunologically sensitive to mold, actual results from skin reactivity tests indicate the prevalence of mold sensitivity is approximately 4%. Low skin reactivity results may be an artifact of the limited variety of mold genera, species, and serotypes that are included in mold extracts used for allergy testing. These extracts are not standardized and vary in composition and potency. Consequently, skin test results

Table 6.5 Selected Fungal Groups that Cause
Allergenic Illness

Zygomycetes	Deuteromycetes	Deuteromycetes (cont.)
Rhizopus	*Acremonium*	*Paecilomyces*
Mucor	*Alternaria*	*Penicillium*
Absidia	*Aspergillus*	*Phoma*
Myxomycetes	*Aureobasidium*	*Sporotrichium*
Filago	*Botrytis*	*Stemphylium*
Stemonitis	*Candida*	*Torula*
	Cephalosporium	*Trichoderma*
Ascomycetes	*Cladosporium*	
Chaetomium	*Curvularia*	**Basidiomycetes**
Claviceps	*Dresclera*	*Agaricus*
Erisyphe	*Epicoccum*	*Boletus*
Eurotium	*Fusarium*	*Coprinus*
Microsphaera	*Gliocladium*	*Polyporus*
	Helminthosporium	*Puccinia*
	Monilia	*Tilletia*
	Nigrospora	*Ustilago*

designed to determine sensitivity to mold have limited reliability. There is evidence to suggest that sensitivity to mold enhances symptoms associated with other allergens.

2. *Allergic fungal sinusitis*

This is a mold-induced disease of the paranasal sinuses. It may include both noninvasive and invasive (penetrates tissue) forms. It commonly occurs in atopic (i.e., allergy-prone) individuals who have rhinitis, nasal polyps, or asthma. In the noninvasive form, sinus membranes thicken and produce an exudate that contains mucinous plugs, eosinophils, and cellular debris; recurrent sinusitis symptoms do not respond to antibiotic therapy. Genera reportedly associated with allergic fungal sinusitis include *Aspergillus, Curvularia, Alternaria,* and *Bipolaris.*

3. *Allergic bronchiopulmonary mycosis*

This ailment results from immune reactions to fungi that have colonized respiratory airways. It is characterized by recurrent pulmonary edema (fluid production), chronic swelling and destruction of affected bronchial tubes, fibrosis, and elevated serum immunoglobin levels. A reduction in total lung volume occurs, reducing the infected individual's breathing capacity.

Aspergillus fumigatus is the most common fungal species associated with allergic bronchiopulmonary mycosis. *Candida, Curvularia,* and *Helminthosporium,* as well as other species of *Aspergillus,* have also been reported to cause this illness.

Individuals with this disease almost universally have an atopic history. It affects approximately 0.5 to 1% of asthmatics and 10 to 15% of individuals with cystic fibrosis.

4. *Asthma*

Asthma can be induced by exposure to mold allergens. As with chronic allergic rhinitis, the prevalence rate of mold-induced asthma is unknown. Because of the limited number of well-prepared diagnostic extracts that are available to physicians, mold sensitization is difficult to confirm. It is likely that fungal species that cause chronic allergic rhinitis also cause asthma.

5. *Hypersensitivity pneumonitis*

Hypersensitivity pneumonitis, as indicated previously (Chapter 5), can be induced by exposure to a variety of organic dusts. Such dusts include fungal spores as well as mycelial fragments. Best known are cases associated with specific occupations. These include malt worker's lung associated with *Aspergillus,* cheese worker's lung associated with *Penicillium,* sequoiaosis associated with *Aureobasidium,* and outbreaks of hypersensitivity pneumonitis associated with ventilation systems.

C. Nonallergenic illness

1. Fungal MVOCs

Because of their ubiquitous presence in mold-infested buildings, fungal MVOCs have been the subject of scientific speculation related to their potential role in causing nasal irritation and stuffiness. Reports linking fungal MVOCs to illness symptoms have been anecdotal, and studies to test the effects of fungal MVOCs on humans have yet to be conducted. Limited laboratory studies indicate that MVOCs produced by *Paecilomyces variotii* are ciliatoxic (adversely affect the hair-like structures, i.e., cilia, in the respiratory tract that are responsible for removing foreign particles). In other studies, MVOCs from *Penicillium* sp. and *Trichoderma veride* increased ciliary beat frequency in respiratory airway cells of animals.

It is likely that fungal MVOCs do have significant biological effects. At present there is little scientific data to indicate that such biological effects occur at the relatively low exposure concentrations present in mold-infested buildings.

2. Mycotoxicoses

Mycotoxin intoxication or poisonings have historically been associated with contaminated livestock feed and human foodstuffs. Outbreaks of mycotoxicoses in animals are associated with climatic or seasonal conditions favoring fungal growth on grain and other crops.

Mycotoxins have a broad range of toxic potentials. Substances such as saratoxin H and cyclochorotine have LD_{50}s (dose required to kill 50% of animals under test) of <1 mg/kg (1 ppm w/w) and are extremely toxic. Other mycotoxins have LD_{50}s as high as 800 mg/kg.

The primary route of mycotoxin exposure in humans is ingestion. However, there is increasing evidence that airborne mycotoxin exposure may be important (particularly to certain mycotoxins). Inhalation exposure may cause more severe toxic responses for a given dose.

Mycotoxins are very large molecules and, as a consequence, not volatile. Therefore, mycotoxin exposure must result from inhalation of particulate-phase substances. Exposure may occur when mold spores that have high mycotoxin concentrations are inhaled. Such spores have aerodynamic diameters in inhalable/respirable ranges.

Mycotoxins have a variety of adverse health effects. The mycotoxin ochratoxin A, for example, causes both kidney and liver damage; liver damage is caused by exposure to aflatoxin B, which is also a potent liver carcinogen. Ochratoxin A is produced by several different species of *Aspergillus* and *Penicillium*. Mycotoxins citroviriden and verruculogen, produced by *P. verrucossen*, are neurotoxic. Mycotoxins patulin (produced by *Aspergillus* and *Penicillium* sp.) and saratoxin H (produced by *Stachybotrys chartarum*) cause lung hemorrhaging in animals.

Of particular concern are the mycotoxins saratoxin H and aflatoxin B, produced in the spores of *S. chartarum* and *A. flavus* and *parasiticus*, respectively. Concerns have also been expressed relative to mycotoxin exposure associated with *A. versicolor*.

Stachybotrytoxicosis has been associated with the handling of heavily infested hay, straw, grain, textiles, etc. Reported symptoms include cough, rhinitis, burning sensation in the mouth and nasal passages, dermal rashes, severe pharyngitis, and bloody nasal excretions. Workers cleaning up *S. chartarum*-infested debris in a home reported severe respiratory and skin symptoms.

A number of mycotoxins are immunotoxic; i.e., they have immunosuppressive or other immunomodulating effects. These include trichothecenes, gliotoxin, and aflatoxin. The trichothecenes are potent inhibitors of protein synthesis, preferentially attack lymphoid tissues, and increase susceptibility to infection.

Stachybotrys chartarum infestation of gypsum board and ceiling tiles in indoor environments is relatively common. As a consequence, potential exposure to airborne *Stachybotrys* spores containing potent mycotoxins may be a special public health concern. Recent investigations conducted by the Centers for Disease Control (CDC) in Cleveland, OH, and Chicago, IL, tentatively identified *S. chartarum* mycotoxin exposure as the potential cause of hemosiderosis, a frequently fatal hemorrhagic disease of the lungs of infants. Animals exposed to high concentrations of *S. chartarum* spores have also developed similar lung hemorrhaging, suggesting a potential causal association. Exposure to high concentrations of *S. chartarum* is not likely to be common since it produces large, initially "sticky" spores which settle out rapidly. Airborne concentrations are usually low unless a source of infestation is disturbed. Based on internal and external reviews of the data, in the year 2000 CDC concluded that, because of methodological flaws, a possible association between acute pulmonary hemorrhage/hemosiderosis and exposure to *S. chartarum* had not been proven.

3. Fungal glucans

Limited studies on the biological and health effects of fungal glucan exposures have been conducted on laboratory animals. Different forms of glucans have different biological properties. On initial exposure, they stimulate immune and inflammatory responses in animals and humans. On prolonged exposure, inflammatory cell numbers are reported to decrease. The clearance of xenobiotic particles from lung tissue and respiratory airways may be impaired because glucans interfere with normal functioning of respiratory system macrophages.

A few studies have been conducted to determine potential relationships between airborne glucan levels and symptoms among building occupants. In one study, a significant relationship was observed between glucan levels and two respiratory symptoms, nasal irritation and hoarseness; in a second study, only throat irritation was observed to be significantly related. These are consistent with human glucan challenge studies which have shown an increase in nose and throat irritation, nasal congestion, and increased airway responsiveness.

V. Mold infestation — risk factors

Infestation of building materials by a variety of mold species depends on conditions that favor the germination and growth of viable mold spores and hyphae. These conditions include nutrient sources, viable mold spores/hyphae, temperature conditions in the range of 40 to 130°F (5.5 to 55°C), and moisture (R.H. ≥70% or the presence of liquid water on surfaces) for 24 hours or more.

Sources that provide the minimum nutritional needs of mold spores abound in building spaces and structures. These include wooden structural materials and furnishings and a variety of paper products, textiles, etc. Mold growth will even occur on the surface of materials that provide no nutrients, such as steel, plastic, synthetic fibers, etc. In such cases, mold utilizes deposited organic dusts or other natural organic substances as nutrients. These dusts allow mold to grow on fiberglass filters and ductwork and synthetic organic textiles (such as those used in carpeting). Mold also grows on metal and plastic furniture that has been subject to human handling (hand oils/skin scales apparently serve as the nutrient source).

A variety of studies indicate that in the absence of liquid water, high relative humidity (≥70%) is needed for germination of mold spores and subsequent optimal infestation conditions on substrate surfaces. Once infestation has taken place, mold growth can continue at lower relative humidity, particularly for xerophilic species (e.g., *Aspergillus* spp.). High relative humidity (if not wetted materials) is required for both the germination and optimal growth of hydrophilic species (e.g., *Stachybotrys*).

Initial mold infestation and subsequent mold growth is favored in building environments with moisture problems. These include persistent high relative humidity, episodic condensation on or in building surfaces, water intrusion through the building envelope, and flooding.

A. High relative humidity

High relative humidity (≥70%) is not uncommon in building environments, particularly residences. It occurs episodically during rainy or foggy weather. Such conditions in many cases are transient and not sufficiently persistent to cause mold infestation. However, buildings, most notably residences, constructed on sites that have been historically wet and have not been provided with adequate drainage, experience chronic elevated humidity. Other sites are subject to seasonal high water tables, which cause wetness in crawlspace, basement, and slab substructures. This wetness is associated with vaporization and subsequent convective movement of water vapor into building interiors, causing high relative humidity in building spaces. In some cases, water may seep into supply air ducts in building slabs and then be vaporized and distributed into building spaces by warm/hot forced air. Water vapor is often drawn into building interiors through leaky return air ducts in houses constructed on crawlspaces.

1. Wet building sites

Wet building sites affect both the absolute humidity (actual amount of water vapor a volume of air contains, i.e., mass per unit volume, or vapor pressure) and relative humidity (percent water in air compared to the maximum it can hold at a given temperature) in buildings constructed atop them. In many cases the relative humidity may be high though the absolute humidity is moderate. At a given absolute humidity, relative humidity increases as temperature decreases. As a consequence, buildings maintained at relatively low temperatures (e.g., vacation homes) and materials in unheated basements (commonly around 55°F, 13°C) are potentially subject to mold infestation. Significant mold growth may also occur on soil in crawlspaces because of wet surfaces and/or high relative humidity. Mold spore dispersal from crawlspace infestation can result in elevated indoor airborne mold concentrations when spores are entrained in leaky return air ducts. Similarly, elevated airborne mold concentrations can occur in living spaces from moldy basements which serve as utility areas for forced air heating and/or cooling systems. Mold spores become entrained in leaky return air ducts and/or the leaky housing of the furnace or air conditioning unit.

2. Occupant practices

Elevated relative humidity may also occur in residential buildings as a result of occupant behavior and practices. When thermostats are set back during sleeping hours, an increase in relative humidity occurs, corresponding to the temperature decrease. Thermostat setbacks, used to reduce energy costs, may result in mold and mildew growth on cool surfaces. Occupants (particularly "empty nesters") often close doors and supply air registers in bedrooms as well as other rooms to reduce heating costs. Since room temperatures and air circulation are reduced, this practice often has the unintended consequence of causing mold infestation.

In many countries where winter temperatures are cool but not cold, residential structures may not be heated at night. Interior temperatures in such houses may drop to 35 to 40°F (3 to 5°C). On wet to moderately wet sites, such houses may experience nighttime relative humidity ≥70%. Consequently, mold infestation of building materials and furnishings such as draperies, mattresses, and carpeting may be common. In a study conducted by the author and colleagues in a typically damp region of southeastern Australia, 19 mold-infested mattresses were identified in a population of 40 houses.

3. Water vapor-generating sources

Relative humidity in buildings can be influenced by a variety of factors. These include the use of vapor-generating sources such as swimming pools, spas, steam baths, large fish tanks, and decorative recirculating fountains, pools, waterfalls and streams. The higher the surface area and temperature of vapor-generating sources, the higher the absolute and relative humidities. High relative humidity may also occur in localized areas such as bathrooms. In the absence of local exhaust ventilation, it occurs episodically with extensive bath and shower use. Consequently, mold growth in showers and on bathroom walls is common.

4. Poor air circulation

Inadequate air circulation within building structures may increase the risk of mold infestation. This is the case in localized areas such as closets and walls obstructed by furniture. Poor air circulation is only a problem if combined with nutrient sources (such as shoes in a closet), elevated relative humidity, and/or a cold surface. When furniture obstructs walls, it prevents heat transfer to wall surfaces; wall cooling causes localized high humidity with subsequent infestation of wall and adjacent furniture surfaces. Poor air circulation occurs in homes that are not heated by forced air systems, in closed-off rooms, and where furniture or other large objects are placed too close to walls.

B. Cold floors

In the absence of adequate insulation, floor surfaces in residences constructed on slabs or crawlspaces are often colder than room air above them. These cold floors may be uncomfortable to occupants as a result of thermal asymmetries; they may also result in high localized relative humidity, which may cause, in the worst case, condensation, or more commonly, an optimum environment for the infestation of jute-backed carpeting or soiled all-synthetic carpeting. Mold infestation of carpeting is common in some countries and climatic regions where interior moisture levels are high and floor surfaces are cool or cold. Mold infestation of carpeting in basements is common because floor surface temperature is typically the same as that of the ground with which the basement slab is in contact (55°F, 13°C).

C. Condensation

When air becomes saturated it reaches its maximum potential to hold water at a given temperature; it then has a relative humidity of 100% and condensation will occur. The temperature at which condensation occurs (i.e., dew point) differs for air that contains different moisture (vapor) levels or absolute humidities. The higher the absolute humidity, the higher the temperature at which condensation takes place. In the psychometric chart in Figure 6.7, a volume of air that has a dry bulb temperature of 75°F (23.5°C) will condense water at approximately 66°F (18.5°C) if the absolute humidity is 0.013 lbs H_2O vapor/lb dry air; condensation will occur at approximately 60°F (15.5°C) if the absolute humidity is 0.011 lbs H_2O vapor/lb dry air.

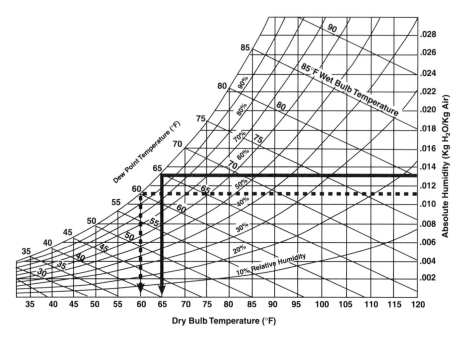

Figure 6.7   Psychrometric chart.

Condensation can occur on any surface if its temperature is below the dew point of the overlying air. The colder the surface, the more likely that condensation and subsequent mold infestation will result (assuming the surface has a sufficient nutrient base).

Condensation on interior building surfaces, within wall cavities, and even within attic areas is relatively common in residential buildings in climatic zones that require seasonal space heating. Condensation problems may occur on or in exterior corners, at exterior wall/roof intersections, on thermal bridges, inside wall cavities, in attic and ceiling areas, and on windows.

1.   Exterior corners

Exterior corners are common locations for both condensation and mold infestation. Exterior corners are colder than other surfaces because they are subject to poor air circulation as well as to wind blowing through corner assemblies (which, due to framing practices, are poorly insulated). They also have a relatively large surface area for heat loss. Wind velocity increases as it flows around corners. Interior surfaces will be cooled significantly if wind enters the corner assembly and blows through it. This airflow is distinct from infiltration.

2.   Exterior wall and roof intersections

Cooler interior surfaces with the potential for condensation may occur where exterior walls intersect roofs. There is often less ceiling insulation present, and that which is there is often compressed, resulting in even greater heat

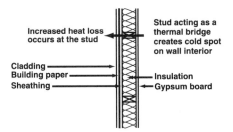

Figure 6.8 Thermal bridge. (From Lstiburek, J. and Carmody, J., *Moisture Control Handbook*, Van Nostrand Reinhold (John Wiley & Sons), New York, 1992. With permission.)

loss. The movement of wind through soffit vents may result in reduced insulation effectiveness and even blow the insulation inward. Cold spots in such areas contribute to condensation and mold infestation.

3. Thermal bridges

Localized cooling of interior building surfaces may occur on thermal bridges. These are areas of relatively high heat conductivity in building envelopes. Examples include uninsulated window lintels, edges of concrete floor slabs, and, most commonly, wood studs in exterior framed walls of residences and other buildings where insulation is installed in wall cavities. Since the wood stud has a higher thermal conductivity than insulation, it provides a pathway for heat loss and localized cold spots on exterior walls. These cold spots can result in high localized relative humidities and potential sites for mold growth. Thermal bridges are illustrated in Figure 6.8.

4. Wall cavities

Condensation occurs when warm, moist, interior air moves into wall cavities where it contacts cold surfaces. Such concealed condensation is common in poorly constructed, poorly insulated walls. However, it may also occur in well-constructed walls (even with vapor barriers). In cool/cold climates, the first condensing surface is typically the interior surface of the exterior non-insulating sheeting (typically plywood or oriented-strand board). In warm/hot climates, the first condensing surface is the exterior side of the interior gypsum board or finish material. If the interior is being cooled, warm, moist outdoor air condenses on contact with interior finishing materials, resulting in mold growth on the back of such materials as vinyl wallpaper.

5. Plumbing

Condensation commonly occurs on uninsulated plumbing lines and fixtures. Uninsulated cold water lines in a relatively humid environment can result in considerable "sweating" and subsequent wetting of adjacent materials. Sweating occurs on toilet fixtures as well. Such sweating may cause both structural damage and mold infestation.

6. Windows

Single-pane windows are common condensation sites. Because glass is relatively thin and has high thermal conductivity, condensation on interior glass surfaces may occur during cold weather. If extensive, water will "puddle" on sashes, sills, and frames, and may even enter wall cavities. Repeated wettings increase the potential for mold growth and subsequent occupant exposure to mold. The degree to which such condensation problems occur depends on interior moisture loads, the heat loss potential of windows, outdoor temperature, and occupant heating practices. In the seasonally cool climate of southern coastal Australia, where use of central heating is relatively limited and internal moisture levels are high, window moisture condensation is a common wintertime occurrence. Such condensation typically results in mold infestation of adjacent drapery material.

7. HVAC system operation

Condensation can occur on building surfaces and materials when heating, ventilation, and air-conditioning systems are not operated properly. This has often occurred in older school buildings during summer months in the Midwest and other relatively humid regions after the installation of air conditioning. During daytime hours, these systems are operated to provide comfortably cool conditions for building occupants; they are subsequently shut down after work hours. Under high dew point (75 to 78°F, 24 to 25°C) ambient weather conditions, outside air passively enters the building where it comes into contact with surfaces below the dew point (72 to 74°F, 22 to 23.5°C). Puddling may occur on floor and other surfaces. Delamination of floor tile is common, as is mold infestation of carpeting. In schools, significant mold infestation occurs on extensively handled books and other materials. This problem may occur in new school buildings as well, but has been much more common in older schools.

D. Water intrusion

Wet building surfaces and other materials that can serve as nutrient sources provide an excellent environment for mold infestation; if wetness is of sufficient duration, it results in structural deterioration. Intrusion of water from rain or snow melt is a common problem in both residential and nonresidential buildings. Risk factors for the penetration of water through the building envelope include poor construction practices; deterioration and inadequate maintenance of roofing, flashing, and cladding; and storms that may cause minor to severe damage to roofing and cladding materials. Evidence of water intrusion in residential buildings has been shown to be a significant risk factor for elevated total airborne mold levels.

1. Construction practices

A variety of construction practices contribute to water intrusion problems. These include poor installation of roofing, flashing, or cladding; inadequate caulking of breaks in the building envelope such as around windows and doors (though most people in the U.S. believe that the reason one caulks windows and doors is to reduce energy losses, the primary reason is to prevent or minimize the penetration of rain water); and use of flat roofs on nonresidential buildings (such roofs are at unusually high risk for water intrusion episodes). Water leaks through rubberized membranes of flat-roofed school buildings in the U.S. are widespread, as can be seen by the numerous water-stained, and often mold-infested, tiles in drop ceilings. In some cases, water-stained ceiling tiles are associated with other moisture problems such as leaky air handling units, condensate from chilled water lines, and plumbing leaks.

Water intrusion through the building envelope in residential structures typically occurs through poorly installed or maintained flashing around chimneys and exhaust structures in roofing, damaged roofing material, poorly caulked windows and doors, and, somewhat surprisingly, brick and stone veneer cladding. In the latter case, rain penetrates brick veneer through tiny holes in mortar, cracks in bricks, and by capillary action through both mortar and bricks. Mortar is relatively porous; bricks, less so. Water penetration can be expected on the weather side of such buildings during wind-driven rains, with subsequent movement of water vapor toward building cavities by sun-affected thermal/physical forces.

Good construction practice requires that brick veneer be provided with an approximate $3/4''$ (19 mm) gap between its interior surface and adjacent wall materials so that water that enters the cavity can trickle down to the base where it can flow outward through functioning weep holes. A direct pathway for moisture to wet wall materials is provided when masons do not remove the excess mortar on the inside of brick or stone veneer. In other cases, water may move by capillary action on metal ties from the brick veneer wall to other components of wall structures. In many cases, masons fail to include weep holes in brick veneer, or these holes are obstructed by excess mortar which accumulates at the interior base of the wall during construction. Weep holes provide an avenue of egress for liquid water from behind brick and stone veneer. They also provide a mechanism whereby air can naturally enter the cavity behind veneer to remove moisture by convection.

On building sites with high shrink/swell-capacity soils, brick/stone veneer wall cladding may develop settling cracks as it ages. These become larger and more numerous with time. Not surprisingly, building age is a significant risk factor for mold infestation, elevated indoor mold levels, and health complaints.

E. Plumbing leaks and flooding

Water damage to buildings may occur as a result of small plumbing leaks, as well as more infrequent episodes in which plumbing breaks cause major floods in buildings. Small plumbing leaks are common and result in localized water damage and often limited mold infestation (Figure 6.9). Small plumb-

Figure 6.9 *Stachybotrys* and other mold species growing on ceiling tile.

ing leaks without timely maintenance may result in significant water damage
and mold infestation problems.

Floods associated with plumbing failures are not uncommon in buildings. They may arise as a consequence of breaks in potable or mechanical
system water lines, or sewage water backups. Significant wetting of building
materials may result, depending on the amount of water released or the
extent of sewage backup. The probability that significant mold infestation
will occur also depends on how quickly building facilities personnel/home-
owners dry the wetted materials. The window for such drying is believed
to be 24 to 36 hours; i.e., if flooding cleanup and material drying can be
accomplished within 24 to 36 hours, little to no mold infestation should
occur. Sewage backups pose problems of both water damage, mold infestation, and potential problems of odor and contamination of building materials
with fecal bacteria.

Overflow of sinks and bathtubs (typically associated with inadequate
supervision of children) is one of the most common flooding problems in
dwellings. Another common problem is poorly constructed and/or maintained shower facilities, wherein water penetrates wall cavities through inadequately grouted tile or windows. Poor use of shower curtains is also a minor
cause of water damage to bathroom floors.

Small floods may occur in mechanically ventilated buildings when cooling coil units freeze in air handling units as a consequence of cold outdoor
temperatures.

Severe flooding in buildings typically occurs as a consequence of cata-
strophic events. Most notable are structural fires that are extinguished with
water, and natural disasters when rising water enters buildings. In such
cases, structural damage and mold infestation are often very severe. These
events represent worst-case scenarios of building mold infestation, associated exposure and health risks, and mitigation requirements.

F. Other sources of indoor mold contamination

Mold contamination of indoor spaces can occur even in the absence of high relative humidities, water intrusion, or flooding. Since high mold levels occur outdoors, indoor spaces can be contaminated by outdoor mold species when windows and doors are open, particularly during warmer months and the autumn season of leaf and other plant decay. Such mold spores are deposited on indoor surfaces and, on resuspension, affect indoor airborne levels for some time after additional spores no longer intrude into indoor spaces in any significant way.

Indoor airborne mold levels and subsequent human exposure can also occur when mold-infested materials are brought into building spaces. Sources may include infested furniture, clothing, firewood, dried plants, books, etc.

Though there is little evidence to indicate that live plants cause significant indoor mold contamination, they are likely to be localized sources of phylloplane fungi when dead leaves and other plant parts undergo decomposition. Soil also has the potential to be a source of airborne mold.

Readings

Burge, H.A., The fungi, in *Biological Contaminants in Indoor Environments,* Morey, P.R., Feeley, J.C., and Otten, J.A., Eds., ASTM STP 1071, American Society for Testing and Materials, Philadelphia, 1990, 136.

Chmel, H., Fungal infections in the immunocompromised host: clinical syndromes and diagnosis, in *Fungal Infections and Immune Responses*, Murphy, J.W., Friedman, H., and Bendinelli, M., Eds., Plenum Press, New York, 1993, 405.

Etkin, D.S., *Biocontaminants in Indoor Environments, Indoor Air Quality Update*, Cutter Information Corp., Arlington, MA, 1994.

Federal–Provincial Committee on Environmental and Occupational Health, *Fungal Contamination in Public Buildings: A Guide to Recognition and Management*, Health Canada, Ottawa, 1995.

ISIAQ Guideline, *Control of Moisture Problems Affecting Biological Indoor Air Quality,* International Society of Indoor Air Quality and Climate, Milan, Italy, 1996.

Kurup, V.P. and Fink, J., Fungal allergy, in *Fungal Infections and Immune Responses*, Murphy, J.W., Friedman, H., and Bendinelli, M., Eds., Plenum Press, New York, 1993, 393.

Levetin, E., Fungi, in *Bioaerosols*, Burge, H.A., Ed., CRC Press/Lewis Publishers, Boca Raton, 1995, chap. 5.

Lstiburek, J. and Carmody, J., *Moisture Control Handbook. Principles and Practices for Residential and Small Commercial Buildings*, Van Nostrand Reinhold, New York, 1994.

Madelin, T.M. and Madelin, M.F., Biological analysis of fungi and associated molds, in *Bioaerosols Handbook*, Cox, C.S. and Wathes, C.M., Eds., CRC Press/Lewis Publishers, Boca Raton, 1995, chap. 14.

Maroni, M., Siefert, B., and Lindvall, T., *Indoor Air Quality — A Comprehensive Reference Book*, Elsevier, Amsterdam, 1995, chaps. 6, 31.

Sorenson, W.G., Mycotoxins: toxic metabolites of fungi, in *Fungal Infections and Immune Responses*, Murphy, J.W., Friedman, H., and Bendinelli, M., Eds., Plenum Press, New York, 1993, 469.

Spengler, J.D., Samet, J.M., and McCarthey, J.F., Eds., *Indoor Air Quality Handbook*, McGraw-Hill Publishers, New York, 2000, chaps. 45, 46.

Questions

1. What do the terms mold and mildew actually mean?
2. What is the significance of mold reproductive processes and potential human exposures?
3. Why is the probability that humans will be exposed to mold spores so high?
4. Mold spores, on release, quickly lose their viability. Why? What significance, if any, does this loss of viability have on human exposures?
5. How do indoor and outdoor airborne mold types and levels compare?
6. Distinguish between hydrophilic and xerophilic mold species.
7. Define water activity. What is its significance in mold growth?
8. Fungi are classified into one of five major divisions. What is the relative role of each of these groups in relationship to human exposures and related health effects?
9. What are fungal MVOCs? What significance, if any, do they have?
10. What are mycotoxins? How and why are they produced?
11. How can one be exposed to high mycotoxin concentrations?
12. What are fungal glucans? What is their source?
13. Why are airborne mold levels different when they are measured by culturable/viable and total mold spore sampling techniques?
14. Characterize airborne mold concentrations in houses; mechanically ventilated buildings.
15. Why are mold infestations an indoor air quality concern?
16. What allergenic/immunologic diseases are caused by exposure to mold?
17. Why are exposures to *Stachybotrys chartarum* during infancy a public health concern?
18. Under what conditions may you be exposed to *Stachybotrys chartarum*?
19. What health concerns are associated with exposure to fungal glucans?
20. What are the two primary risk factors for mold infestation of building materials?
21. How can conditions in a crawlspace affect indoor mold levels?
22. Describe factors that contribute to condensation in building structures and wall cavities.
23. What is water intrusion? How does it contribute to mold growth?
24. What is the dew point? What is its significance relative to mold-related building concerns?

chapter seven

Problem buildings

Buildings that serve nonresidential purposes, such as office complexes, a large diversity of retail and commercial units, and institutional buildings (schools, universities, hospitals, day-care centers, and convalescent and retirement homes), experience a variety of complaints that may be associated with poor indoor air/indoor environment quality. When a building is subject to complaints sufficient to convince management to conduct an indoor environment (IE) investigation, it is often characterized as a "problem" or "sick" building. Health complaints have been described as being due to building-related illness, or tight building/sick building syndrome.

I. Building illness concepts

A. Building-related illness

The terms, "building-related illness" (BRI) or "specific building-related illness" (SBRI), are used to characterize cases in nonresidential, nonindustrial buildings wherein causal factors for illness symptoms and complaints have been convincingly identified. Building-related illness is characterized by unique symptoms that may be accompanied by clinical signs, laboratory confirmations, and identifiable contaminants. Included in BRI/SBRI are nosocomial (hospital-acquired) infections, hypersensitivity diseases (hypersensitivity pneumonitis, humidifier fever, asthma, and chronic allergic rhinitis), Legionnaires' disease, fiberglass dermatitis, and toxic effects associated with high exposures to carbon monoxide (CO). It could also include formaldehyde (HCHO). Formaldehyde, however, produces symptoms that are often indistinguishable from classical IAQ-type symptoms.

The term building-related illness could be more generally applied. If illness symptoms (no matter the cause, known or unknown) can be shown to be associated with a building or indoor environment, they are, in fact, building-related.

B. *Work-related illness and symptoms*

The concept of work-related illness or symptoms as distinct from those that are building-related have not yet been distinctly described in the scientific literature. In many instances, illness symptoms may be associated with specific work activities rather than exposures to components of building environments. These would include eyestrain, headache, fatigue, and muscle ache associated with working with video-display terminals and keyboards; upper respiratory and skin symptoms associated with handling carbonless copy paper; illness associated with exposure to glutaraldehyde in medical and dental offices; latex allergy associated with using latex gloves in medical and dental offices; severe mucous membrane irritation from ammonia emanating from blueprint machines; illness associated with wet-process photocopiers, laser printers, and spirit duplicators; and neurotoxic symptoms due to solvent vapor exposures associated with printing, silk screening, painting, etc.

In such cases, exposures are directly associated with a localized work activity and not with contaminants in the general building environment. However, if activity-related contaminants migrate and affect others, symptoms may be better described as building-related.

In many problem building investigations, skin symptoms are reported and assessed within the context of an IAQ/IE problem. Skin symptoms are typically caused by direct contact with irritant substances and materials. Causal agents are, in most cases, unlikely to be airborne. As a consequence, most IE reports of skin symptoms are likely to be work-, rather than building-, related.

C. *Sick building syndrome*

The term, "sick building syndrome" (SBS), has been historically used to define a spectrum of subjective illness symptoms associated with building/work environments with which a specific causal agent or agents cannot be identified. Panels of major organizations, i.e., the World Health Organization (WHO), the Commission of European Communities, and the American Thoracic Society, have attempted to define the apparent phenomenon of SBS. Definitions overlap to some degree, but also define the nature of the phenomenon quite differently. For illustrative purposes, "sick building syndrome" is described here within the context of the WHO definition.

Sick building syndrome has been defined by WHO on the basis of frequently reported symptoms and complaints. These include: (1) sensory irritation of the eyes, nose, throat; (2) neurotoxic or general health problems; (3) skin irritation; (4) nonspecific hypersensitivity reactions; and (5) odor and taste sensations. Sensory irritation is described as pain, a feeling of dryness, smarting, stinging irritation, hoarseness, or voice problems; neurotoxic/general health problems such as headache, sluggishness, mental fatigue, reduced memory, reduced concentration, dizziness, intoxication, nausea, vomiting, and tiredness; skin irritation such as pain, reddening, smarting, itching sensations, or dry skin; nonspecific hypersensitivity reactions such as runny nose or eyes, asthma-like symptoms among nonasthmatics; and odor and taste sensations such as changed sensitivity of olfactory and taste senses, or unpleasant odor and taste.

In defining SBS, the WHO panel concluded that: (1) the major symptoms are mucous membrane irritation of the eyes, nose, and throat; (2) symptoms should appear especially frequently in individual buildings or parts thereof; (3) a majority of occupants should report symptoms; and (4) there should be no evident symptom relationship to occupant sensitivity or excessive exposures.

The WHO characterization of SBS appears to be based on the theory that SBS complaints of a sensory nature occur as a consequence of the nonspecific irritation or overstimulation of trigeminal nerves (responsible for the common chemical sense) in mucous membranes. Trigeminal nerves respond to chemical odors, producing sensations of irritation, tickling, or burning. Exposure to many different chemicals produces similar responses.

Within this context, a WHO committee has suggested that indoor air contains a complex of sensory stimuli that produces irritant responses not specific to individual contaminant exposures. As a consequence, no single contaminant is likely to be responsible for SBS. Reactions of the "referred pain" type may take place (i.e., headaches that may be due to the irritation of trigeminal nerves). Following absorption of contaminants on nasal mucosa, upper respiratory symptoms would occur as a result of numerous subthreshold stimulations.

As defined by WHO, SBS is a phenomenon in which high prevalence rates of illness symptoms occur in buildings with no single apparent causal factor responsible. This concept of SBS suggests that reported symptoms are due to collective exposure to a variety of chemical substances present at low concentrations.

The concept of SBS was defined in the early 1980s at a time when there was little understanding of causal or risk factors for illness symptoms in building occupants. It was also defined at a time when ventilation rates used in buildings were relatively low, and emissions from various building materials, furnishings, finishes, etc., were high. Since that time, building ventilation rates have increased and emissions from materials have decreased. As a result, our early understandings of SBS as a unique phenomenon are less applicable today.

Scientific studies conducted over the past two decades suggest that a so-called "sick building syndrome" may not in fact exist. SBS-type symptoms reported in any individual building are likely to be multifactorial in origin, i.e., a variety of exposures occurring at the same time may be responsible for the reported symptoms.

D. Sick/tight/problem buildings

Outbreaks of illness symptoms with high prevalence rates in northern European and North American buildings in the late 1970s and early- to mid-1980s led investigators to conclude that such buildings were "sick." Other buildings where no complaints were reported were thought to be "healthy." Under the WHO characterization, a sick building was distinguished from a normal one by the prevalence of symptoms, i.e., in a sick building a large percentage of occupants report symptoms. Based on this characterization, WHO concluded that 30% of new buildings in the early 1980s were sick buildings. The term, "sick building," is still widely (and loosely) used by laypersons to describe buildings subject to health-related indoor air quality/indoor environment (IAQ/IE) complaints.

The terms, "tight building" and "tight building syndrome," were used in the late 1970s and early 1980s when it was widely believed that SBS-type phenomena were due to the implementation of energy conservation measures in the design, construction, and operation of buildings. These terms were unfortunately simplistic and wrongly described the true nature of building-/work-related health and comfort complaints in buildings.

Buildings vary significantly in symptom prevalence rates that may be associated with the building and work environment. A high prevalence of symptoms may result in complaints to building management requesting that an investigation be conducted. Such buildings can best be described as "problem buildings." The term, "problem building," is an appropriate characterization of any building subject to complaints, whether complaints are limited to a few individuals or involve a much larger building population. Because of difficulties inherent in defining a sick building and the negative connotation this term conveys to both building occupants and managers/owners, the term, "problem building," better describes an indoor environment in which there are building-related health, comfort, and odor complaints.

II. Field investigations

An apparent relationship between building/work environments and occupant illness complaints was initially determined from building investigations conducted by governmental agencies and private consultants providing industrial hygiene or IAQ/IE services.

Field investigations are conducted at the request of building owners. They vary considerably in methodologies employed, training and experience of those conducting the investigation, and success in identifying potential causal factors.

A. NIOSH investigations

Field investigations of >1000 problem buildings have been conducted in the U.S. since 1978 by health hazard evaluation teams of the National Institute of Occupational Safety and Health (NIOSH). These three-member investiga-

Table 7.1 Frequency of Reported
Symptoms in NIOSH Building
Investigations

Symptom	% of buildings
Eye irritation	81
Dry throat	71
Headache	67
Fatigue	53
Sinus congestion	51
Skin irritation	38
Shortness of breath	33
Cough	24
Dizziness	22
Nausea	15

Source: From Wallingford, K.M. and Carpenter, J., *Proc. IAQ '86: Managing Indoor Air for Health and Energy Conserv.,* American Society of Heating, Refrigerating and Air-Conditioning Engineers, Atlanta, 448, 1986. With permission.

tive teams are comprised of an epidemiologist, industrial hygienist, and HVAC system engineer or technician. Summary reports have been published periodically. Buildings investigated have included schools, universities and colleges, health-care facilities, and private offices.

In many NIOSH investigations, symptom complaints were subjective and not attributable to a specific causal agent. Reported symptoms have included headache; eye, nose, throat, and skin irritation; fatigue; a variety of respiratory symptoms such as sinus congestion, sneezing, cough, and shortness of breath; and, less frequently, nausea and dizziness. The frequencies of reported symptoms in several hundred building investigations are summarized in Table 7.1. In a large percentage of cases (>50%), occupants reported eye irritation, dry throat, sinus congestion, headache, and fatigue. The former three are described as mucous membrane symptoms; the latter two as general (or neurotoxic) symptoms.

Major problem types identified in over 500 NIOSH building investigations are briefly summarized in Table 7.2. Inadequate ventilation was an IAQ/IE concern in >50% of buildings investigated. Inadequate ventilation was determined by reference to a guideline value of 1000 ppmv carbon dioxide (CO_2). Other ventilation problems included poor air distribution and mixing, draftiness, pressure differences among building spaces, and filtration problems caused by inadequate maintenance.

Indoor air quality problems due to indoor sources included exposures associated with office equipment, e.g., methanol from spirit duplicators, butyl methacrylate from signature machines, and ammonia and acetic acid from blueprint machines. Other contamination problems included misapplied pesticides, boiler additives in steam humidification units, combustion gases from cafeterias and laboratories, and cross-contamination between building zones.

Table 7.2   Problem Types Identified in NIOSH Building Investigations

Problem type	Buildings investigated	%
Contamination from indoor sources	80	15
Contamination from outdoor sources	53	10
Building fabric as contaminant source	21	4
Microbial contamination	27	5
Inadequate ventilation	280	53
Unknown	68	13
Total	529	100

Source: From Seitz, T.A., *Proc. Indoor Air Qual. Internatl. Symposium: The Practitioner's Approach to Indoor Air Qual. Investig.,* American Industrial Hygiene Association, Akron, 163, 1989. With permission.

Outdoor sources of indoor contamination included entrainment/reentry problems associated with motor vehicle exhaust, boiler flue gases, rooftop and building side exhausts, dusts and solvents from road and parking lot asphalt work, and gasoline vapors infiltrating basements or sewage systems.

Contamination associated with building products and materials included HCHO emissions from urea–formaldehyde-bonded wood products, fiberglass particles eroded from duct liners, organic solvents from adhesives, and PCBs from fluorescent light ballast failure.

Microbial contaminants were identified as the major cause of complaints in approximately 5% of NIOSH investigations. Hypersensitivity pneumonitis associated with high levels of exposure to spores of fungi or thermophilic actinomycetes was the major health problem in buildings with microbial contamination.

NIOSH health hazard evaluations represent a significant resource of documented building investigations. NIOSH investigations differ in quality from many early investigations conducted without benefit of a systematic investigative protocol. NIOSH reports provide a general overview of the type of problems observed by field staff. They are not likely to be representative of the frequency of problems found in U.S. buildings because the building population is biased toward institutional buildings and to buildings with problems that are likely to be more difficult to identify and resolve. NIOSH investigations are often conducted when other government investigators or private consultants have failed to identify and resolve reported problems. The relatively high percentage of building cases with hypersensitivity pneumonitis is likely due to NIOSH expertise in this area.

III. Systematic building investigations — symptom prevalence

Field investigations have served to initially identify and define the nature of problem building phenomena. However, they have limited scientific use-

Table 7.3 Symptom Prevalence Rates (%) Among Employees of Danish Municipal Buildings

	Prevalence Rate (%)	
	Males (N = 1093–1115)	Females (N = 2280–2345)
Symptoms		
Eye irritation	8.0	15.1
Nasal irritation	12.0	20.0
Blocked, runny nose	4.7	8.3
Throat irritation	10.9	17.9
Sore throat	1.9	2.5
Dry skin	3.6	7.5
Rash	1.2	1.6
Headache	13.0	22.9
Fatigue	20.9	30.8
Malaise	4.9	9.2
Irritability	5.4	6.3
Lack of concentration	3.7	4.7
Symptom Groups		
Mucous membrane irritation	20.3	
Skin reactions	4.2	
General symptoms	26.1	
Irritability	7.9	

Source: From Skov, P. and Valbjorn, O., *Environ. Int.*, 13, 339, 1987. With permission.

fulness due to the inherent bias involved in conducting investigations/studies of buildings subject to occupant health and comfort complaints. In addition, many building investigations have been conducted relatively unsystematically.

Systematic epidemiological studies have been carried out in problem and noncomplaint buildings in order to assess symptom prevalence rates and potential risk factors that may be associated with symptoms or symptom reporting rates.

Major systematic cross-sectional epidemiological building studies have been conducted in Denmark, the United Kingdom, Sweden, the Netherlands, and the U.S. Studies have differed in symptom prevalence assessment methodology, building types evaluated (commercial office, governmental, schools), and complaint status (complaint vs. noncomplaint). Symptom prevalence rates among male and female employees in 14 noncomplaint Danish municipal buildings are summarized in Table 7.3. Note that 20+% of females in these putatively nonproblem buildings reported symptoms of nasal irritation, headache, and fatigue. Prevalence rates of 20 and 26%, respectively, were reported for the two symptom groups, mucous membrane irritation and general symptoms (headache, fatigue, malaise). Based on this study, illness symptoms associated with building/work environments occur at significant rates even when no complaints have been previously reported.

Table 7.4 Symptom Prevalence Rates (%) Among
Occupants in 11 Swedish "Sick" Office Buildings

Symptom	Total mean Prevalence (%)	Range
Eye irritation	36	13–67
Swollen eyelids	13	0–32
Nasal catarrh	21	7–46
Nasal congestion	33	12–54
Throat dryness	38	13–64
Sore throat	18	8–36
Irritative cough	15	6–27
Headache	36	19–60
Abnormal tiredness	49	19–92
Sensation of getting a cold	42	23–77
Nausea	8	0–23
Facial itch	12	0–31
Facial rash	14	0–38
Itching on hands	12	5–31
Rashes on hands	8	0–23
Eczema	15	5–26

Source: From Norback, D., Michel, I., and Widstrom, J., *Scand.
J. Work Environ. Health,* 16, 121, 1990. With permission.

A systematic cross-sectional epidemiological study was conducted in 11 Swedish office buildings presumed to be "sick buildings" due to occupant complaints. Reported symptom prevalence rates are summarized in Table 7.4. High prevalence rates (>30%) were reported for eye irritation, nasal congestion, throat dryness, sensation of getting a cold, headache, and abnormal tiredness. In these "sick buildings," rates appear, on average, to be considerably higher than those observed in Danish noncomplaint municipal office buildings. It must be noted, however, that the Swedish study included all symptoms reported in a 6-month period. In the Danish study, symptom prevalence rates were limited to symptoms that occurred often or always and resolved when leaving the building.

It is notable here to compare results of Danish and Swedish studies to an intensive study conducted in the headquarters building of the U.S. Environmental Protection Agency (USEPA) in 1989. This building had been the focus of occupant complaints (including litigation) of poor air quality, and subject to considerable notoriety because of the irony of the situation and failure of USEPA, NIOSH, and a host of private consultants to identify and mitigate the causes of USEPA staff complaints.

The USEPA headquarters complex comprised three buildings. These included Waterside Mall (WM) (a large building divided into sectors), and two smaller buildings, Crystal City (CC) and Fairchild (FC). Prevalence rates for IAQ-type and respiratory/flu-like symptoms are shown in Table 7.5. Prevalence rates of symptoms reported to occur often or always which

Table 7.5   Symptom Prevalence Rates (%) in USEPA Headquarters Buildings

Symptoms	N =	WM sectors						Building (avg.)		
		1	2	3	4	5	6	WM	CC	FC
		772	600	400	500	435	223	3070	445	407
IAQ-type										
Headache		14	13	18	19	16	18	16	11	16
Runny nose		7	9	9	10	8	8	8	9	7
Stuffy nose		15	13	16	21	16	16	16	17	15
Dry eyes		14	15	21	18	13	20	17	12	15
Burning eyes		9	10	13	11	9	10	10	8	11
Dry throat		8	9	15	12	8	14	10	7	9
Fatigue		12	15	17	17	12	15	15	14	11
Sleepiness		13	14	18	17	14	20	15	19	13
Respiratory/Flu-like										
Cough		4	5	6	6	4	2	4	5	4
Wheezing		1	1	1	2	1	2	1	1	2
Shortness of breath		1	2	3	3	3	2	2	1	2
Chest tightness		1	1	3	2	2	2	1	1	0
Fever		4	0	0	1	1	5	1	1	0
Aching muscles/joints		3	4	5	5	4	6	4	4	2

Note: N denotes number of persons in a sector or building. WM = Waterside Mall; CC = Crystal
City; FC = Fairchild.

Source: From Fidler, A.T. et al., *Proc. 5th Internatl. Conf. Indoor Air Qual. Clim.*, Toronto, 4, 603,
1990.

resolved on leaving the building varied from 7 to 21%. Though having been
publicly labeled as an archetype "sick building," the prevalence rates for
building/work-related symptoms were in the same range as noncomplaint
Danish municipal buildings.

Systematic building studies have shown that building environ-
ment-/work-related health complaints occur in all buildings surveyed,
regardless of their complaint status. They also show a broad range of prevalence rates among buildings and variation in symptom prevalence.

IV. Work performance and productivity

Symptoms characteristic of IAQ-type complaints are not life-threatening.
They are relatively minor in their seriousness and in most cases do not
constitute a significant health concern. They are best described as "quality
of life" symptoms. Their effects, however, may not be without consequence.
Concerns have been expressed about the potential for IAQ-type symptoms
to result in decreased productivity by decreasing work performance and
increasing absenteeism. If poor IAQ did decrease productivity, it would
impose potentially significant economic costs on employers of affected building occupants.

A limited number of studies have been conducted to assess the relationship between IAQ and worker productivity. Studies employing subjective ratings of productivity and IAQ indicate that perceived productivity decreases when symptom prevalence rates increase. Quantitative studies have been limited, and no quantitative relationship between IAQ and productivity in office buildings has been reported to date.

V. SBS-type symptom risk factors

As previously suggested, occupant symptoms associated with office, commercial, and institutional building environments appear to have a multifactorial origin. In addition, a variety of factors have been identified which, either directly or indirectly, contribute to increased symptom prevalence or reporting rates. These include personal characteristics of occupants, psychosocial factors, tobacco smoke, building environmental conditions and furnishings, office materials and equipment, and individual contaminants.

A. Personal characteristics

A variety of personal characteristics of building occupants have been evaluated as potential contributing factors to SBS-type symptom prevalence/reporting rates in systematic building studies. These have included gender, age, marital status, atopy, and lifestyle factors such as smoking, alcohol consumption, coffee consumption, exercise, use of contact lenses, etc.

Gender and allergic history have been reported to be major risk factors for SBS-type symptom prevalence/reporting rates. Mixed results have been reported for tobacco smoke.

1. Gender

In systematic building studies, females consistently report SBS-type symptoms at rates 2 to 3 times that of males. These differences are evident in the Danish municipal building study (Table 7.3). Gender differences have also been reported among children, with parents reporting more symptoms in females in problem schools. These differences appear to start at an early age and increase with age. It has been suggested that females may be more sensitive to environmental influences or be more aware of physical symptoms. In the former case, studies have shown that females have a more responsive immune system and are more prone to mucosal dryness and facial erythema than males. Differential illness perceptions and treatment responses between males and females appear to be universal across all population groups. In developed countries, females in general report more physical symptoms, take more prescribed medication, and visit physicians more frequently. Changes in social roles involving stresses associated with combining work and family responsibilities have also been suggested as contributing factors in increased illness in adult females.

British scientists have proposed that gender differences in symptom prevalence may be due to the tendency of males to underreport symptoms. This proposition is based on observations that there are no gender differences in symptom reporting rates among office workers who (1) report extreme dissatisfaction with their work environment, (2) have considerable control over their work, (3) work <6 hours/day, and (4) have worked in the same office environment for more than 8 years. In controlled exposure studies to irritant chemicals, objective measurements of inflammatory responses of mucous membranes indicate that males and females are equally affected. Nevertheless, males report fewer symptoms. This may be due to cultural conditioning, i.e., males may perceive symptom reporting to be an admission of personal weakness.

Females are likely to be exposed to environmental and psychosocial factors in building/work environments that are different from those of males. Females in office environments tend to perform clerical work; males are more often supervisors. A clerical worker has a lower social status and compensation level. Clerical work also includes unique exposures to office materials and equipment.

Female/male symptom reporting differences may be a significant factor in building management decisions to have an IAQ/IE investigation conducted to identify and resolve complaints. In general, based on the author's experience, when complaints are preponderantly from females, male managers/owners are less likely to take them seriously. This exacerbates "the problem" and makes it more difficult for investigators to resolve.

2. Allergic history

In most systematic epidemiological studies of SBS-type symptom prevalence in nonresidential buildings, a self-reported history of atopy (genetic predisposition to common allergens) has been strongly associated with increased symptom prevalence rates. This has particularly been the case for mucous membrane (eyes, nose, throat, sinuses) inflammatory symptoms. Individuals with atopy are highly allergic and experience symptoms of chronic allergy that overlap the spectrum of SBS-type symptoms.

There are several possible reasons why self-reported allergy is an apparent risk factor for SBS-type symptoms. These include the possibility that allergic individuals are more sensitive to irritant chemicals than average individuals. Because of their experience with chronic allergy and/or asthma, atopic individuals (including males) may be preconditioned to report symptoms when they actually occur. There is also increasing evidence that exposure to common allergens in building environments may be responsible for a portion of symptom prevalence rates.

The apparent relationship between self-reported allergy/asthma and SBS symptom prevalence has been evaluated by Danish scientists using an objective measure of atopy (specific IgE blood tests for common allergens such as pollen, dust mites, and mold). Though self-reported allergy was strongly correlated with SBS symptom prevalence, there was no correlation between positive specific IgE and SBS symptoms. Paradoxically, the increased prevalence of work-related SBS symptoms was mainly associated with individuals reporting a history of asthma or allergy but having a negative IgE test. Apparently, individuals who report SBS symptoms tend to interpret such symptoms as allergies.

B. *Psychosocial phenomena and factors*

Building/work environments are characterized by behavior dynamics among occupants and between occupants and building management. Human behavior has often been cited as the principal cause of complaints.

1. *Mass psychogenic illness*

Prior to the present era of increased scientific understanding of problem buildings and causes of occupant complaints, investigators often failed to identify any causal factor in complaint investigations. As a consequence, outbreaks of illness with high prevalence rates were diagnosed and reported in the medical literature as having been caused by psychological factors. These were variously described as mass hysteria, hysterical contagion, epidemic hysteria, psychosomatic illness, epidemic psychogenic illness, and mass psychogenic illness. Common to all such reports were (1) a sudden onset of illness (2) the perception by investigators that illness problems became worse as a consequence of verbal and visual contact among those affected, (3) high prevalence rates among females, and (4) investigators were unable to identify a causal agent of either an infectious or toxic nature. Notably, a very high percentage (circa 75%) of such episodes were reported in school buildings.

Though terms such as *mass psychogenic illness* and its synonyms have been widely used and carry an aura of medical authority, these putative phenomena were based on anecdotal reports of field investigators. There is no credible scientific evidence that mass psychogenic illness is a real phenomenon.

In most instances, problem buildings have a characteristic dynamic related to the behavior of individuals with health/comfort/odor complaints, with various levels of individual and group emotion. The extremity of emotion varies with circumstances involved. These include, for example, the suddenness of problem onset, occupant perception of risks or threats to them personally, an inability to convince building management that a problem exists, and individual lability.

In buildings where there is no dramatic onset of symptoms or sudden awareness of "toxic" chemical odors, occupants conduct their daily activities unaware that symptoms and discomfort they are experiencing are related to their building/work environment. This is apparent in systematic studies of noncomplaint buildings which, nevertheless, show relatively high building/work-related symptom prevalence rates. In such buildings and those

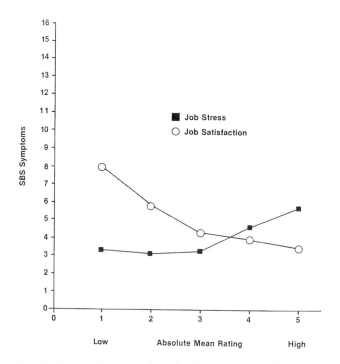

Figure 7.1 Relationship of job stress/satisfaction ratings and SBS symptom prevalence. (Courtesy of Hedge, A., Cornell University, personal communication.)

described as problem buildings, an awareness develops among some occupants, which is communicated to others. A "contagion" of awareness may consequently develop that is construed as psychogenic since it develops as a consequence of increasing occupant-to-occupant communication.

2. Psychosocial factors

A number of psychosocial factors have been significantly associated with SBS-type symptom prevalence. These include job function; dissatisfaction with supervisors, colleagues, physical environment, and/or job; quantitative work demands; job stress; conflicting roles; perceived degree of control over environmental conditions; and occupant density.

The most significant consistent relationships between SBS-type symptom reporting rates and psychosocial variables have been job stress and job satisfaction or dissatisfaction. The relationships between job stress, job satisfaction, and symptom prevalence can be seen in Figure 7.1. Symptom reporting rates can be seen to increase with increasing stress (particularly perceived high stress) and decrease when occupants report high job satisfaction.

3. Seasonal affective disorder

Seasonal affective disorder (SAD) is a recurring mood-changing phenomenon characterized by depression-type symptoms, with onset in late fall, continuing during winter, and resolving in spring. It is thought to result from body biochemical changes associated with decreases in solar intensity and photoperiod. Building occupants with high SAD scores have significantly higher symptom prevalence rates, which decrease as the winter season ends and spring begins.

4. Significance of psychosocial factors and SBS-type symptoms

The most scientifically plausible explanation of the observed relationship between psychosocial factors and SBS-type symptoms (other than potential effects of SAD) is the effect of the former on symptom reporting rates. It can be anticipated that individuals experiencing work-related stress are more likely to report symptoms (when they occur), and those experiencing little or no stress are less likely to report symptoms. Similarly, individuals reporting high job satisfaction may be less likely to report symptoms even when present. The effect of psychosocial factors may be indirect (affecting reporting rates) rather than direct, as is often supposed.

C. Tobacco smoking

Studies that have evaluated passive smoking as a risk factor for SBS-type symptoms have consistently shown a relationship between office workers' perception of exposure to environmental tobacco smoke (ETS) and increased symptom prevalence. Several studies have shown weaker relationships between objective measurements and symptom prevalence. These studies indicate that exposure to ETS may increase the rate of SBS-type symptom reporting. Danish studies assessing satisfaction or dissatisfaction with air quality using trained panels indicate that about 25% of reported dissatisfaction with perceived air quality in office buildings is due to ETS.

Exposure to ETS may have an indirect effect. Because of subjective annoyance with ETS, it is probable that those exposed to it will report increased SBS-type symptoms.

Most studies evaluating the relationship between ETS and SBS-type symptoms have been conducted in northern Europe, where tobacco smoking rates are high and common in nonresidential buildings. In North America, smoking restrictions are in place in restaurants and a large majority of nonresidential, nonindustrial buildings. As a consequence, ETS is unlikely to be a risk factor for building-related illness symptoms in most North American nonresidential buildings.

D. Environmental factors

Environmental conditions in building spaces at any given time are due to a number of physical variables. These include air temperature, relative humidity (R.H.), air movement, ventilation, lighting, noise, vibration, and electrical and magnetic phenomena. Many of these environmental factors have been evaluated relative to their potential role in contributing to health and/or comfort complaints in buildings.

Factors that have significant effects on human comfort, such as temperature, ventilation, air flow, and in some cases R.H., are mechanically controlled by environmental/climate control systems. Increasingly, new buildings are being designed to provide year-round climate control, and older buildings are being retrofitted to do so. Systems which provide heating, cooling, and ventilation are described by the acronym HVAC. As indicated in Chapter 11, many factors associated with the design, construction, maintenance, and operation of HVAC systems have the potential to cause, or contribute to, building-related health and comfort complaints.

Human comfort is affected by a number of physical factors, the most important of which are thermal conditions, characterized by air temperature, R.H., air movement, and the radiant effects of indoor surfaces. Relative humidity and air velocity may have effects on comfort that are independent of thermal effects.

1. Thermal conditions

Occupants of buildings who report illness symptoms and dissatisfaction with air quality work in environments where a number of coexisting factors influence their sense of well-being and personal comfort. Dissatisfaction with temperature and other factors that affect thermal comfort in buildings is common. This can be seen from results of systematic studies conducted in the USEPA headquarters and Library of Congress buildings (Table 7.6). Dissatisfaction with environmental conditions was relatively high, with temperature and lack of air movement being particularly significant sources of dissatisfaction in the Waterside Mall building of the USEPA headquarters complex. Waterside Mall occupants also reported higher prevalence rates of SBS-type symptoms.

A number of epidemiological studies have attempted to evaluate work-space temperature and SBS-type complaints. Varied results have been

Table 7.6 Occupant Dissatisfaction with Environmental Conditions in USEPA Headquarters Buildings and the Library of Congress

Location	Environmental parameter (% dissatisfied)		
	Temperature	Relative humidity	Air movement
Library of Congress	39	26	43
USEPA buildings			
Waterside Mall	57	36	53
Crystal City	40	37	48
Fairchild	40	33	41
Total	53	36	49

Source: From Selfridge, O.J., Berglund, L.G., and Leaderer, B.P., *Proc. 5th Internatl. Conf. Indoor Air Qual. Clim.*, Toronto, 4, 665, 1990.

reported. Initial statistical analysis often shows strong correlations between increasing temperatures and increasing symptom prevalence. On more intensive analysis, temperature usually is shown to vary colinearly with other factors which investigators determine to be more directly related to SBS-type complaints. However, increasing symptom prevalence with increasing temperature, independent of other factors, has been reported. Several studies have shown significant increases in symptom prevalence/reporting rates at temperatures >73.5°F (22°C).

Factors which affect the thermal balance (thermal comfort) of the human body include air temperature, mean radiant temperature, air velocity, and R.H. Relative humidity affects thermal comfort because it affects evaporative heat loss from the body. With increasing R.H., water vapor pressure gradients between the skin and surrounding air decrease, resulting in reduced evaporation of perspiration and therefore reduced heat loss. Air velocity affects heat loss by convection; higher velocities result in greater convective heat loss.

Local thermal discomfort occurs when parts of the body are subjected to different thermal heat gain or loss conditions. An example of such a phenomenon is when significant vertical differences in air temperature exist between head and feet, e.g., when floor temperature is warmer or cooler than air temperature.

Local thermal discomfort also occurs as a result of excessive air velocity (drafts) or radiant temperature asymmetries. Radiant field asymmetries are a common cause of thermal comfort complaints. When an exterior wall is warm, it radiates heat toward an individual who absorbs it. When it is cool, heat is readily lost from that portion of the body facing it; as a consequence, an individual may experience a localized chilling effect. Radiant asymmetries are noticeable at different seasons on north and south faces of buildings. Seasonal air temperatures, air velocities, and radiant asymmetries associated with optimal thermal comfort are summarized in Tables 7.7 and 7.8.

2. Humidity

In addition to influencing thermal comfort, relative humidity can affect mucous membranes. In low-humidity, winter-time conditions, low R.H. (<30%) appears to be a risk factor for SBS-type symptoms. Humidification to attain a range of 30 to 40% R.H. has been reported to reduce symptom prevalence/reporting rates.

3. Air flow and air movement

Air movement (or lack thereof) in building spaces may directly affect human comfort, perceived air quality, and even symptoms. When air movement is perceived as excessively cool, it is described as a draft and may be thermally uncomfortable. Uncomfortable drafts may result from air leakage through the building envelope during cold weather conditions or during the cooling season when chilled air flows from diffusers at high velocity onto individuals

Table 7.7 Thermal Comfort Requirements During Summertime Conditions Under Light, Mainly Sedentary Conditions

Air temperature between 23 and 26°C.
Vertical air temperature difference between 1.1 m and 0.1 m above floor less than 3°C.
Mean air velocity less than 0.25 m/s.

Source: From Fanger, P.O., *Proc. 3rd Internatl. Conf. Indoor Air Qual. Clim.,* Stockholm, 1, 91, 1984.

Table 7.8 Thermal Comfort Requirements During Wintertime Conditions Under Light, Mainly Sedentary Conditions

Air temperature between 20 and 24°C.
Vertical temperature difference between 1.1 m and 0.1 m (head and ankle level) <3°C.
Surface temperature of floor between 19 and 26°C.
Mean air velocity <0.15 m/s.
Radiant air temperature asymmetries from windows or other cold vertical surfaces:
  <10°C in relation to a small vertical plane 0.6 m above floor.
Radiant temperature asymmetry from a warm or heated ceiling:
  <5°C in relation to a small horizontal plane 0.6 m above the floor.

Source: From Fanger, P.O., *Proc. 3rd Internatl. Conf. Indoor Air Qual. Clim.,* Stockholm, 1, 91, 1984.

in its trajectory. Warm, dry air flows, on the other hand, have the potential to cause discomfort by drying the mucous membranes of the eyes.

Building occupants commonly complain of inadequate air movement in building spaces. This suggests that occupants perceive that air circulation is not adequate and insufficient ventilation is being provided.

E. Office materials and equipment

A variety of materials and equipment used in the work environment have been implicated as potential risk factors for causing SBS-type illness symptoms.

1. Paper products

Paper products such as carbonless copy paper (CCP), photocopied/printed bond, and green-bar computer printing paper have been associated with building-/work-associated illness symptoms. Of these, CCP has been the most extensively studied.

a. Carbonless copy paper. Carbonless copy paper, also known as NCR (no-carbon-required) paper, is a pressure-sensitive product designed to provide multiple copies of forms or documents without the use of carbon paper. The principle of CCP or NCR paper is to reproduce information typed in or written on the top sheet of 2- to 5-part forms. The image is reproduced as a consequence of chemical reactions between colorless color formers and developing compounds. In 3-part CCP forms, the top sheet (CB) is coated on the back side with color formers contained within microcapsules; the

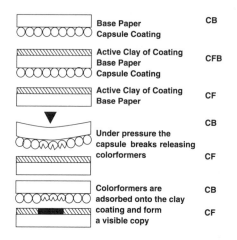

Figure 7.2 Carbonless copy paper and copy formation. (From Murray, R., *Contact Derm.*, 24, 321, 1991, © Munksgaard Int. Publ. Ltd. With permission.)

middle sheet (CFB) is coated on the front with a color developer and on the back with color formers; the third sheet (CF) is coated on the front with color developers (Figure 7.2). The microcapsules break when the CCP form is subjected to the pressure of a pen or pencil or struck by type, releasing color formers which react with color developer on the page beneath, producing legible copies.

Microcapsules are a few micrometers in diameter. The outer layer, made of gelatin, gum arabic, or carboxymethylcellulose, encloses a mixture of oils, dyes, and diluents. When microcapsules are broken, this mixture is transferred to the developer on the underlying page. Microcapsules vary in composition, reflecting different manufacturing formulations. These have included hydrogenated terphenyls mixed with aliphatic hydrocarbons, diacyl ethanes, alkyl naphthalenes, chlorinated paraffins, and alkyl benzenes. Starch or other binders are used to adhere microcapsules to the base paper.

Color developers are based on clays, phenolic resins, or salts of aromatic carboxylic acids. Clay developers have been dominant in Europe, phenolic resins in the U.S. and Japan. The developer coating contains binding substances that are not reactive with clay or other binders. Commonly used binding substances include styrene–butadiene rubber, acrylic latex, and water soluble polymers such as carboxymethyl cellulose, polyvinyl acetate, or polyvinyl alcohol.

Color-forming chemicals used have included crystal violet lactone, benzoyl leuco methylene blue, rhodamine blue lactone, Michler's hydrol of paratoluene sulfonate, indoyl red, phthalide red, phthalide violet, spirodipyranes, and fluorans.

Base CCP paper is purchased by companies who print custom forms for individual business customers. They apply printing inks, which include pigments and a variety of solvents, to the top page and, in whole or in part, to other copy pages. Printers use desensitizing inks to deactivate the developer coating to prevent printed materials from appearing on all copy forms.

   i. Complaint/exposure investigations. A number of complaint/exposure investigations of symptom outbreaks in office workers handling large quantities of CCP forms in building environments have been conducted in the last two decades. Reported symptoms have included itchy hand rashes, headache, swollen eyelids, burning throat and tongue, fatigue, thirst, burning sensation on the face and forearms, nausea, eye irritation, dry throat, sore/dry burning eyes, facial rash, burning sensation in the mouth, itching eyes and nose, and hoarseness.
   Symptom development associated with handling CCP appears to depend on individual sensitivity (approximately 10% of individuals working with CCP report symptoms) and the number of contacts, or the amount of CCP handled, per day by an individual office worker. This relationship can be seen for symptoms of rash, itching, and mucous membrane irritation in Figure 7.3.
   A number of investigators have attempted to identify response patterns to chemical components of CCP used in office environments, with mixed results. Components identified as potential causal agents of one or more symptoms have included free formaldehyde (HCHO); melamine HCHO; resorcin; the dye, paratoluene sulfonate; alkyl phenol novalac resin; terphenyls; and desensitizers. In response to health concerns associated with these materials, some manufacturers have changed CCP formulations. Consequently, substances such as Michler's hydrol of paratoluene sulfonate, HCHO, and alkyl phenyl novolac are generally no longer used in CCP manufacture.
   Cross-sectional epidemiological studies have identified handling CCP as a risk factor for SBS-type symptoms. Handling CCP was observed to be a risk factor for (1) mucous membrane and general symptoms (Danish

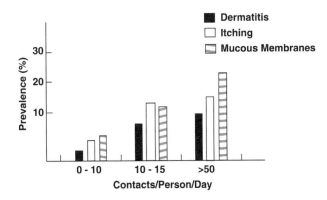

Figure 7.3 Carbonless copy paper exposure and symptom prevalence. (From Menne, T. et al., *Contact Derm.*, 7, 72, 1981, © Munksgaard Int. Publ. Ltd. With permission.)

Municipal Building Study), and (2) eye/nose/throat irritation, tight
chest/difficulty breathing, and fatigue/sleepiness (California Healthy Building Study). Both studies showed a relationship between the number of
contacts or contact hours with CCP per day and symptom prevalence.

There is significant evidence from epidemiological and exposure studies
and complaint investigations that CCP has the potential to cause a variety
of symptoms, including SBS-type symptoms. It has, however, proven very
difficult to isolate chemicals that may be causing reported symptoms and
definitively link symptoms to specific CCP materials. This may be due to
the fact that (1) components differ in manufacturing formulations, (2) additional chemical substances are introduced by printing companies using base
CCP paper, (3) there are a large number of commercially available compounds used in CCP papers, (4) compounds are released and new compounds produced when CCP is used, (5) formulations change with time,
and (6) exposed individuals often handle a diverse mix of CCP products.

Exposure to CCP does not appear to be associated with contaminants
released into the general air environment of buildings. There is limited
evidence to indicate that symptoms are due to dermal contact or very localized (breathing zone) exposure.

b. Other papers. Several studies have indicated that office exposures
to a variety of paper types cause symptoms in users. These include skin
symptoms from carbon paper, and mucous membrane and general symptoms associated with handling large quantities of ordinary bond paper.

Extensive studies of emissions from electrostatically copied, and laser-
and matrix-printed paper have been conducted. Results are summarized in
Table 7.9. Since many of the reported compounds were also found in toner
products, they may have originated from both the paper and toner. The most
commonly observed compounds were benzene, 1-butanol, toluene, hexanal,
1-butyl ether, ethyl benzene, *m*-, *p*-, and *o*-xylene, styrene, 2-phenylpropane,
ethyl toluene isomers, benzaldehyde, diethylbenzene isomers, and 2-ethyl-
hexyl acrylate.

The significance of such emissions on the health of individuals handling
large quantities of these papers is not known. Studies associated with handling bond paper (which is likely to have been copied or used for printing)
and studies with laser printers and electrostatic copiers suggest that such
exposures may contribute to SBS-type symptom complaints.

2. Office equipment

Exposure to irritating substances may occur as a consequence of using office
equipment such as electrophotographic printers, a variety of duplicators,
laser printers and copiers, microfiche, blueprint and signature machines, and
computers. Individual pieces of office equipment may represent significant
point sources, with exposure primarily to individuals working with them or
to those nearby. Since chemicals may be transferred to paper surfaces,

Table 7.9 Volatile Organic Compounds Emitted from Photocopied (A-F), Laser Printed (G-I) and Matrix-Printed (J,K) Paper

Compound	A	B	C	D	E	F	G	H	I	J	K
Hexane	x										
1,1-Dichloro-1-nitroethane	x										
Benzene	x	x+	x+	x+	x+	x+	x	x+	x	x	x
Octene (isomer)	x										
Pentanal	x				+						
Trichloroethene	x										
1-Butanol	x+	x+		x	x+	x+	x+	x	x+		x
Toluene	x+	x+	x+	x+	x+		x+	x+	x		x
Pyridine					x+						
4-Methyl-2-pentanone		x+		x+							
Hexanal	x	x	x	x	x	x	x	x	x	x	
C₄-Cyclohexane isomers				+	x+		+		+		
1-Butyl ether		x+		+	x	+	x+	x+	x+	x	x
m- and *p*-Xylene	x+	x+	x+	x+	x+	x+	x+	x+	x+	x	x
o-Xylene	x+	x+	x+	x+	x+	x+	x+	x+	x+	x	x
Styrene	x+	x+	x+	x+	x+	x+	x+	x+	x+	x	x
1-Butyl acrylate		x+			x+		x+				
2-Phenylpropane	x	x+	x	x+	x+	x+	x+	x+	x+	x	x
3-Heptanol			x+								
1-Phenylpropane	x	x+	x+	x+	x+	x+	x+	x+	x+	x	x
Ethyl toluene (isomers)	x	x+	x+	x+	x+	x	x+	x+	x+	x	
3-Ethoxy-3-ethyl-4,4-dimethylpentane		x+									
1-Butyl methacrylate	x+						+				
Benzaldehyde	x			x+	x+	x+		x+	x+	x	x
Diethylbenzene isomers			x	x			x+	x+	x		x
2-Ethyl-1-hexanol				x+	+		+	x+	x+		x
2-Ethylhexyl acetate			+	+				x+	+		
2,2-Azo-*bis*-isobutyronitrile	+	x+		x+					x+		
2-Ethylhexyl acrylate	x+	x+				x+	x+	x	x		
Methylbiphenyl		+							x		

Note: x = detected in paper; + = detected in toner powder.

Source: From Wolkoff, P. et al., *Indoor Air*, 3, 118, 1993. With permission.

printed/copied paper may be a secondary point source of exposure to those who handle it.

Wet-process photocopiers, electrostatic photocopiers, laser printers, and possibly computers, may cause exposures that result in SBS-type complaints. Therefore, they are of particular concern.

a. Wet-process photocopiers. Wet-process photocopiers are widely used in Canada and, to a lesser extent, the U.S. They emit large quantities of C_{10}–C_{11} isoparaffinic hydrocarbons, and therefore have the potential to cause significant indoor contamination by these volatile organic compounds (VOCs). Because of high emission rates and their significant contribution to total VOC levels, they are believed to be potential contributors to SBS-type symptom prevalence rates in some buildings. Several complaint investigations have implicated wet-process photocopiers as a cause of a variety of irritation-type symptoms.

b. Electrostatic copying machines and laser printers. Electrostatic, or xerographic, copiers have been in use for over four decades. They are the most common copy machines used in offices. Because of the high voltages used, they can be a significant source of ozone (O_3). Emissions vary widely and newer products have been designed to reduce O_3 emissions. Electrostatic copiers also emit toner particles. Toner powder consists of carbon black and a resin that adheres carbon black to paper; particle size is 10 to 20 μm. Because of the electrostatic charge imparted on toner particles, they are deposited on surfaces near operating machines.

Laser or electrophotographic printers also appear to be a source of contaminants to which office workers may be exposed. Laser printers use a high-voltage generator that charges the surface of a continuously rotating photoconducting drum. The drum surface is exposed to a scanning laser beam that discharges the photoconducting surface selectively (in accordance with the pattern to be printed). The toner is charged with an electrostatic potential and is attracted to the drum surface where a negative image is produced. Paper is brought into contact with it and toner is transferred by a second high-voltage generator. The image is made fast, or fused, by thermal or chemical means. In the latter case, laser printers use a mixture of Freon and acetone, which are emitted from the machine. Emissions of a variety of compounds have been associated with laser printers employing heat fusion. These emissions appear to be toner decomposition products. Unlike electrostatic copiers, laser printers do not produce significant quantities of O_3 or carbon black. Toner particles are so highly charged that they are quickly deposited on machine surfaces.

As indicated above, emissions from heat-fusing laser printers appear to be primarily associated with toner resins and decomposition products. One of the more common laser printer toners is described as consisting of 10% carbon black, 5% charge control agents (derived from diphenyl hydrozones), and 85% styrene–acrylate copolymer binder.

Vapor emissions from electrostatic copiers include residual monomers from toner resins and decomposition products including (1) unstable ozonides, diperoxides, and epoxides and (2) oxygenated hydrocarbons such as lower aldehydes, ketones, and organic acids.

A variety of vapor-phase substances are emitted from toners used in electrostatic and laser printers. These include solvent residues (benzene,

Table 7.10 Volatile Organic Compound Emissions from Computer/Video
Display Terminals

2,6-*bis* (1,1-dimethyl)-4-methyl phenol	Ethylbenzene
n-Butanol	Heptadecane
2-Butanone	Hexanedioc acid
2-Butoxyethanol	4-Hydroxy benzaldehyde
Butyl 2-methylpropyl phthalate	3-Methylene-2-pentanone
Caproloactum	2-Methyl-2-propenoic acid
Cresol	Ozone
Decamethyl cyclopentasiloxane	Phenol
Diisoctyl phthalate	Phosphoric acid
Dimethylbenzene	2-*tert*-butylazo-2-methoxy-4-methyl-pentane
Dodecamethyl cyclohexasiloxane	Toluene
2-Ethoxyethyl acetate	Xylene

Source: From Brooks, B.O. and Davis, W.F., *Understanding Indoor Air Quality*, CRC Press, Boca Raton, 1991. With permission.

toluene, xylene, octene, C_4-cyclohexones, 1-butanol, butyl acetate, 2-ethoxy-ethanol); monomers (styrene and acrylate esters); monomer impurities; coalescent agents; and monomer or polymer oxidation products (benzaldehyde, acetophone). Human exposure to HCHO was observed to be significantly increased in a simulated office environment, although the cause of this increase was unknown. Investigators speculated that it may have been associated with O_3 reactions with VOC emissions from toner powders or VOCs in air, or thermal decomposition of toners.

Several epidemiological studies have shown a relationship between photocopying/printing and SBS-type symptoms. Increased symptom prevalence was associated with increased use of photocopy equipment or number of pages photocopied. A case of "laser printer rhinitis" has been reported and confirmed in challenge studies.

    c. Computers/video-display terminals. Several cross-sectional epidemiological studies have implicated working with video display terminals (VDTs)/computers as a risk factor for SBS-type symptoms. Most notably, there appears to be a significant relationship between hours worked with VDTs (> 6 hours) and symptom prevalence rates. This may be due to exposure to substances emitted from VDTs/computers, particularly when they are new (Table 7.10). Emissions may occur from electronic components, adhesives, and plastic covers.

Working with VDTs/computers for long hours has significant ergonomic effects characterized by neuromuscular discomfort, eye strain/fatigue, and general stress. Such factors can contribute to overall stress experienced by office clerical workers and, as a result, increase SBS-type symptom reporting rates.

F. Building furnishings

Furnishings in nonresidential buildings include floor and wall coverings, workstations, office dividers, and furniture items such as desks and chairs.

Furnishing materials can be, in many cases, significant emission sources of VOCs, semivolatile VOCs (SVOCs), and aldehydes such as HCHO. A variety of concerns have been expressed relative to building furnishings, their emissions and effects on air quality, and contributions to health complaints.

1. Floor coverings

A number of investigators have implicated floor coverings as a potential cause of health complaints in problem building environments or as a risk factor for SBS symptoms in general. Floor coverings used to furnish office and institutional buildings are of two major types, textile carpeting and vinyl tile. Both (1) are used in large quantities and have large surface-to-volume ratios; (2) are composite materials, i.e., they are made from a variety of components or substances responsible for their physical properties; (3) require use of a bonding agent to effect adhesion to a substrate surface; and (4) require periodic cleaning. Because of these factors, floor coverings have the potential to adversely affect air quality and potentially contribute to building-related health complaints.

Effects may be direct or indirect. In the former case, floor coverings may be the source of contaminants that affect air quality and the health of building occupants. In the latter case, associated products may cause a variety of contaminant problems. These include (1) emissions of irritant chemicals from bonding agents used to apply floor covering, plasticizers, waxes/waxing compounds, shampoos/cleaning agents and (2) reservoir effects involving VOCs, SVOCs, and organic dust which, on exposure, elicit immunological/allergic responses (see Chapter 9). Soiled carpeting may serve as a medium for microbial growth and a potential source of air contamination by mold, bacteria, and organic dust.

a. Carpeting.   Wall-to-wall textile carpeting is widely used in North American and European office, commercial, and institutional buildings. It is attractive, conveys a sense of warmth, absorbs sound, and reduces undesirable reverberation.

Textile carpeting is produced in a large variety of types or grades depending on desired applications. Commercial/industrial grades used in nonresidential buildings are characterized by short piles, dense weaves, soiling resistance, and durability. Textile fibers may be attached to backing materials that are (1) woven natural or synthetic polyolefin fiber materials or (2) rubber/latex. Carpets, as indicated previously, are "composite" products with various manufacturing requirements (e.g., fiber preparation and weaving, dyes, solvents, mordants, biocides, bonding agents, and stain and soil resistance). Consequently, textile carpeting can be expected to be a potentially significant source of VOC (and possibly SVOC) emissions, particularly when the product is new.

Emission studies have been conducted on carpet products as well as adhesives in response to building occupant complaints that the installation of new carpeting was responsible for health problems. Carpet emissions

Table 7.11 Volatile Organic Compounds
Identified in Emissions from Carpeting

Benzene	Ethylmethylbenzenes
4-Phenylcyclohexene	Trimethylbenzenes
Ethanol	Chlorobenzene
Carbon disulfide	Chloroform
Acetone	Benzaldehyde
Ethyl acetate	Styrene
Ethylbenzene	Undecanes
Methylene chloride	Xylenes
Tetrachloroethene	Trichloroethene
Toluene	Phenol
1,1,1-Trichloroethane	Dimethylheptanes
1,2-Dichloroethane	Butyl benzyl phthalate
Hexanes	1,4-Dioxane
Octanal	Pentanal
Acetaldehyde	Methylcyclopentane
Methylcyclopentanol	Hexene

Source: From Bayer, C.W. and Papanicolopolous, C.D.,
Proc. 5th Internatl. Conf. Indoor Air Qual. Clim., Toronto,
3, 713, 1990.

include a large variety of VOCs and a limited number of SVOCs, which vary among products and manufacturers. An example of VOC emissions from textile carpeting manufactured in the U.S. in the late 1980s can be seen in Table 7.11. Emissions included a variety of aliphatic and aromatic hydrocarbons, aldehydes, esters, alcohols, and chlorinated hydrocarbons. Among these was 4-phenylcyclohexene or 4-PC, a by-product of the manufacture of styrene–butadiene latex used to bond fibers to backing, and the source of what has been described as the typical odor of carpeting. Exposure to 4-PC was suggested to be the cause of health complaints associated with new carpeting. However, little is known of 4-PC's toxicity and effects on humans.

Carpeting is bonded to floor substrates using adhesives that vary in composition. Adhesives used in the late 1980s emitted VOCs at levels that were several orders of magnitude higher than those emitted from the carpeting it was designed for.

In response to USEPA's Carpet Initiative (see Chapter 13), carpeting and adhesive manufacturers in the U.S. have voluntarily (since 1990) instituted a program of emission testing and production of "low-emission" products. As a consequence, total emissions of VOCs from carpet and adhesive materials have decreased dramatically, including emissions of 4-PC.

Significant associations between SBS-type symptom prevalence and textile carpeting has been reported in five of six major cross-sectional epidemiological studies. These were the studies of Danish municipal buildings, Swedish primary schools, Danish schools, Canadian office buildings, and California office buildings. In Danish municipal buildings, textile floor covering was significantly related to mucous membrane symptoms. In Swedish primary schools, there was an increased prevalence of eye and upper respiratory system symptoms, face rash, headache, and abnormal tiredness among children in school buildings with wall-to-wall carpeting compared to those with hard floor covering. In Danish schools, asthmatic children were reported to have more severe asthmatic symptoms in carpeted buildings. A significant association between SBS-type symptoms and carpeting was observed in both Canadian and California office buildings. However, no relation between carpeting and SBS-type symptoms was observed in a Dutch study of office buildings.

In addition to these epidemiological studies, Danish studies of human exposures to carpet emissions under controlled laboratory chamber conditions have shown significant symptom responses among asthmatic and non-asthmatic subjects after a 6-hour exposure to rubber-backed nylon carpeting. These increased symptoms included head feeling heavy, tiredness, eye irritation, itchy/runny/congested nose, and facial itching/burning.

Carpeting serves as a sink for a variety of contaminants. These include VOCs and SVOCs that are emitted from carpeting and other sources which are adsorbed on, and released by, carpeting. Fleecy material such as carpeting has a large specific surface area (total surface area per weight or volume). As such, carpet has a large capacity for adsorbing a variety of VOCs and SVOCs.

In addition to VOCs and SVOCs, carpeting serves as a reservoir for organic particles. These include human skin scales, pet danders, insect and mite excreta and body parts, plant materials, viable and nonviable mold spores/hyphal fragments, and bacteria. One of the strongest correlations between illness symptoms in the Danish municipal building study was with the macromolecular organic fraction (MOD) of surface dust and individual volatile compounds associated with it. Both mold and bacteria are likely to grow on organic dust found on carpets and other fleecy surfaces. However, at this time, there is little evidence that such dust and the presence of carpeting significantly increased airborne exposure to culturable/viable mold or bacteria. Carpeting is a major reservoir for allergenic substances such as cat, dog, and dust mite allergens. Though commonly present, dust mite allergen levels in nonresidential buildings are low (<1 µg/g) compared to residential buildings. As such, exposures are likely to cause, at worst, a minimum risk of sensitization in nonsensitized individuals and symptom initiation in those who are. Significant levels of both cat and dog allergen have been found in carpeting of elementary and other schools. Cat allergen in carpeted schools appears in concentrations sufficient to cause both sensitization and symptoms in sensitive individuals. Though dog allergen levels are high, there are as yet no exposure guidelines available to evaluate the significance of dog allergen in floor dust samples. It is worth noting that, in the Swedish school study described above, symptoms were associated with older, soiled carpets and mostly resolved when carpeting was removed.

b. Vinyl floor covering. Compared to textile carpeting, vinyl floor covering has received relatively little attention as a risk factor for SBS-type symptoms. Only a few complaint investigations have been conducted. However, a variety of substances are emitted from vinyl flooring. These include benzyl and benzal chloride, 2,2,4-trimethyl-1,3-pentanediol diisobutyrate (TXIB), dodecene, methyl phenol, ethylhexyl acrylate, 2-butanone, toluene, trichloroethane, cyclohexanone, butanal, hexanol, benzaldehyde, phenol, benzyl butyl ether, decanol, and the plasticizers, dibutyl phthalate and diisobutyl phthalate. The health significance of exposure to vinyl floor covering emissions is unknown. Danish human chamber exposure studies, however, indicate that vinyl floor covering has the potential to cause a variety of SBS-type symptoms. Exposure to plasticizers has been suggested as a potential contributing factor to asthma.

G. Exposure to specific vapor- and particulate-phase contaminants

Exposure to specific vapor- and particulate-phase contaminants has been evaluated as a potential risk factor for SBS-type symptoms. These include HCHO, TVOCs, biological contaminants, and airborne particles (dust). Although they have yet to be evaluated in a systematic way, vapor-phase human bioeffluents are suspected of contributing to SBS-type symptoms. The potential relationships between individual exposure to HCHO, TVOCs, and biological contaminants, and SBS-type symptom responses in nonresidential buildings, have been discussed in previous chapters. The reader is directed to Chapter 4 for discussions related to HCHO and TVOCs and Chapters 5 and 6 for biological contaminants. The discussion here is limited to human bioeffluents and dust (airborne and surface) as risk factors for SBS-type symptoms.

1. Human bioeffluents

Contaminants generated by humans, described as bioeffluents, have long been a major IAQ/IE concern which has focused primarily on odor and comfort.

Since metabolically produced CO_2 is emitted at high rates by humans, levels in occupied buildings are commonly greater than those outdoors. Because of these concentrations and ease of measurement, CO_2 has served as an indicator of total bioeffluent levels and a crude indicator of ventilation adequacy (see Chapters 8 and 9). Carbon dioxide in buildings may vary from the approximate ambient (outdoor) concentrations of 360 to 370 ppmv to >4500 ppmv. The wide range of CO_2 levels observed in buildings is due to differences in occupant density and ventilation rates. Other bioeffluents reported in indoor air include acetone, acetaldehyde, acetic acid, alkyl alcohol, amyl alcohol, butyric acid, diethyl ketone, ethylacetate, ethyl alcohol, methyl alcohol, phenol, and toluene. Substances such as toluene, and possibly phenol, may be secondary bioeffluents, as their exhalation may be associated with previous inhalation exposure. A systematic characterization of bioeffluent emissions from humans has yet to be conducted. There is, however, indirect evidence that humans produce such gases as ammonia, methane, hydrogen sulfide, and compounds that have pheromonal properties.

It is conventional wisdom among IAQ/IE investigators that there is a relationship between bioeffluent levels and human comfort. This belief or perception is based on personal experience in conducting investigations of high-occupant-density, poorly ventilated buildings. Such environments "feel" oppressive, with even a sense of nasal astringency.

Despite these perceptions, systematic building studies have been unable to establish a relationship between CO_2 levels (the most abundant bioeffluent) and illness symptoms. Several studies, however, have observed relationships between perceived air quality and CO_2. It appears that bioeffluents, as reflected in CO_2 concentrations in buildings, are related more to odor and comfort concerns than health effects.

Scientific evidence to support a causal association between human bioeffluents and comfort, and possibly health effects, is limited. There are, nevertheless, biologically plausible mechanisms to explain why bioeffluents may contribute to such effects. Exposure to human pheromones is suggested here as a potential contributing factor.

Pheromones are volatile or semivolatile substances produced by animals to elicit specific behavioral responses in individuals of the same species. In many cases, these are associated with sexual behavior and reproduction. Pheromonal effects have been reported to be nearly universal in social animals, including primates. Typically, male odors/scents in such species serve to mark territory, assert dominance, repel rivals, attract females, and synchronize female sexual cycles. Synchrony of female menstrual cycles is the best known pheromonal response in humans. Experimental studies with the putative male pheromones, androstenol and androstenone, found in male sweat, urine, and semen, have shown that they stimulate avoidance behavior among males, and attraction of females. The avoidance response in males may play a spacing function. In a poorly ventilated space with high male density, pheromonal concentrations would be expected to be relatively high, with little opportunity to elicit pheromonally induced behavior. It has been proposed that such a sensory overload may be responsible for symptoms of headache and fatigue commonly reported in poorly ventilated buildings. However, studies are needed to confirm the presence of pheromones in indoor air, assess the relationship between occupant density and ventilation, and determine whether they may cause symptoms because normal pheromonally induced behavior cannot be expressed. Females produce different pheromones than males; at present, little is known of their structure and potential to cause behavioral effects.

2. Dust

Though concentrations of particulate-phase matter are rarely assessed in problem building investigations, there is increasing evidence that exposure to airborne dust, or dust on interior building surfaces, is a risk factor for

SBS-type symptoms. The term, "dust," includes a broad range of particulate-phase materials, which vary in size, chemical composition, type, and source.

Airborne dust concentrations may be reported as total suspended particles (TSP), respirable suspended particles (RSP), ultraviolet particulate matter (UVPM), particle numbers, or concentrations of specific particle fractions such as man-made mineral fibers. TSP particles represent the largest range of particle sizes measured, including those that are respirable (<3.0 μm, RSP), inhalable (<10.0 μm), and not inhalable (10 to 100 μm). Respirable suspended particles are generally considered to have the greatest biological significance since they can be deposited in lung tissue. Larger particles (3 to 10 μm) tend to be deposited in the upper respiratory passages where they have the potential to cause irritation or allergic/inflammatory responses. Though UVPM particles are in the respirable size range, they differ somewhat from RSP concentrations since the measurement technique is dependent on light scattering/absorption properties of particulate matter.

Cross-sectional epidemiological studies to determine whether relationships between exposure to airborne dust and SBS-type symptoms exist have reported mixed results. Several studies have shown significant relationships between symptoms and airborne dust concentrations measured as TSP and RSP. Indirect evidence to demonstrate a causal relationship is available from breathing-zone filtration studies, which have shown decreases in workplace symptoms with decreases in airborne dust concentrations.

A number of studies have observed significant relationships between SBS-type symptoms and surface dust concentrations. Building cleaning studies have shown significant reductions in SBS symptom prevalence rates after the implementation of major, systematic building surface cleaning efforts.

Surface dusts may cause symptoms on resuspension as a result of toxic effects, irritation, or immunological (allergenic) mechanisms. Potential toxic or irritation effects may be increased as a result of the adsorption of gases or vapors. Personal exposure measurements indicate that individual airborne dust exposure may be 3 to 5 times greater than those predicted from area sampling (used in most systematic epidemiological studies). It appears that individuals create their own "dust cloud" as surface dust is resuspended during activities. Indoor activities increase airborne concentrations in the very coarse (12.5 to 25 μm) and very fine (0.05 to 0.4 μm) size ranges; particles in the 0.8 to 6.4 μm range appear to be least affected.

The effect of human activities on airborne particle concentrations can be seen in day-care center studies. Day-care centers subject to significant floor and surface dust cleaning had average airborne particle concentrations of 59 μg/m³ (range 32 to 98 μg/m³) compared to 96 μg/m³ in control day-care centers (range 42 to 204 μg/m³). Notably, in both experimental and control cases, supply air concentrations were <20 μg/m³ and averaged 5 to 6 μg/m³. These studies indicate that a large percentage of airborne dust may be associated with resuspension of surface dust by human activity; increased cleaning reduces airborne dust concentrations.

Readings

ACGIH, 1984. *Evaluating Office Environmental Problems,* American Conference of Governmental Industrial Hygienists, Cincinnati, 1984.

Cone, Z.E. and Hodgson, M.J., Eds., Problem buildings: Building associated illness and the sick building syndrome, *Occupational Medicine. State of the Art Reviews,* Hanley and Belfus, Philadelphia, 1989, pp. 575–592, 593–606.

Godish, T., *Sick Buildings. Definition, Diagnosis and Mitigation,* Lewis Publishers, Boca Raton, 1995.

Maroni, M., Siefert, B., and Lindvall, T., Epidemiology of principal building-related illnesses and complaints, in *Indoor Air Quality. A Comprehensive Reference Book,* Elsevier, Amsterdam, 1995, chap. 19.

O'Reilly, J.T. et al., *A Review of Keeping Buildings Healthy: How to Monitor and Prevent Indoor Environmental Problems,* John Wiley & Sons, New York, 1998.

Spengler, J.D., Samet, J.M., and McCarthy, J.F., Eds., *Indoor Air Quality Handbook,* McGraw-Hill Publishers, New York, 2000, chaps. 3, 4, 28, 53, 54, 64-68.

Wallace, L., Indoor particles: a review, *J. Air Waste Mgt. Assoc.,* 46, 98, 1996.

Weekes, D.M. and Gammage, R.B., The practitioner's approach to indoor air quality investigations, *Proc. Indoor Air Qual. Symp.,* American Industrial Hygiene Association, Fairfax, VA, 1989, pp. 1–18, 163–171.

Woods, J.E., Cost avoidance and productivity in owning and operating buildings, in *Problem Building-Associated Illness and the Sick Building Syndrome,* Cane, J.E. and Hodgson, M.J., Eds., *Occupational Medicine State of the Art Reviews,* Hanley & Belfus, Philadelphia, 753, 1989.

Questions

1. How do the concepts of building-related illness and sick building syndrome differ?
2. What is the relevance of seasonal affective disorder (SAD) in conducting problem building investigations?
3. What is the significance of psychosocial factors in reporting rates of SBS-type symptoms?
4. Characterize sick building syndrome.
5. What is the difference between a "sick" building and a problem building?
6. Characterize indoor air quality/SBS-type symptoms.
7. Is working with carbonless copy paper an indoor air quality problem? Explain.
8. How is inadequate ventilation a risk factor for indoor air quality complaints?
9. Describe the relationship between SBS-type symptom prevalence rates, gender, and atopy.
10. What is mass hysteria? Is it a real phenomenon?
11. How is exposure to ETS a risk factor for SBS-type symptoms among non-smokers and smokers?
12. How is dissatisfaction with thermal conditions related to indoor air quality concerns?
13. In studies of the USEPA headquarters building, what was unique?
14. Describe factors that affect thermal comfort.
15. Working with office copy machines and copy paper may be a risk factor for SBS-type symptoms. Explain why this may be the case.

16. How may working with personal computers or video display terminals contribute to SBS-type symptom reporting?
17. What role may floor coverings have in indoor air quality complaints and illness symptoms?
18. What are human bioeffluents and how may they contribute to indoor air quality complaints?
19. How may surface dust be related to SBS-type symptom prevalence rates?
20. How does indoor air quality in an office building affect the performance of clerical workers?

chapter eight

Investigating indoor environment problems

As indicated in previous chapters, built environments are subject to a number of potentially significant indoor air quality/indoor environment (IAQ/IE) problems that may cause acute symptoms, long-term health risks, discomfort, or odor. With the exception of severe cases of acute illness or unpleasant odors, most problems (or potential problems such as radon) go unrecognized. In the case of residential and nonresidential buildings, a need to conduct an IAQ/IE investigation develops only after occupants become aware that health and other problems may be associated with home or work environments. Awareness development is rapid when an odor problem or some type of physical discomfort occurs. In most cases, building occupants do not suspect a causal relationship between acute and chronic illness symptoms (which characterize classical air quality-related symptoms) and their building environment. High prevalence rates of sick building syndrome (SBS)-type symptoms in noncomplaint buildings suggest that most individuals so-affected do not realize that their building/work environment is in any way responsible.

I. Awareness and responsibility

A. Residential buildings

Illness symptoms associated with exposure to formaldehyde (HCHO), combustion by-products such as carbon monoxide (CO), inhalant allergens, and long-term health risks associated with elevated radon levels are major IAQ/IE problems in residential environments. When such problems occur, it is the individual homeowner's/lessee's responsibility to recognize that a problem exists and seek professional assistance to identify and resolve it.

Table 8.1 Factors Indicating Health Problems May Be Associated with Exposures to Contaminants in Residential Environments

Symptoms diminish in severity when building is ventilated by opening windows.
Symptoms diminish in severity or resolve completely when occupants are away from home for several days, and recur upon returning.
Symptoms show a seasonal pattern, that is, associated with the heating/cooling season, building closure conditions, or operation of heating/cooling appliances.
Similar symptoms occur in several or more building occupants.
Symptoms are more severe in individuals who spend the most time at home.
Symptoms develop after moving into a new home (not necessarily a new house).
Residential environment subject to severe moisture and/or mold infestation problems.
Symptoms experienced by visitors.

In our society an individual experiencing the relatively minor symptoms of headache, fatigue, and mucous membrane irritation seeks to achieve relief by using over-the-counter or prescription medication. The emphasis is on symptom relief rather than identifying and mitigating causal factors. Such a symptom amelioration approach assures the problem will persist.

How may homeowners/lessees develop an awareness that health problems may be due to contaminant exposures in their home environments? Awareness, in most cases, comes as a matter of chance — in the form of national or local news reports, internet sites, conversations with acquaintances, or physician suggestions.

Awareness requires some degree of education on the part of homeowners/lessees, and physicians (if medical assistance is sought). Factors that suggest illness symptoms may be associated with one's home environment are summarized in Table 8.1. Listed factors represent common sense epidemiological observations. They reflect exposure/response relationships that are either simply helpful or essential in determining a causal relationship between building environments and persistent health problems.

An individual must be exposed to concentrations sufficient in magnitude to experience symptoms caused by gas/particulate-phase contaminants of either a chemical or biological origin. Exposure is related both to the concentration of causal substances and to duration. As a consequence, exposures and illness symptoms can be expected to diminish when a building is ventilated, and when individuals are away from home for a period of time. On the other hand, individuals who spend the most time at home commonly experience more severe symptoms.

As indicated in Table 8.1, other factors also suggest a potential causal relationship with one's home environment. Multiple individuals experiencing similar symptoms indicate common exposures, as do visitors reporting similar symptoms. Houses that experience high moisture levels are known to be at special risk for mold infestation and high dust mite populations. Significant HCHO exposures have occurred, and in some cases continue to occur, in mobile homes.

For individuals experiencing IAQ/IE-related illness problems in their homes, it is, in most cases, very difficult to make the associations that seem so logical in Table 8.1. Nevertheless, some association must be made before the problem can be professionally investigated and mitigation efforts recommended.

Longer-term health risks like those associated with radon and environmental tobacco smoke (ETS) require an educated public that recognizes the need for conducting radon testing in their homes, implementing mitigation measures when elevated radon levels occur, and minimizing exposure to tobacco smoke.

B. Nonresidential buildings

Occupants of nonresidential office, commercial, and institutional buildings must also develop an awareness that illness symptoms may be associated with their building environment. Awareness of potential IAQ/IE problems in such buildings is more likely to occur because: (1) of the larger population base and potential for interaction among occupants, (2) these buildings are prone to a larger variety of problems than residential buildings, (3) building thermal and ventilation conditions are less under the control of occupants, (4) many buildings are poorly ventilated or have poor thermal control, which may contribute to vocal occupant dissatisfaction, and (5) of "odor" problems. Odor complaints often trigger investigations of unrelated illness symptoms.

Investigations of IAQ/IE concerns are commonly conducted in nonresidential buildings because the factors described above increase the probability that IAQ and other environmental complaints will be reported to building management.

The relationship between health symptoms and one's building/work environment is more clear cut than is the case for residential environments. In most instances, symptoms resolve within a few hours after leaving the building/work environment and begin anew within an hour or two after beginning work the next day. Symptoms typically do not occur over weekends and during vacations. Symptom prevalence is often high as well, with 15+% of building occupants reporting symptoms in building investigations and questionnaire studies.

II. Conducting indoor environment investigations

The task of investigating building-related health, comfort, and sometimes odor complaints falls to a variety of local, state, and federal public or occupational health agencies, and increasingly, private consultants. Typically, residential complaints are investigated by local and state public health agencies. Nonresidential complaints are more commonly investigated by staff of private consulting firms. Health hazard evaluation teams from the National Institute of Occupational Safety and Health (NIOSH) conduct investigations of building environments on request when workers in schools, office buildings, etc., are involved.

The primary goal of conducting building investigations is to identify and mitigate IAQ/IE problems and prevent their recurrence. Successful conduct of building investigations requires that investigators have (1) extensive knowledge of buildings and building systems, (2) a broad understanding of the nature of IAQ/IE problems and factors that contribute to them, (3) an understanding that IAQ/IE investigations have both political and technical dimensions, (4) knowledge of environmental testing procedures and their limitations, and (5) knowledge of investigative protocols and their application to conducting successful IAQ/IE investigations.

A. Residential investigations

Residential IAQ/IE problems are, in theory, easier to diagnose than those that occur in nonresidential buildings. In many, but not all, cases they are more easily resolved. Residential structures are smaller and more simply designed, and activities that occur within them, and equipment and materials used, are fewer and less diverse. The design and operation of mechanical heating/cooling systems are also less complex than in nonresidential buildings.

To successfully conduct an investigation of residential IAQ/IE complaints, an investigator must know what the most common problems are, various aspects of housing construction, the location and operation of heating and cooling systems, aspects of human behavior, and what can go wrong in heating/cooling system operation and building maintenance. Unfortunately, many of the tools required to conduct a successful residential IAQ/IE investigation are only acquired by an investigator after many years of conducting investigations. Though dwellings are simple structures, each has its own unique construction and renovation history as well as history of occupancy. With the possible exception of many manufactured houses, IAQ/IE investigations will rarely be the same for any two houses.

1. Investigative practices

There are three basic approaches to conducting residential IAQ/IE investigations. These can be described as (1) an *ad hoc* or "seat of the pants" approach, (2) conducting air testing only, and (3) a systematic approach that is designed to identify and resolve problems.

a. Ad hoc approach. The *ad hoc* approach is used by investigators who have had little or no experience in responding to homeowner complaints. Such investigations are conducted by personnel in small public health departments where few resources are available and where housing investigations are often of the "nuisance" type. They are also conducted on occasion by private consulting personnel who have little experience. Such investigations have a limited probability of identifying and mitigating IAQ/IE problems.

b. Air testing. At a somewhat higher plane are investigations conducted by local and state public health departments in which the primary focus is air testing. Such investigations are done in response to a home-owner/lessee request to determine by air testing what might be wrong with their home. Air testing is usually limited to a few well-defined contaminants for which methodologies and equipment are readily available, e.g., CO, HCHO, and mold, and which have a history of being a potential cause of contamination and health problems in dwellings.

Generic air testing, as an investigative protocol, has significant limitations. Low contaminant levels based on one-time sampling are often interpreted as indicating that a problem does not exist. Such interpretations ignore the often episodic nature of contaminant emissions and concentrations as well as seasonal variations that occur with CO, HCHO, and mold. Air testing results are often compared to guideline values that may not be sufficiently health-based or protective of sensitive populations. Air testing tends to be hampered by the "magic number" syndrome: levels above guideline values are unsafe; levels below them are safe. Unfortunately, safe or acceptable levels of exposure that protect the most sensitive or vulnerable populations are less clear cut than guideline values, which are often based on what can be reasonably achieved.

c. Systematic approaches. To successfully conduct an IAQ/IE investigation in a residential environment, it is essential that the investigator approach the problem in a systematic manner. This includes pre-site-visit information gathering, an on-site investigation and occupant interview(s), conducting air/surface dust testing when appropriate, and evaluating potential causal factors and mitigation requirements when the on-site investigation has been completed.

Pre-site-visit information gathering is typically conducted in a phone interview with an adult building occupant (preferably the female head of house). This limited interview should be designed to elicit information on the perceived nature and history of the problem; symptom types and patterns; house type, construction, age, and recent changes and renovations; previous investigations and results; and any mitigation efforts. Information gathered in the phone interview may be used in initial hypothesis formation and in suggesting air testing and environmental sampling needs.

The on-site investigation typically provides information essential to successfully diagnosing an IAQ/IE problem. The on-site investigation should include a careful inspection of both the interior and exterior of the building. This includes basement and crawlspace (and, in some cases, attic as well).

The investigator should be cognizant of any distinctive or unusual odors that may indicate the nature of the problem or factors contributing to it. These include chemical odors, pesticides, new carpeting and other new materials, mold odor, etc. The investigator should be able to recognize materials, equipment, etc., that may be a source of health-affecting contaminants.

These would include large volumes of pressed wood products bonded with urea–formaldehyde resins (e.g., particle board, hardwood paneling, medium-density fiber board [MDF]); malfunctioning space and/or water heaters; mold-infested materials; hobbies/crafts (e.g., silk screening, stained glass, etc.); pets; insect infestations; recent pesticide applications; lead-based paint; etc.

The inspection should include an evaluation of any structural problems such as water-damaged interior/exterior materials, rotting timbers, damaged gutters, cracks/holes in brick veneer, and wet basements/crawlspaces. It should include a site evaluation as well. Site conditions of note include moderate to heavy shade, poor site drainage as evidenced by ponding after rains, moss growth, frost heaving, capillary wicking on substructure walls, etc.

Heating/cooling system appliances, including hot water heaters, should be inspected, as well as associated flue systems and supply and cold air duct systems. Evidence of flue gas spillage (in the absence of CO measurements) can be determined from condensation stains and corrosion on draft hoods and flue pipes as well as the design and assembly of flue pipes.

The location of furnace/air conditioner/blower fans and return air ducts is important in conducting a building investigation. The presence of such systems in musty basements or wet crawlspaces provides a pathway for the transport of mold spores from an infested source to spaces throughout the home. Ductwork in slab-on-grade houses should be inspected (by opening supply air registers) to determine whether water entry occurs, if insect/dust contamination is present, and what materials ducts are made of.

During the on-site investigation, an intensive interview should be conducted with an adult occupant to better define the nature of health problems experienced as well as to gain additional information on various factors observed during the inspection of the interior and exterior of the home. This interview may provide information on aspects of the inspection that might require more detailed evaluation.

Air/environmental testing may be conducted to confirm and elucidate the nature of the problem. Air testing for HCHO is desirable if evidence indicates that occupants have been experiencing symptoms consistent with the presence of significant HCHO sources. Carbon monoxide testing is appropriate if symptoms are characteristic of CO exposures or there is visible evidence of flue gas spillage. Air testing for mold using both culturable/viable and total mold spore sampling is desirable if the building environment has been subject to moisture problems with or without evident mold infestation. The use of a portable flame ionization detector (FID) to determine sources of methane would be appropriate if a sewer gas problem appears to exist without an evident source. Surface dust sampling and monoclonal antibody testing for dust mite, pet, and cockroach allergens may be appropriate in cases where health histories are suggestive of inhalant allergens. A role exists for air testing and environmental sampling in conducting residential building investigations. That role is to confirm a hypothesis or to more fully evaluate the nature of a problem.

After the on-site inspection, occupant interview, and air testing/environmental sampling, the investigator should evaluate all information obtained during the investigation. During this evaluation process, he/she should determine whether symptoms/health problems/complaints are consistent with observations made during the inspection, as well as with results of air/environmental sampling.

2. Diagnosing specific residential indoor environment problems

Though the nature of residential IAQ/IE complaints varies, only a relatively few contaminants are responsible. These include biological contaminants such as mold and allergens produced by dust mites, pets, and insects; HCHO; CO; pesticides; lead dusts; sewer gases; and, increasingly, soot produced by candle burning. Complaints may include significant health effects or be of a nuisance nature. Diagnostic criteria used to evaluate residential indoor environment problems are summarized below.

a. Biological contaminants. Biological contaminants, as indicated in Chapters 5 and 6, are the major cause of allergy, asthma, and recurring sinusitis in tens of millions of North Americans annually. These ailments have characteristic symptoms and clinical findings that can be used to identify potential causal agents. Individuals affected often have a family history of allergy or asthma and test positive to specific allergens in standard allergy testing.

Building diagnoses associated with allergens is best conducted in consultation with a physician trained in allergy or immunology. When such consultation is impractical, the investigator should use professional judgment in evaluating the potential cause of allergy/asthma/sinusitis among occupants of a residence.

Risk factors that can be used to evaluate biological contaminants as potential causes of allergy/asthma/sinusitis associated with residential environments are summarized in Table 8.2.

b. Formaldehyde. Fortunately, HCHO is less likely to cause IAQ-related health complaints today than at any time in the past three decades. Because of changes in the use of construction materials and improvements in products bonded with urea–formaldehyde resins, indoor HCHO concentrations (even in many new dwellings) are relatively low and are unlikely to cause health problems. Despite this, HCHO may cause symptoms in sensitive individuals in environments such as new mobile homes constructed with urea–formaldehyde-bonded wood products and homes with new wood cabinetry or furniture. Factors that suggest HCHO exposures may be responsible for reported health complaints include: (1) symptoms characterized by eye and upper respiratory system irritation, headache, and fatigue; (2) symptoms more severe on warm, humid days; (3) potent HCHO-emitting sources present; and (4) HCHO levels determined under near-optimum testing conditions (closure, humidity >50%, temperature 22 to 25°C, moderate outdoor conditions) ≥0.05 ppmv.

Table 8.2   Risk Factors for Biological Contaminant-Associated IAQ/IE-Related
Health Problems

Contaminant	Risk factor
Mold	Obvious active/past mold infestation on building materials
	Building history of water damage
	Wet building site
	Musty odors
	Culturable/viable airborne mold test results (uninfluenced by outdoor mold sources) >1000 CFU/m³
	Sample results dominated by one or several mold genera
	Total mold spore counts >10,000 S/m³
	Positive allergy tests
Dust mites	Damp/moist interior building conditions
	High (>2 µg/g) mite allergen levels in floor dust samples
	Positive allergy tests
Pet danders	Presence of pets in or near home
	House history of pets (without pets necessarily being present)
	Positive allergy tests
	High allergen levels in floor dust samples (cat ≥1 µg/g)
Other allergens	Evidence of organisms present
Cockroach, birds, rodents, crickets, spiders	Positive allergy tests
	High allergen levels in floor dust samples (cockroach ≥2 U/g)

c. Carbon monoxide.   Flue gas spillage and associated exposures to CO and other combustion by-products commonly occur in residences. Such problems have been reported in residences of all ages; they are, however, more likely in older buildings. Carbon monoxide exposures are characterized by symptoms of headache, extreme fatigue, sleepiness/sluggishness, and even nausea. Risk factors include the presence of combustion appliances that show some evidence of malfunction (loose flue pipes, flue pipe condensation staining and/or corrosion); symptoms associated with the heating season; measured CO levels >20 ppmv; carboxyhemoglobin (COHb) levels >2% in nonsmokers not exposed in the workplace; and high CO levels emanating from supply registers (indicating cracked heat exchangers).

Extreme care must be taken in diagnosing a CO problem. Flue gas spillage is, by its very nature, episodic. As such it is not uncommon to measure very low CO levels in a residence even though a CO exposure problem exists. Blood tests for COHb may be desirable when a building occupant reports symptoms. As CO exposures are common in small industrial environments, workplace exposures should also be evaluated in the context of COHb test results.

d. *Pesticides.* As indicated in Chapter 4, a variety of pesticide exposure problems occur in residences. These may be due to indoor application of pesticide products or passive transport of pesticides from the outdoors. Acute pesticide-caused symptoms are typically associated with applications within the home. Diagnosis of an exposure problem requires a knowledge of recent pesticide usage and the type used. Symptoms may be due to active compounds, inert ingredients, or both. Because of low vapor pressures of pesticidal compounds, air testing will generally not show significantly elevated concentrations. Nevertheless, it may be desirable to conduct air testing in response to a homeowner's request or to confirm that air levels are not excessive. The level of indoor contamination and potential for future exposure may be determined from floor dust or surface wipe samples. High pesticide residue concentrations in such samples may indicate the need for significant remediation measures.

e. *Lead.* Exposure to lead-contaminated dust and soils is the major cause of pediatric lead exposure and poisoning. Building investigations associated with elevated blood lead levels and frank symptoms of lead poisoning are commonly conducted by public health personnel and, in some cases, private consultants. It is desirable for investigators to have obtained records and reports of blood lead tests and physician diagnoses and recommendations. The investigator should conduct a complete residential risk assessment for potential lead exposure using the protocol described for investigation of elevated blood lead levels in the 1995 HUD (Department of Housing and Urban Development) guidelines for lead in housing. In conducting such investigations, all potential sources of lead are taken into consideration. These include lead-based paint, lead-contaminated dusts and soil, water, lead associated with hobbies and crafts, lead brought home from work environments, lead-containing ceramics or glassware, home remedies, candles with lead-containing wicks, etc.

f. *Miscellaneous nuisance problems.* Sewer gas odors are the most common IAQ/IE nuisance problems experienced in residences. Sewer odors are usually associated with dry sink/drain traps or the absence of drain traps in air-conditioning condensate drain lines connected to sewer lines. Sewer gas problems are easily resolved by locating all sink/drain traps that lead to sewer lines and filling them with water. Air testing with a portable flame ionization detector (FID) is desirable when dry traps cannot be located easily.

An increasingly important nuisance problem in North American homes is soot deposition on wall and ceiling surfaces associated with the frequent burning of candles and incense. Soot deposition occurs on surfaces with differential temperature conditions (e.g., around wall heating units, on wall surfaces with thermal bridges [see Chapter 6]). Soot and wick fragments can be identified microscopically.

B. *Nonresidential investigations*

The investigation of IAQ/IE complaints in nonresidential office, commercial, and institutional buildings is a much more difficult task than investigating residential problems. Nonresidential buildings represent much more complex environments in terms of the larger populations of individuals involved; a greater diversity of contaminants and potential exposure sources; mechanical systems that provide heating, cooling, ventilation, and sometimes steam humidification; lack of individual control over thermal and ventilation comfort; interpersonal dynamics between occupants and building management; and the often-multifactorial nature of health complaints in such buildings. However, principles employed in conducting residential and nonresidential investigations are, for the most part, similar.

Investigative techniques and protocols used in problem building investigations reflect the knowledge and experience of individuals conducting them. They also reflect the availability of resources of government agencies and private consultants that provide such services, as well as the resources of those requesting services on a fee basis.

Early problem building investigations were conducted on an *ad hoc* basis. They were usually limited to brief discussions with building management, a building walk-through, and a few simple screening air tests. Increasingly, building investigations conducted by state government agencies and consulting companies that specialize in conducting IAQ/IE investigations have become more systematic.

The purpose of an IAQ/IE investigation is, in theory, to identify and resolve complaints in a way that prevents them from recurring. However, building managers/owners may see the problem in different terms. They may view it in the context of occupant complaints only; i.e., the focus of their concern may be to mollify those who complain rather than identify and resolve the actual cause or causes. Consequently, they may request that a government agency or private consulting company conduct air testing to demonstrate to occupants their "good faith" in responding to occupant concerns. Such investigations are usually limited to providing screening measurements. Since contaminant levels in screening measurements are rarely above guideline values (with the exception of CO_2), air testing alone is often unsuccessful in identifying and resolving building-related problems. It is not uncommon for building managers to conclude from such testing that an IAQ/IE problem does not exist. The type of services that a building manager/owner requests is discretionary, whether it makes technical sense or not.

1. *Investigative protocols for problem buildings*

A number of systematic protocols have been developed for conducting problem building investigations in the U.S., Canada, and northern Europe. In the

U.S., NIOSH has developed a protocol to serve the needs of its health hazard evaluation teams. It employs a multidisciplinary approach, with an investigative team consisting of an industrial hygienist, an epidemiologist, and a technical person familiar with the operation and maintenance of building mechanical systems. The NIOSH protocol, like most others, utilizes a multi-stage approach, with more intensive investigative efforts when the cause of occupant complaints cannot be easily identified.

NIOSH and the USEPA have jointly developed a model investigative protocol for in-house personnel. It is a multistage or phased approach that emphasizes information gathering; a building walk-through; and an evaluation of ventilation systems, contaminant sources, and pollutant pathways. It gives limited attention to air testing, suggesting that such testing may not be required to solve most problems, and that test results may be misleading. Exceptions to this are routine tests to determine ventilation adequacy and thermal comfort parameters, such as temperature, humidity, and air movement. A flow diagram with suggested activities in conducting an in-house investigation is illustrated in Figure 8.1. This investigative protocol is included in a more expansive USEPA *Building Air Quality Manual*, which is available from the USEPA (www.epa.gov) and Government Printing Office. It includes discussions of factors that can be used to prevent IAQ problems; diagnosis of IAQ problems; mitigation measures; and appendices that describe common IAQ measurements, HVAC system operation and IAQ concerns, moisture and mold problems, and a number of forms and check-lists that are useful in conducting IAQ investigations. USEPA has also developed a kit (*Tools for Schools*) somewhat along the lines of the *Building Air Quality Manual*, which is designed to assist facilities managers in improving and maintaining good air quality in schools. Like the *Building Air Quality Manual*, it includes a number of useful checklists.

Two other investigative protocols have been developed for use in the U.S. The AIHA (American Industrial Hygiene Association) protocol was designed for use by industrial hygienists (professionals who have historically conducted air testing and safety evaluations of industrial workplaces). Because of their experience and training, industrial hygienists most commonly conduct problem building investigations. Because of an initial over-emphasis on air testing and reference to occupational health standards, the AIHA protocol emphasizes investigation rather than air testing. It is also multiphasic; that is, it becomes more intensive in its conduct when problems are not easily identified and resolved.

The Building Diagnostics protocol is utilized by a number of private consulting firms. Because of its engineering origin, it emphasizes the evaluation of building system performance, in contrast to identifying specific causal factors. It assumes that if a building and its systems are performing as designed, or meet generally accepted performance criteria for comfort, such as standards recommended by the American Society of Heating, Refrigeration, and Air-Conditioning Engineers (ASHRAE), most (≥80%) building occupants will be satisfied with building air quality.

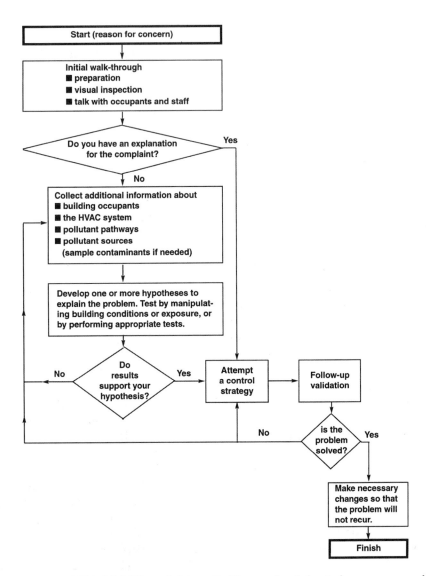

Figure 8.1 USEPA/NIOSH model investigative protocol for in-house personnel. (From USEPA/NIOSH, *EPA/400/1-91/003, DHHS Publication No. 91-1141*, 1991.)

The Building Diagnostics approach places primary emphasis on evaluation of the performance of the building and its systems and the quantitative assessment of airborne contaminants and a variety of environmental parameters. It eschews information gathering from occupants and uses air testing only in the final stages of an investigation.

Investigative protocols have also been developed by European investigators. These include the essentially similar Danish Building Research Institute (DBRI) and Nordtest (Nordic Ventilation Group) protocols. They contain five and eight investigative stages, respectively, and are designed to system-atically evaluate factors that are likely to be problems, with more intensive technical investigations involving surveys of buildings and occupants, fol-lowed by simple, and, if needed, more complex environmental measure-ments. DBRI and Nordtest protocols describe specific inspection criteria to be used in conducting investigations, and tables that describe environmental and contaminant risk factors for health and comfort complaints in buildings. These tables are provided for illustrative purposes (Tables 8.3 and 8.4).

Investigative protocols briefly described above represent efforts to stan-dardize procedures for conducting building investigations by a variety of public agencies, professional groups, and private consultants. A high degree of success in identifying and resolving IAQ/IE problems should be the measure of the value of an investigative protocol. There is, unfortunately, no evidence to indicate the relative success of individual protocols in iden-tifying and resolving IAQ/IE problems since systematic follow-up studies of the efficacy of recommended mitigation measures have not been con-ducted and reported. Because of differences in how investigations are con-ducted, success rates in solving IAQ/IE problems are likely to be quite variable. However, because of relatively high prevalence rates for ventila-tion-related problems, such as inadequate outside air (high CO_2 levels), cross-contamination, re-entry, and entrainment, these problems are likely to be identified in most systematic investigations.

2. Generalized investigative protocol

A generalized investigative protocol is presented here that combines ele-ments common to most protocols and elements which, in the author's expe-rience, are essential to successfully identify and resolve IAQ/IE-related com-plaints. It includes multiple stages of investigation: pre-site-visit information gathering, on-site inspection of the building environment, assessment of occupant symptoms and complaints, assessment of HVAC system operation and maintenance, assessment of potential contaminant sources, and envi-ronmental measurements.

a. Pre-site-visit information gathering. Before beginning an on-site inspection, it is important to obtain information that will facilitate its con-duct. Such information gathering can be accomplished by means of a phone interview with building managers/owners or others who may be familiar with the problem.

It is desirable to obtain the building manager's/owner's perception of the problem to gauge their view of occupant complaints and commitment to identifying and resolving complaints. Important initial requested infor-mation should include: (1) both the general and, if available, specific nature of complaints (e.g., comfort vs. health complaints, general IAQ symptoms vs. cases of severe respiratory illness); (2) relative time period when com-plaints began; (3) coincidental events (e.g., building renovation, introduction of new furnishings, manifestation of an odor problem, etc.); (4) general

Indoor environmental quality

Table 8.3 Environmental Risk Factors for Health and Comfort Complaints in Buildings

Factor	Level of risk			Effects
	Low	Medium	High	
Air temperature, °C	21–23	21–22 23–24	<20 >24 >26 (summer)	Draft, cold, hot, dryness, SBS
Daily temperature rise, °C	<2	2–3	>3	Dryness, SBS
Air velocity, m/s	<0.15	0.15–0.20	>0.20	Drafts
Noise, dBA				
Average	<60	60–65	>65	Noise complaints
Background	<35	35–40	>40	General symptoms
Low frequency noise, dB	<20	20–25	>25	General symptoms
Lighting				
General	Suitable	Suitable	Poor	General symptoms, eye complaints
Individual	Yes	No	No	General symptoms, eye complaints
Glare	No	Control	Glare	General symptoms, eye complaints
Contrast	Good	Control	Too much/little	General symptoms, eye complaints
Static electricity, kV	<1	1–2	>2	Complaints-shocks
Ventilation/person, L/s	>14	14–8	<8	Bad air, SBS
Ventilation/area, L/s/m²	>2	2–1	<1	Bad air, SBS

	Natural exhaust, well-working supply system	Supply with heating or cooling	Humidification, badly monitored	SBS, allergy
Ventilation system				
Size of room; number of workplaces/room	<3	3–7	>7	General symptoms
Office machinery & processes	In separate room		Several in same room	SBS
Cleaning program	Thorough (daily) (3–4/week)	Suitable	Unsuitable, superficial (<2/week)	SBS, allergy
Cleaning, tidiness, accessibility	Good	Average	Poor	SBS, allergy
Floor coverings	Medium hard	Carpet	Carpet >10 yrs service	SBS, allergy
Fleece factor, m^2/m^3	<0.35	0.35–0.70	>0.70	SBS, allergy
Shelf factor, m/m^2	<0.2	0.2–0.5	>0.5	SBS, allergy
Odor assessment, decipols	<0.8	0.8–1.5	>1.5	Bad air
Moisture damage	None	Minor, short duration	Major, long duration	SBS

Source: From Kukkonen, E. et al., *Nordtest Report NT Technical Report 204,* Helsinki, 1993.

Table 8.4 Contaminant Risk Factors for Health and Comfort Complaints in Buildings

Contaminant	Level of risk			Effects
	Low	Medium	High	
Formaldehyde,				
mg/m^3	<0.05	0.05–0.10	>0.10	Mucous membrane
ppmv	0.04	0.04–0.08	>0.08	irritation, SBS
VOCs, mg/m^3				SBS
Ozone,				
mg/m^3	<0.05	0.05–0.10	>0.10	Mucous membrane
				irritation
ppmv	<0.03	0.03–0.05	>0.05	
Hydrogen chloride,				
mg/m^3	<1.4	1.4–4	>4	Mucous membrane
				irritation
ppmv	<1	1–3	>3	
Nitrogen dioxide,				
mg/m^3	<0.2	0.2–0.5	>0.5	Mucous membrane
				irritation, asthma
ppmv	<0.1	0.1–0.3	>0.3	
Carbon monoxide,				
mg/m^3			>10	General symptoms
ppmv			>9	
Carbon dioxide,	<700	700–1000	>1000	Stale air
ppmv				
Mineral fibers,				
Air, f/m^3	<200	200–1000	>1000	Mucous membrane
				and skin irritation
Surfaces, f/m^2	<10	10–30	>30	
Bacteria in air,				Allergy, SBS,
CFU/m^3				respiratory
				complaints
Fungi in air,				Allergy, respiratory
CFU/m^3				complaints
Tobacco smoke	None	Sometimes	Constant	Eye irritation, SBS
Dust (air), mg/m^3	<0.1	0.1–0.3	>0.3	SBS, mucous
				membrane irritation
Floor dust, g/m^2	<0.2	0.2–0.5	>0.5	SBS
Bacteria in floor	$<6 \times 10^3$	$6–10 \times 10^3$	$>10 \times 10^3$	SBS ?
dust, CFU/g				
Fungal spores in	<1000	1000–3000	>3000	SBS ?, allergy
floor dust, CFU/g				
Macromolecular	<1	1–3	>3	SBS
organic dust, mg				
MOD/g dust				
Dust mites				
Allergen/g dust	<100 ng	100–2000 ng	>2000 ng	Allergy, asthma
Mites/g dust	<5	5–100	>100	

Source: From Kukkonen, E. et al., *Nordtest Report NT Technical Report 204*, Helsinki, 1993.

description of the building/work environment; (5) nature of ventilation system management; (6) previous investigations/environmental testing and their results; and (7) any manager/owner interview/complaint documentation. Depending on the information obtained in this preliminary phase, the investigator may develop one or more prospective hypotheses as to potential causes of reported complaints. Such information may be used to plan the investigation and determine environmental sampling needs.

Pre-site-visit information gathering may be used to provide building managers with problem-solving recommendations on a self-help basis (as is done by NIOSH staff in their investigative protocol). Of particular importance is whether the ventilation system is being operated properly and whether thermal comfort needs are being addressed.

Pre-site-visit information gathering can be conducted systematically with note-taking or by using a checklist. A checklist is a valuable tool, lest the investigator fail to request desired information before the site visit.

b. On-site investigation. The on-site investigation is the primary means by which IAQ/IE problems are identified and resolved. The site investigation should include a preliminary meeting with the building manager/owner and facility staff to obtain detailed information on the nature of the problem (the building and its use and operation), and an investigator/management agreement as to the scope of the investigation.

After the initial meeting with the building manager/owner and other personnel, the investigator should conduct a walk-through inspection to (1) ascertain the layout of the building and the nature of activities conducted by its occupants, and (2) identify potential sources of contaminants and problems through sensory means. The latter may include, for example, inadequate ventilation (e.g., human odor, a sense of stuffiness), solvents, known or unknown odors, excessive/inadequate air movement, and/or thermal discomfort. The presence of identifiable odors is an important tool in identifying potential causes of occupant complaints. The walk-through investigation should also include a walk around the building (in less densely urbanized areas) to identify where building intakes, loading docks, and possible outdoor sources may be located, and to assess the potential for entrainment and re-entry (see Chapter 11).

During the walk-through inspection, the investigator attempts to identify potential contaminant sources that may be responsible for the reported problem(s). Source assessments are of a qualitative nature. Contaminant/source problems may have been identified in previous investigations, known from experience, drawn from the reports of others, or have been the subject of intensive research investigations. As a consequence, investigators should have a sense of the kind of indoor environment/health complaints and other problems that could occur in a building and how such problems should be evaluated. What one considers to be a problem will, of course, significantly affect the outcome of an investigation.

Table 8.5   Guide for the Initial Characterization
of Complaints

Nature of complaints/symptoms
Site/organ affected (e.g., respiratory)
Severity
Duration
Associations
Treatment/confirmations
Timing of complaints
Long-term (continuing, periodic, seasonal, weekly, daily)
Short-term (isolated events)
Location of affected and nonaffected groups
Numbers affected
Demographics of occupants with and without complaints
Age
Gender
Employment status

Source: From McCarthy, J.F. et al., in *Indoor Air Pollution: A Health Perspective,* Samet, J.M. and Spengler, J.C., Eds., The Johns Hopkins University Press, Baltimore, 1991, 82. With permission.

Though not commonly conducted by many IAQ/IE investigators, it is essential that occupants reporting complaints be surveyed by in-person interviews (preferred) or by the use of standardized questionnaires. Interviews can be of a general nature or structured (preferred). Ideally, individuals interviewed should include those who have complained and those who have not. A guide for the initial characterization of complaints is summarized in Table 8.5. Complaints may be of a health, comfort, or odor nature. It would not be unusual during the course of an interview for an individual to report apparent building-related symptoms, discomfort with thermal conditions, and unpleasant odors associated with the building environment.

Interviews with complainants often elicit information that helps define the nature of health complaints, their onset and time-dependent variation, their occurrence among specific individuals (e.g., those working with large quantities of paper), and their possible relationship to work activities or other factors that may indicate a potential causal association. In many cases, occupant interviews are sufficient to identify risk factors and exposures that are responsible for major complaints. An extensive building investigation may not, as a consequence, be necessary.

In other circumstances, results of occupant interviews may not reveal a clear-cut pattern or the occupant population might be too large for an interview assessment of the problem. In such cases, it may be desirable to administer questionnaires to all building occupants or to individuals in areas of the building where complaints have been reported.

Questionnaires are used in many IAQ/IE investigations. They are often designed to obtain occupant demographic information, prevalence rates of illness symptoms that may be associated with the building/work environment, and perceptions of the building environment and work conditions. Questionnaires may be self-administered or administered by investigators. Self-administered questionnaires are more commonly used because they minimize staff time required.

Questionnaires have advantages and limitations. The major advantage is that they standardize data collection from a large spectrum of building occupants who have complained and those who have not. As such, they can provide an overview of the nature of the problem, relative symptom prevalence, occupant satisfaction/dissatisfaction with the building/work environment, etc. However, because of their often generalized nature, they may fail to collect vital information about the nature of the problem that may be provided by direct occupant interviews. They also have the disadvantage of being too impersonal. Part of problem resolution involves listening to those who have reported complaints to building management; interviews often convey investigator and building management concern for problems occupants are experiencing.

A variety of questionnaires have been designed for the conduct of building investigations. Two questionnaires are included at the end of this chapter. The first (Appendix A) is used in conducting NIOSH health hazard investigations of problem buildings. It is designed to be used for either structured interviews or to be self-administered. The second (Appendix B) is a more extensive questionnaire developed by the Danish Building Research Institute. As can be seen, it is more detailed and structured. It is designed to collect demographic information, recently experienced symptoms, and histories relative to allergy and asthma, as well as physical and psychosocial factors in the work environment. It is far more comprehensive than the NIOSH questionnaire and reflects an epidemiological assessment of risk factors for illness and other IAQ/IE complaints.

   c.   Assessment of HVAC system operation and maintenance.   Because of the high probability (≥50%) that a problem building will be experiencing some type of ventilation deficiency, many building investigations will assess, to some degree, the operation and performance of HVAC (heating, ventilation, and air-conditioning) systems. The evaluation of HVAC systems and their potential role in contributing to occupant comfort and health complaints requires that the investigator have a technical understanding of the design, operation, and maintenance of HVAC systems, as well as familiarity with problems that occur as a result of system deficiencies. A limited discussion of ventilation system design, operation, and performance may be found in Chapter 11.

A variety of HVAC system-associated problems have been linked to health and comfort complaints. These include: (1) insufficient outdoor air flow rates needed to control bioeffluent levels; (2) migration of contaminants from one building space to another (cross-contamination); (3) re-entry of building exhaust gases; (4) entrainment of contaminants generated outdoors; (5) generation of mineral fibers and black fragments of neoprene from duct liner materials in air-handling units (AHU) and supply and return air ducts; (6) microbial growth in condensate drip pans, wet duct liner, and AHU filters; (7) steam humidification with toxic corrosion-inhibitor additives; (8) inadequate dust control; (9) inadequate control of temperature, relative humidity, and air velocity; and (10) inadequate ventilation air flows into building spaces due to system imbalances.

Inadequate outdoor air flows to maintain bioeffluents at acceptable levels may be associated with a variety of factors. These include: (1) provision of outdoor air for ventilation not included in building design and construction; (2) HVAC systems not designed/selected to provide adequate outdoor air flows, i.e., they were underdesigned for occupant/building needs; (3) building or space occupant capacity increased beyond the original building design; (4) malfunction, obstruction, or deliberate disabling of outdoor air system dampers; (5) system-operating practices that minimize the provision of outdoor air to conserve energy and reduce costs; and (6) reduced air flows due to poor maintenance of filters, fans, and other HVAC system components.

It is important that the problems described above be evaluated when conducting problem building investigations. Checklists for evaluating HVAC system operation and performance are included in both USEPA's *Building Air Quality Manual* and *Tools for Schools* kit. Because of their extensive nature, they are not reproduced here. Readers should consult one or both of these documents for their personal use of these checklists.

The investigator should obtain information about the nature, operation, and maintenance of the HVAC system from building facilities personnel familiar with it. Subsequently, a walk-through inspection should be conducted to locate HVAC system components and determine pathways of air flow. The inspection should include all AHUs, heating/cooling elements, fans, filters, supply air ducts and diffusers, return air plenums and ducts, and outside air intakes. The inspection should include an assessment of potential problems. These include clogged condensate drip pans and growth of microbial slime, and system design and malfunction problems such as overloaded/collapsed filters, nonfunctioning dampers, etc.

In comprehensive HVAC system assessments, outdoor air ventilation rates are measured using flow hoods (preferable with small intakes), pitot tubes, or other airflow measuring devices. Carbon dioxide levels are measured in building interiors to determine whether they conform with guideline values, particularly when direct measurement of ventilation airflows is not possible or practical. Other factors evaluated may include occupancy and space use patterns, differences in air flow in problem and nonproblem areas (using a flow hood), and the potential for air stratification and poor air distribution.

Assuming that there is little or no infiltration due to pressure imbalances, ventilation rates can be determined by the use of instruments that measure airflow through outside air intakes. The preferred instrument for this task is a flow hood. Unfortunately, many outside air intakes are too large for the practical use of flow hoods for such measurements. As a consequence, outdoor airflows must be determined by means of pitot tubes or other velocity measuring devices over an often uneven flow surface. Because of these limitations, outdoor airflows are usually determined indirectly using one or more techniques, e.g., CO_2 and tracer gas measurements, and conducting enthalpy balances.

i. CO_2 techniques. Outdoor airflow rates or effective ventilation rates can be determined from measurements of CO_2 in return, supply, and outdoor air. The percentage of outdoor air can be determined from Equation 8.1.

$$\text{outdoor air } (\%) = \frac{C_R - C_S}{C_R - C_O} \times 100 \tag{8.1}$$

where C_S = ppmv CO_2 in supply or mixed air in the AHU
C_R = ppmv CO_2 in return air
C_O = ppmv CO_2 in outdoor air

This outdoor air percentage can be converted to an outdoor airflow rate using Equation 8.2.

$$\text{outdoor airflow rate (CFM, L/s)} = \frac{\% \text{ outdoor air} \times \text{total airflow (CFM, L/s)}}{100} \tag{8.2}$$

The total airflow rate may be the volumetric flow rate into a room, zone, AHU, or HVAC system determined from actual measurements.

Outdoor airflow rates can be determined by graphical methods. Assuming that peak CO_2 levels occur at the time of measurement, ventilation rates in CFM/person can be determined from Figure 8.2. A peak CO_2 level of 800 ppmv would be equal to a ventilation rate of 20 CFM/person or approximately 10 L/s/person.

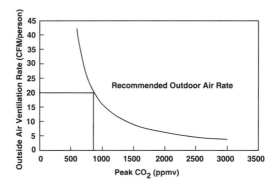

Figure 8.2 Outdoor air ventilation requirements as a function of peak indoor CO_2 concentrations. (From Salisbury, S.A., in Weekes, D.M. and Gammage, R.B., Eds., *Proceedings Indoor Air Quality International Symposium: The Practitioner's Approach to Indoor Air Quality Investigations*, American Industrial Hygiene Association, Fairfax, VA, 1990, 87. With permission.)

Use of Figure 8.2 to obtain ventilation rates requires the measurement of peak CO_2 levels. Depending on the actual building air exchange rate, peak CO_2 levels may occur within a few hours of early morning occupancy to upwards of 4 to 5 hours later (the latter under poorly ventilated building conditions).

ii. Thermal balance.   Percent outdoor air can also be determined from temperature measurements of outdoor air, return air, and mixed air measured in the AHU box where outdoor and recirculated air are mixed. In the latter case, access to the mixing box is essential, and such access may not be easily achieved in small AHUs.

The percent of outdoor air can be calculated from Equation 8.3.

$$\text{outdoor air } (\%) = \frac{T_R - T_M}{T_R - T_O} \qquad (8.3)$$

where T_R = temperature of return air
T_M = temperature of mixed air
T_O = temperature of outdoor air

Percent outdoor air can be converted to a volumetric flow rate using Equation 8.2.

The thermal balance method is limited to assessing airflow through the HVAC system and does not take into account air exchange associated with infiltration (as do CO_2 techniques).

iii. Tracer gases.   Outdoor air ventilation rates can, in theory, be determined by the use of tracer gases such as sulfur hexafluoride (SF_6). Their effective use is limited to small buildings or building spaces where tracer gases can be easily injected, mixed with building air, and sampled.

Tracer gas use is based on the mass balance assumption that the change of tracer gas concentrations in building spaces is a function of the amount of tracer gas introduced minus the amount removed by mechanical ventilation, exhaust systems, and exfiltration.

After the introduction of a tracer gas and mixing with air in building spaces, the decay or decrease of tracer gas levels with time is exponentially related to the outdoor air ventilation rate. From measurements of tracer gas levels over time, air exchange rates (ACH) can be calculated using Equation 8.4.

$$n = \left(\frac{\ln \dfrac{C_o}{C_t}}{T} \right) \qquad (8.4)$$

where C_o = initial tracer gas concentration (ppmv)
C_t = tracer gas concentration at time, t (ppmv)
T = hours
n = air exchange rate/hr (ACH)

The ventilation rate per person can be calculated by multiplying the air exchange rate by the volume of the building space and dividing by the number of individuals present.

Use of tracer gases such as SF_6 to determine building ventilation rates is technically more demanding than using CO_2 and thermal balance techniques. It requires systems for tracer gas injection, sampling, and analysis. It cannot be used to measure air exchange rates in large buildings. An alternative to the SF_6 method for large buildings is to monitor CO_2 levels during and after building occupancy. There are two limitations to the use of CO_2 as a tracer gas: (1) CO_2 measurements must be made after working hours and (2) CO_2 is often not uniformly distributed in building spaces.

d. Environmental measurements. Environmental measurements are commonly conducted in problem building investigations. Protocols developed by NIOSH for health hazard evaluations, and the USEPA/NIOSH protocol for in-house personnel, de-emphasize contaminant measurements, reasoning that in most cases no contaminant will have a sufficiently high concentration to explain reported symptoms. Carbon dioxide is the exception. It is measured in most building investigations to determine the adequacy of ventilation.

Most investigators will conduct environmental measurements. In the early stages these will typically include CO_2, temperature, and relative humidity. Significant quantitative contaminant measurements are usually reserved for advanced stages of an investigation, when the walk-through inspection indicates what potential causal agents may be, or when initial efforts fail to identify the problem. In theory, it would be desirable to conduct measurements of only those contaminants that have a reasonable chance of being a potential cause of complaints based on symptom types, patterns, and potential sources present. These could include CO, HCHO, TVOCs (total volatile organic compounds), airborne mold, respirable particles, and surface dust. Surface dust would be analyzed for common allergens associated with cats, dogs, etc., or a variety of semivolatile organic compounds.

3. Interpretation of investigation results

Once a problem building investigation has been completed, the investigator has the task of evaluating the data collected, assessing the problem in both technical and political terms, and formulating hypotheses that may be tested by comparison to IAQ/comfort guidelines and by the implementation of mitigation measures.

Table 8.6 Thermal Performance Criteria

Parameter	Guideline
Operative temperature (winter)	20.3–24.4°C (68.5–76°F) (at 30% RH)
Operative temperature (summer)	22.8–26.1°C (73–79°F) (at 50%)
Dew point	1.7°C (>35 F) (winter); 16.7°C (62°F) (summer)
Air movement	9.1 MPM (≤30 FPM) (winter); 15.2 MPM (≤50 FPM) (summer)
Vertical temperature gradient	Shall not exceed 2.8°C (5°F) at 10.2 cm (4 in.) and 170.1 cm (67 in.) levels
Plane radiant symmetry	10°C (<18°F) in horizontal direction; 5°C (<9°F) in vertical direction

Source: From Woods, J.E. et al., in *Design and Protocol for Monitoring Indoor Air Quality*, ASTM STP 1002, Nagda, N.L. and Harper, J.P., Eds., American Society for Testing and Materials, Philadelphia, 1989, 80. With permission.

a. IAQ/comfort guidelines.   Interpretation of environmental measurements is invariably a difficult task. As a consequence, most investigators use some form of IAQ/comfort guidelines. The use of guidelines reflects a desire on the part of investigators to determine whether measured values are within or exceed those generally accepted in the professional community. By using guidelines, investigators have some degree of confidence that a reasonably accurate interpretation of sampling data will be made. Guideline values for acceptable thermal environments have been developed by the American Society of Heating, Refrigerating and Air-Conditioning Engineers (ASHRAE) (Table 8.6) and for indoor contaminants by the World Health Organization (WHO) and other bodies. Guideline values for HCHO are summarized in Table 8.7. Note the considerable range of acceptable levels of HCHO in indoor air. It is notable that concentrations of 0.08 to 0.12 ppmv

Table 8.7 Guidelines for Acceptable Formaldehyde Levels in Indoor Air

Agency/government	Permissible level (ppmv)	Status
HUD–USA	0.40 target	Recommended
ASHRAE–USA	0.10	Recommended (withdrawn)
California Dept. of Health–USA	0.10 (action level) 0.05 (target level)	Recommended
Canada	0.10 (action level) 0.05 (target level)	Recommended
Sweden	0.20	Promulgated
Denmark	0.12	Promulgated
Finland	0.12	Promulgated
Netherlands	0.10	Promulgated
Germany	0.10	Promulgated
Italy	0.10	Promulgated
WHO	0.08	Recommended

that represent the large majority of guideline values in Table 8.7 are above exposure concentrations that have been linked to HCHO-related symptoms in epidemiological studies.

Used properly, guidelines can serve as useful tools for investigators to evaluate the probability that building-related health complaints are related to contaminants measured. As such, they should not be used as "magic numbers" that divide safe and unsafe levels of exposure. They are reference values from which judgments can be made relative to the potential role of specific contaminants in the observed problem and what mitigation measures should be implemented. These judgments should include a consideration of consensus-based guidelines and recent scientific literature. The investigator should also use good sense, particularly when the facts of the exposure strongly indicate that the contaminant in question is the cause of the complaints.

The use of guideline values poses other concerns. These include the unavailability of guidelines for some contaminant exposures, limitations of environmental testing based on one-time sampling, and conduct of contaminant measurement simply because guidelines exist. In the absence of guidelines, attempts to relate symptoms to exposures should be consistent with toxicological and epidemiological evidence. For example, it makes good scientific sense to consider solvent exposures when symptoms/health complaints are of a neurotoxic nature.

Exposures to gas and particulate-phase contaminants vary significantly over time. As indicated previously, there is a risk that the results of one-time sampling may be misinterpreted (when compared to guideline values) if little or no consideration is given to the potential range of contaminant levels that may occur or have occurred. Results are more likely to be misinterpreted when one-time sampling results are in the low end of their range of variation.

b. Hypothesis formulation/testing. Based on information gathered from building managers and occupants, the on-site investigation, and environmental measurements, the investigator should be able to formulate one or more hypotheses that might identify the cause or causes of complaints which can be tested by implementation of mitigation measures. Such testing should involve a quantitative assessment of symptom/complaint prevalence and targeted environmental measurements. These results should be compared to those obtained in the original investigation. Though such a follow-up assessment is technically desirable, building owners/managers do not commonly request them after investigators report results of their initial assessment of the cause(s) of complaints and their control recommendations.

4. Political considerations

Most problem building investigations have both technical and political dimensions. Many investigators have a technical background and therefore, by their education, training, and experience, are relatively well suited to conduct problem building investigations. On the other hand, most investigators are poorly prepared to appropriately navigate the political tempest (or its potential) that characterizes many problem building situations.

Most problem building investigations are conducted in response to complaints from occupants who have no control over their environment. When one or more occupants have complaints that they believe are associated with the building, they must first inform, and then attempt to convince, building management that a problem exists so that managers/owners take appropriate action to resolve complaints.

It is common for managers/owners to be skeptical of the fact that symptoms/health problems may be due to contaminants in the building environment. This skepticism is based on a variety of factors, including: (1) the preponderance of complainants are female employees, (2) building managers themselves do not perceive that any air quality/indoor environment problem exists, (3) building managers/owners often view the problem in political terms and thus hope that with time it will go away, and (4) building managers/owners have no technical understanding of the issues and are uncertain how to proceed. The political temperature may rise as a disbelieving building management fails to respond to occupant concerns and may be aggravated by occupant perceptions of management indifference or hostility. Such political differences between occupant complainants and building managers/owners are unfortunately very common, if not the norm.

The problem building investigation itself has its genesis in the dynamics of human behavior. It may have come about as a result of a genuine management/owner concern for the health and welfare of occupants or it may have originated from a desire by managers/owners to mollify occupants by having an investigation conducted. In the latter case, managers/owners see the problem primarily as a political problem rather than the technical problem with the building environment that it may actually be. The political dimensions of the problem are usually unknown to investigators prior to the conduct of the investigation. In many cases, building managers/owners prefer that there be little or no contact between investigators and occupants to minimize what they perceive as political problems that may result.

It is, nevertheless, critical to the successful conduct of an investigation that occupants be interviewed or surveyed in order to obtain information on symptoms and potential exposure patterns that are vital to correctly diagnosing and resolving IAQ/IE problems. Occupant interviews in themselves serve an important political (public relations) function. Occupants get to tell "their story" to a neutral or sympathetic third-party technical expert who is the manager's/owner's representative. Such interactions convey a sense that someone cares, a sense that is essential in solving the political dimensions of a problem. Occupant confidence in the performance of the investigator is vital to the successful resolution of building environment problems.

The investigator must educate both occupants and building management as to what constitutes a building/working environment problem and what does not (for the latter, e.g., reported cancer clusters involving very disparate cancer types and organ systems). The investigator must educate building owners/managers as to the causes of building environment complaints and how best to resolve them.

5. Diagnosing specific nonresidential IAQ/IE and other building-/work-related problems

The successful diagnosis and resolution of building environment complaints requires that the investigator be familiar with the most common causes of complaints and how to resolve them. It is important to realize that in most problem building environments there is no single causal agent present that can be uniquely identified as responsible for the spectrum of health concerns reported by building occupants. Reported health symptoms are likely to be multifactorial in their genesis. As a consequence, it is unlikely that a problem building investigation will resolve all health/IAQ/IE concerns. Nevertheless, it is possible to diagnose and address the major problems present. These include inadequate ventilation, cross contamination/entrainment/re-entry problems, exposure to work materials and equipment, excessive concentrations of volatile organic compounds (TVOCs), HCHO exposures, surface dust, and hypersensitivity pneumonitis.

a. Inadequate ventilation. As indicated earlier in this chapter, inadequate ventilation has a variety of causes. It is characterized by occupant complaints of headache, fatigue, and a lack of air movement. It is recognized by occupant use of fans, excessive CO_2 levels (>1000 ppmv) at peak occupant capacity, and noticeable human odor on entering a building or building space.

b. Cross contamination/entrainment/re-entry. Along with ventilation system deficiencies, this group of related problems is a common cause of complaints in buildings. They are often characterized by odors that are out of place, e.g., odors from printing solvents contaminating other areas because of pressure imbalances, motor vehicle exhaust odors drawn through outdoor air intakes or by infiltration, boiler exhausts and cooking odors that re-enter the building due to poor siting of exhausts and intakes, and pressure imbalances. Not uncommonly, reports of odor are associated with building-related health complaints that are only coincidental.

c. Work materials/equipment. Some employees report health complaints that uniquely affect them. As such, they do not constitute a building-wide IAQ/IE problem. Such complaints are often associated with intensive handling of paper products such as carbonless copy paper (CCP) and photocopied bond paper. Individuals sensitive to CCP typically report a variety of upper respiratory symptoms characterized by hoarseness and, in some cases, laryngitis. Complaints increase with increased CCP contacts. Skin rashes and a chemical taste in the mouth may also be reported. Individuals at risk include billing personnel, medical records clerks, and other individuals who process large numbers of forms (typically >25 to 50+ paper contacts per day). Those reporting very severe symptoms may handle upwards of hundreds of CCP forms per day. Individuals who handle large quantities of photocopy paper, conduct photocopying activities, or use laser printers extensively may be at special risk of developing upper respiratory symptoms.

d. TVOC exposures. A variety of volatile organic compounds (VOCs) may produce combined total VOC (TVOC) exposures during early building occupancy sufficient in magnitude to cause mucous membrane irritation-type symptoms. These TVOC exposures may be associated with solvent-type odors characteristic of new or newly renovated buildings. Such exposures diminish significantly with time and should resolve within several months. It is difficult to diagnose a TVOC exposure building health problem from environmental measurements since TVOCs represent such a broad range of volatile organic compounds. However, TVOC concentrations >1 mg/m^3 may be cause for concern.

e. Formaldehyde exposures. Levels of HCHO in nonresidential buildings are usually relatively low. Elevated levels (>0.05 ppmv) may be associated with large quantities of office furniture whose components are bonded with urea–formaldehyde resins. Classic symptoms, as previously described, are irritation of the eyes and mucous membranes of the upper respiratory system, headache, and fatigue. Both concentrations and symptoms may increase or decrease with changes in temperature and relative humidity.

f. Mold infestation. Various levels of mold infestation occur in problem buildings. These include localized problems associated with leaking roofs, plumbing, and HVAC systems; groundwater seepage into sublevels; moisture penetration of the building envelope; high relative humidity and condensation problems associated with the improper operation of HVAC systems; and infested HVAC system duct liners, condensate drip pans, and filters. Infestation problems may be characterized by mold growth on building materials, furnishings, books, paper, etc., and, in some cases, elevated airborne mold levels. Sampling airborne mold using both culturable/viable and total mold spore sampling methods may be desirable to determine dominant mold types present and their relative abundance. It is important to realize that because mechanically ventilated buildings use recirculated air, even poor quality filters remove mold spores sufficiently so that airborne culturable/viable and total mold spore levels are typically low. Levels above 300 CFU/m^3 and 5000 S/m^3 indicate a potential infestation problem. Regardless of the results of air testing, the presence of significantly infested materials indicates that a potential mold-related health problem may exist.

g. Surface dust. A number of European studies have indicated a relationship between components of surface dust and building-related symptoms. The exact nature of this relationship needs to be better elucidated. Studies in schools indicate that surface dust often contains elevated concentrations of pet, and in some cases cockroach, allergens. Pet allergens (cat and dog) usually originate outside the building environment and are brought in by passive transport. It is likely that some portion of symptoms reported in problem and nonproblem buildings is due to classic allergy symptoms associated with sensitization and subsequent exposure to pet allergens. The potential for such allergen-based problems can be elucidated by allergen testing of surface dust samples and occupant allergy testing. Surface dusts can also contain significant concentrations of phthalate plasticizers, which may contribute to the development of an inflammatory response in the upper airways.

h. Hypersensitivity pneumonitis. Outbreaks of hypersensitivity pneumonitis, as well as individual sporadic cases, occur in some problem buildings. Hypersensitivity pneumonitis can be diagnosed by its classic pattern of flu-like symptoms, which are initiated early in the work week, improve as the week progresses, and begin anew with the new work week or after vacation. Exposures are often associated with contaminated HVAC systems, and outbreaks may occur at the beginning of the heating season as microbial slimes in condensate drip pans dry and aerosolize. Causal agents may include thermophilic actinomycetes and mold species such as *Penicillium* or *Aspergillus,* which produce small spores. Cases may be confirmed immunologically. Potential exposures to causal organisms may be evaluated by conducting airborne sampling by culturable/viable and total mold spore methods for molds and actinomycetes, aggressive sampling in problem areas, surface sampling, and visual inspection of infested materials and surfaces.

Appendix A

NIOSH IAQ Survey Questionnaire

1. Complaints Yes ____ No ____
    (If yes, please check)
    ____ Temperature too cold
    ____ Temperature too hot
    ____ Lack of air circulation (stuffy feeling)
    ____ Noticeable odors
    ____ Dust in air
    ____ Disturbing noises
    ____ Others (specify)

2. When do these problems occur?
    ____ Morning   ____ Daily
    ____ Afternoon   ____ Specific day(s) of the week
    ____ All day   Specify which day(s):_____
    ____ No noticeable trend _____

3. Health Problems or Symptoms
    Describe in **three words or less** each symptom or adverse health effect you experience **more than two times** per week. (Example: runny nose)
    Symptom 1 _____
    Symptom 2 _____
    Symptom 3 _____
    Symptom 4 _____
    Symptom 5 _____
    Symptom 6 _____
    Do **all** of the above symptoms clear up within 1 hour after leaving work? Yes ____ No ____
    If no, which symptom or symptoms persist (noted at home or at work) throughout the week?
    Please indicate by drawing a circle around the symptom number below.
    Symptom: 1   2   3   4   5   6
    Do you have any health problems or allergies which might account for any of the above symptoms? Yes ____ No ____
    If yes, please describe._____
    _____
    _____

4. Do any of the following apply to you?
   ___ Wear contact lenses
   ___ Operate video display terminals at least 10% of the workday
   ___ Operate photocopier machines at least 10% of the workday
   ___ Use or operate other special office machines or equipment
   (Specify) _____

5. Do you smoke? Yes ____ No ____

6. Do others in your immediate work area smoke? Yes ___ No ___

7. Your office or suite number is _____

8. What is your job title or position? _____

9. Briefly describe your primary job tasks. _____
   _____
   _____

10. Can you offer any other comments or observations concerning your office environment? (Optional)
    _____
    _____
    _____
    _____
    _____

11. Your name? (Optional) _____

12. Your office phone number? (Optional) _____

Source: From Gorman, R.W. and Wallingford, K.M., in *Design and Protocol for Monitoring Indoor Air Quality*, ASTM STP 1002, Nagda, N.L. and Harper, J.P., Eds., American Society for Testing and Materials, Philadelphia, 1989, 63. With permission.

Appendix B

Danish Building Research Institute Indoor Climate Survey Questionnaire

Background Factors

Year of birth _____   Occupation_____

Sex   male ____ female ____   How long have you been at your present place
                              of work?

Do you smoke?   Yes ____ No ____   _____ years

Work Environment

Have you been bothered during the last three months by any of the following
factors at your work place?

	Yes, often (every week)	Yes, sometimes	No, never
Draft	____	____	____
Room temperature too high	____	____	____
Varying room temperature	____	____	____
Room temperature too low	____	____	____
Stuffy "bad" air	____	____	____
Dry air	____	____	____
Unpleasant odor	____	____	____
Static electricity, often causing shocks	____	____	____
Passive smoking	____	____	____
Noise	____	____	____
Light that is dim or causes glare and/or reflections	____	____	____
Dust and dirt	____	____	____

Work Conditions

	Yes, often	Yes, sometimes	No, seldom	No, never
Do you regard your work as interesting and stimulating?	____	____	____	____
Do you have too much work to do?	____	____	____	____
Do you have any opportunity to influence your working conditions?	____	____	____	____
Do your fellow workers help you with problems you may have in your work?	____	____	____	____

Past/Present Diseases/Symptoms

	YES	NO
Have you ever had asthmatic problems?	____	____
Have you ever suffered from hayfever?	____	____
Have you ever suffered from eczema?	____	____
Does anybody else in your family suffer from allergies (e.g., asthma, hayfever, eczema?)	____	____

Present Symptoms

During the last 3 months have you had any of the following symptoms?

	Yes, often (every week)	Yes, sometimes	No, never	If YES: Do you believe that it is due to your work environment? Yes	No
Fatigue	____	____	____	____	____
Feeling heavy-headed	____	____	____	____	____
Headache	____	____	____	____	____
Nausea/dizziness	____	____	____	____	____
Difficulty concen-trating	____	____	____	____	____

	Yes, often (every week)	Yes, sometimes	No, never	If YES: Do you believe that it is due to your work environment?	
				Yes	No
Itching, burning or irritation of the eyes	____	____	____	____	____
Irritated, stuffy or runny nose	____	____	____	____	____
Hoarse, dry throat	____	____	____	____	____
Cough	____	____	____	____	____
Dry or flushed facial skin	____	____	____	____	____
Scaling/itching scalp or ears	____	____	____	____	____
Hands dry, itching red skin	____	____	____	____	____
Other	____	____	____	____	____

Further Comments

_____

_____

_____

_____

_____

Source: From *Indoor Climate and Air Quality Problems*, SBI Report 212, Danish Building Research Institute, Aarhus, 1996. With permission.

Readings

ANSI/ASHRAE Standard 55-1981, *Thermal Environmental Conditions for Human Occupancy,* American Society for Heating, Refrigerating and Air-Conditioning Engineers, Atlanta, 1981.

ASHRAE Standard 62-1981R, *Ventilation for Acceptable Indoor Air Quality,* American Society of Heating, Refrigerating and Air-Conditioning Engineers, Atlanta, 1989.

ASTM, *Guide for Inspecting Water Systems for Legionella and Investigating Outbreaks of Legionellosis (Legionnaires' Disease and Pontiac Fever),* American Society for Testing Materials, West Conshohocken, PA, 1996.

Davidge, R. et al., *Indoor Air Quality Assessment Strategy,* Building Performance Division, Public Works Canada, Ottawa, 1989.

Kukkonen, E. et al., *Indoor Climate Problems — Investigation and Remediation Measures,* Nordtest Report NT Technical Report 204, Helsinki, 1993.

Light, E. and Sundell, J., *General Principles for the Investigation of Complaints,* TFII-1998, International Society of Indoor Air Quality & Climate, Milan, Italy, 1998.

O'Reilly, J.T. et al., *A Review of Keeping Buildings Healthy: How to Monitor and Prevent Indoor Environmental Problems,* John Wiley & Sons, Inc., New York, 1998.

Quinlan, P. et al., Problem building associated illness and the sick building syndrome, in *Occupational Medicine, State of the Art Reviews,* Cone, J.E. and Hodgson, M.J., Eds., Hanley and Belfus, Philadelphia, 1989, 771.

Rafferty, P.J., Ed., *The Industrial Hygienist's Guide to Indoor Air Quality Investigations,* American Industrial Hygiene Association, Fairfax, VA, 1993.

Spengler, J.D., Samet, J.M, and McCarthy, J.F., Eds., *Indoor Air Quality Handbook,* McGraw-Hill Publishers, New York, 2000, chaps. 49, 52, 53.

USEPA/NIOSH, *Building Air Quality: A Guide for Building Owners and Facility Managers,* EPA/400/1-91/003, DDHS (NIOSH) Publication No. 91-114, Washington, D.C., 1991.

USEPA, *Tools for Schools,* USEPA, Washington, D.C., 1996.

Woods, J.E. et al., Indoor air quality diagnostics: qualitative and quantitative procedures to improve environmental conditions, in *Design and Protocol for Monitoring Indoor Air Quality,* ASTM STP 1002, Nagda, N.L. and Harper, J.P., Eds., American Society for Testing and Materials, Philadelphia, 1989, 80.

Questions

1. How can you know whether illness symptoms experienced by you or members of your family are associated with contaminants in your home?
2. Who is most at risk for developing illness symptoms associated with residential indoor contaminants? Why?
3. How can one know whether illness symptoms experienced by an individual are associated with the work environment?
4. What is the significance of odor complaints in buildings?
5. What approaches can an investigator use in responding to a homeowner's request for services in conducting a residential indoor environment investigation?
6. Under what circumstances is it desirable to conduct air testing in an IAQ/IE investigation?

7. What advantages are there in conducting occupant interviews in both residential and nonresidential investigations?

8. As an investigator, what factors would indicate that health complaints in a house were due to CO exposures? Be specific.

9. A child in a residence has been diagnosed with symptoms of pesticide poisoning. Air testing has revealed that air concentrations of pesticides are within acceptable limits. Explain this apparent contradiction.

10. Under what circumstances would it be desirable to conduct surface dust sampling and analysis as a part of an indoor environment investigation?

11. Why is it important to obtain symptom information from individuals who report that they are experiencing illness symptoms in a building environment?

12. A child has elevated blood lead levels. You have been asked to conduct a risk assessment of the child's potential exposure. Describe what potential exposures you would consider.

13. What factors would allow an investigator to conclude that formaldehyde exposures in an individual's home were the cause of reported illness symptoms?

14. Describe two nuisance IAQ/IE problems experienced in residences and their causes.

15. Describe political considerations involved in conducting a problem building investigation.

16. Describe procedures you would use in conducting a residential IAQ/IE investigation.

17. Describe procedures you would use in conducting an IAQ/IE investigation in a nonresidential building.

18. What value do consensus guidelines have in evaluating the results of environmental sampling? What are their limitations?

19. What are the objectives of the building walk-through in an IAQ/IE investigation?

20. Describe advantages and limitations of questionnaire use in a problem building investigation.

21. Describe HVAC system-associated problems that may result in health and comfort complaints.

22. How can one determine the percentage of outdoor air in a building environment?

23. Describe risk factors that contribute to the following problems:
    a. Cross contamination
    b. Entrainment
    c. Re-entry

24. What work materials/equipment have been associated with building/work-related health complaints?

25. Under what conditions would exposure to TVOCs result in illness symptoms?

26. What is the significance of surface dusts in contributing to health complaints in buildings?

chapter nine

Measurement of indoor contaminants

Contaminant measurements are made in most, if not all, investigations conducted to evaluate potential causal relationships between illness or illness symptoms and residential and nonresidential building environments. Contaminant measurements may include sampling of airborne concentrations of gas/vapor or particulate-phase substances, sampling of airborne biological contaminants, surface sampling, and bulk sampling of building materials.

Contaminant measurements are made for various reasons. In the case of carbon dioxide (CO_2), they are used to determine the adequacy of ventilation; in other cases they may be used as a screening tool to determine whether target contaminants are within or above acceptable guideline values. The best reason to conduct contaminant measurements in problem buildings is to identify and confirm the presence of contaminants that may be causally associated with reported illness symptoms. In the case of carbon monoxide (CO), measurements of COHb in blood may be used to confirm a CO exposure and its magnitude. Contaminants that may pose long-term health exposure risks (e.g., radon) may also be measured.

I. Measurement considerations

It is important when conducting environmental measurements in indoor environments that investigators are familiar with principles and practices associated with such measurements and conduct these activities with specific objectives in mind. Contaminant concentrations are determined from samples that have been collected.

A. Sampling

In sampling, one attempts to identify or determine the concentration of a substance or substances in a relatively small volume of indoor air or human blood, on a limited surface area, or in a small mass of material. For purposes of contaminant identification and quantification, a sample is assumed to be representative of a larger volume of air (e.g., room), blood, or material surface or mass. This assumption, when used in conjunction with an appropriate sampling protocol, can be expected to provide reasonably reliable measurements that can be used to confidently interpret sampling results.

B. Sampling objectives

Environmental sampling is conducted in indoor environments for a number of reasons. It has, as a consequence, one or more stated or inferred objectives. These may include (1) general or specific measurements requested by a homeowner/building owner/client, (2) routine screening measurements to determine whether major identified contaminants are within guideline values or other acceptable limits, (3) measurements to confirm a hypothesis relative to problem contaminants and health effects that may be associated with such exposures, and (4) measurements to determine the effectiveness of mitigation measures.

1. Requests

Environmental sampling is often requested by building managers/owners of both nonresidential and residential properties. These may be made in response to regulatory requirements (asbestos, and in some cases, lead); as a part of environmental site assessments; or as a condition of a real estate transaction (asbestos, lead, radon). They may also be made in response to problem building complaints in the case of nonresidential buildings and general or specific health concerns expressed by occupants of residential buildings. In the former case, environmental sampling may be requested (1) in response to occupant requests, (2) to demonstrate empathy for occupant concerns, (3) to allay occupant fears by demonstrating that an air/surface contamination problem does not exist, and (4) to identify the potential cause of occupant complaints.

Investigators have different professional responsibilities as they relate to building manager/owner requests. Private consultants are obliged to provide only the services requested and any additional services that may be subsequently agreed upon. Public health and environmental agency staff have an obligation to protect public health. In theory, they have more latitude in conducting investigatory activities beyond simple requests for air or other environmental sampling. In practice, public health/environmental agency investigators acting in a nonregulatory mode generally respect the wishes of building managers/owners relative to the scope of environmental sampling and other building investigation activities.

2. Routine screening

Routine screening is a common sampling objective. Measurements are made without any consideration of the probability that a contaminant is present or the nature of health or other concerns. Common indoor contaminants are measured using sampling instruments and techniques that are readily available, easy to use, and which do not impose an undue financial cost on building owners or public health and environmental agencies.

Screening measurements have been, and continue to be, widely used by homeowners/lessees and other building owners to determine radon levels. Such measurements are designed to identify buildings with high radon concentrations so that owners can implement appropriate mitigation measures to reduce exposure.

Routine screening is an important infection control measure in hospitals. Of particular concern is the maintenance of low airborne levels of *Aspergillus fumigatus* in surgical operating and convalescent patient rooms, as well as oncology, transplant, and AIDS wards. Screening measurements for formaldehyde (HCHO) in urea–formaldehyde foam-insulated (UFFI) houses were conducted by the Canadian government and many homeowners in both Canada and the U.S. in the 1980s; such measurements are rare today. Measurements of HCHO, CO, CO_2, respirable particles (RSP), airborne mold, and nitrogen dioxide (NO_2) are commonly conducted in epidemiological studies and problem building investigations.

On a population basis, such screening has the potential to identify indoor environments that exceed guideline values and, as a consequence, are in need of mitigation measures. Because of its generic nature, routine screening has limited value in identifying specific causal factors responsible for building health complaints.

3. Identifying causal contaminants

Ideally, environmental sampling is conducted to identify and quantify contaminants which, based on information gathered in an investigation, can reasonably be expected to be a potential causal factor in occupant health complaints. In some cases, the targeting of specific contaminants is facilitated by: (1) unique symptomology (CO exposure, hypersensitivity pneumonitis); (2) suggestive evidence that ventilation may be inadequate (human odor, poorly designed/operated heating, ventilation, and air conditioning [HVAC] systems); (3) odor (ammonia, solvents); (4) water-damaged materials and evident mold infestation; and (5) occupant allergy tests that indicate sensitivity to particular allergens.

Nonspecific mucous membrane and general (headache, fatigue) symptoms are commonly reported in many problem building and residential investigations. They cannot easily be associated with unique contaminant exposures. As a result, environmental sampling is unlikely to identify environmental contaminants that may be causal agents.

4. *Evaluating effectiveness of mitigation measures*

Environmental sampling is routinely used to test the effectiveness of radon mitigation measures. This requires that sampling be conducted prior to and after mitigation activities have been completed. Environmental sampling is also conducted to determine the effectiveness of post-abatement cleaning measures for asbestos and lead in buildings, and is increasingly being used to determine the effectiveness of clean-up activities after the abatement of toxigenic mold infestations such as *Stachybotrys chartarum* and *Aspergillus versicolor*. In asbestos abatements, aggressive sampling is conducted to meet clearance airborne asbestos fiber concentrations; it is also a desirable practice in toxigenic mold abatements. Surface sampling is conducted in lead abatements to determine whether clearance guidelines have been achieved.

Environmental sampling has particular value in determining the effectiveness of measures implemented to improve ventilation. Pre- and post-measurements of CO_2 levels are commonly conducted to evaluate the performance of ventilation systems after operation and maintenance changes have been made.

Environmental sampling in conjunction with occupant health and comfort surveys can be used to evaluate the effectiveness of mitigation efforts in reducing symptom prevalence and increasing occupant satisfaction with air quality. Such coordinated pre- and post-environmental sampling and occupant health and comfort surveys are rarely conducted. Those conducted in research studies have, for the most part, not demonstrated significant reductions in symptom prevalence rates and increases in occupant satisfaction with air quality.

C. *Sampling airborne contaminants*

Airborne gas and particulate-phase substances have historically been the major focus of environmental sampling in indoor environments subject to indoor air quality/indoor environment (IAQ/IE) concerns. The conduct of air sampling requires selection of appropriate sampling and analytical procedures including: (1) instrument selection and calibration; (2) sampling location, time, and duration; and (3) number of samples. It also involves sampling and analysis administration and quality assurance.

There are a variety of approaches, sampling instruments, and analytical methods that can be used to conduct measurements of airborne contaminants. Selection of sampling and analytical techniques that provide acceptable performance is, of course, very important. Acceptable performance is ensured by implementing appropriate quality assurance practices. These performance considerations include accuracy, precision, sensitivity, and specificity.

1. *Performance considerations*

a. Accuracy.   Accuracy is the closeness of a value to its true or known value. It can be described by an error value expressed as a percentage. A sampling/analytical procedure may have an accuracy of 95%. This means that repeated measurements indicate that the measured value deviates from the true value by –5%, or 5% < true value. Another procedure may have an accuracy of 110%; i.e., on average the concentration is +10%, or 10% > true value. A good sampling/analytical procedure will have an accuracy within ±10% of the true value.

  b. Precision. Sampling/analytical procedures should also be relatively precise. Precision in the scientific (rather than dictionary definition) sense is the reproducibility of measured results. Precision indicates the relative variation around the mean. It is reported as a ± percentage around the mean of multiple values determined from measuring the same known concentration. It is determined by calculating the coefficient of variation

$$CV = \frac{\sigma}{x}(100) \qquad (9.1)$$

where σ = standard deviation
  x = mean

  Sampling/analytical procedures should have both high accuracy (within ±10% of the true value) and high precision. Acceptable precision values depend on the technique employed. For many instrumental techniques, ±10% is desirable. For techniques such as gas sampling tubes and passive samples, a precision of ±25% is generally acceptable.

  Though accuracy and precision are scientifically well-defined concepts, they are used interchangeably by the lay public as well as by technically trained individuals. As a consequence, it is often difficult to communicate their scientific meaning and relevance in environmental sampling.

  c. Sensitivity. In addition to acceptable accuracy and precision, it is important to use sampling/analytical procedures that are sufficiently sensitive to measure contaminant concentrations expected. Sensitivity is determined from reported limits of detection (LODs) for different sample sizes and durations. The LOD varies with different analytical procedures. It can often be extended (within limits) by increasing the volume of air sampled into/onto sorbing media by increasing sampling duration. This cannot be done on direct-read, real-time instruments.

  d. Specificity. In most, but not all cases, a procedure should be specific for the contaminant under test. This is especially true for gases and vapors. Nonspecific techniques have diminished accuracy when two or more contaminants with similar chemical characteristics are present. Measured concentrations may be higher than they actually are. Such results can be characterized as positive interference. In other cases they will be lower; therefore, interference is negative. Interference with measured concentrations can also occur even when a sampling/analytical procedure has relatively high specificity. This is true for the DNPH–HPLC method used for HCHO and other aldehydes. It is subject to significant negative interference from ozone (O_3).

e. Reference methods. The use of sampling/analytical procedures to conduct sampling for vapor-phase substances should be a relatively conservative one. It is desirable to use reference, approved, or recommended methods that have been systematically evaluated by the National Institute of Occupational Safety and Health (NIOSH), the U.S. Environmental Protection Agency (USEPA), the Occupational Safety and Health Administration (OSHA), or the American Society for Testing and Materials (ASTM). The use of approved methods provides a relative (but not absolute) degree of confidence in the accuracy, precision, and reliability of sampling/analytical procedures being employed.

f. Quality assurance. Contaminant measurements conducted in problem buildings or in systematic research studies need to be quality assured to provide confidence in their accuracy and reliability. Quality assurance procedures include instrument calibration, and, when appropriate, use of field blanks, media blanks, replicate samples, and split or spiked samples. They focus on both field and analytical aspects of contaminant collection and measurement.

i. Calibration. Calibration is a process whereby measured values of air flow and/or contaminant levels are compared to a standard. In the case of air flow, the standard may be primary or secondary, with the latter traceable to the former. Primary standards can be traced directly to those at the National Institute of Standards and Technology (NIST). A gas burette serving as an airflow measuring device is a primary standard; a rotometer (which must be calibrated) is a secondary standard. All dynamic sampling instruments, particularly gas sampling pumps and tube systems, should be calibrated frequently. In common practice, calibration of real-time, direct-read instruments is conducted by using calibration gases that have been prepared to provide sample concentrations within the measuring range of the instrument. Single- and multiple-point calibrations are conducted depending on the user; multipoint calibrations are preferred.

ii. Blank/replicate samples, etc. In collecting samples onto a medium, it is essential that field or media blanks be used. A field blank is a media sample taken to the field, opened, then closed and returned to the laboratory where it is analyzed. Media blanks are samples of liquid media prepared at the same time as samples used in the field or the same lot of solid media. Both are used to adjust environmental sample concentrations for contaminant levels present in unexposed media. Field blanks are used to determine whether contamination occurred as a consequence of sample media being taken into the field. Two field blanks for each of 10 environmental samples are generally recommended.

Replicate samples are often collected to assess the accuracy and precision of analytical results, split samples to compare the performance of different analysts, and spiked samples to assess analytical performance relative to a known sample concentration.

2. Resource limitations

In addition to performance characteristics described above, selection of a sampling/analytical procedure will often be determined by the availability of resources. Despite significant performance limitations, sampling/analytical procedures are often chosen because of their relatively low cost. This is particularly true for gas sampling tubes and, in some cases, passive samplers. Other factors affecting the selection of sampling/analytical procedures include equipment availability, portability, and degree of obtrusiveness.

3. Sampling procedures

Samples of airborne contaminants can be collected and analyzed by utilizing both dynamic and passive sampling procedures.

a. Dynamic sampling. In dynamic sampling, vapor-phase substances are drawn (by means of a pump) at a controlled rate through a liquid or solid sorbent medium or into a sensing chamber. In sampling for particulate-phase substances, air is drawn through a filter, impacted on an adhesive-coated surface, attracted to collecting surfaces by electrostatic or thermostatic processes, or brought into a sensing chamber. The sample volume is determined from the known flow rate and sampling duration. Concentrations can be calculated or read directly from electronic real-time instruments.

Dynamic sampling is conducted using two approaches. In the first, sampling and analysis are discrete events. Sampling is conducted with pumps that collect vapor-phase substances in a liquid medium (absorption) or onto one or more solid sorbents (adsorption) such as charcoal, tenax, silica gel, etc. Exposed sorbent media must be analyzed by a laboratory. Depending on laboratory schedules, results may be available in a matter of days or up to several weeks. In the case of particulate matter samples, gravimetric analysis may be conducted within a day or so after samples have been sent to a laboratory. The relatively long period between sample collection and availability of results is an obvious disadvantage of this type of sampling. For many contaminants, alternatives may not yield acceptable results or may be too expensive.

Dynamic sampling (and analysis) can also be conducted using direct-read, real-time (instantaneous) or quasi-real-time instruments. Such instruments are available for a limited number of common indoor contaminants. Sampling and analysis are combined so that results are immediately available to individuals conducting investigations and testing (providing an obvious advantage).

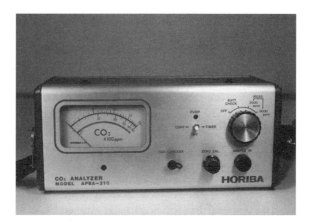

Figure 9.1   Real-time electronic CO_2 monitor.

Real-time sampling/analysis is conducted with electronic direct-read instruments or by gas sampling tubes. A variety of electronic direct-read instruments are commercially available for determining concentrations of gases, vapors, or particulate matter. They are usually pump-driven devices which draw air at a known, low, constant rate into a small chamber where sensors measure specific chemical or physical properties of contaminants under test. Commonly used principles for gas/vapor substances include electrochemistry, photometry, infrared absorption, and chemiluminescence. Commonly used principles for airborne particles include optical techniques that measure light scattering or absorption and piezoelectric resonance. Portable, direct-read, real-time instruments are widely used to measure CO_2, CO, and RSP in indoor air. Electronic instruments have a continuous output of concentration readings, and as a consequence, concentration values can be continuously recorded. An electronic direct-read sampling instrument for CO_2 is illustrated in Figure 9.1.

Gas sampling tube systems are used to measure a variety of contaminants in industrial workplaces and, less commonly, IAQ/IE investigations. They consist of a gas syringe or bellows into which a gas sampling tube designed to detect and quantify a specific gas or gases is inserted. The syringe/bellows is designed to draw a minimum sample volume of 0.1 L. Larger volumes can be utilized by employing multiple pumping actions. Gas sampling tubes are hermetically sealed prior to use. They contain a granular sorbent such as silica gel, alumina, or pumice, impregnated with one or more chemicals that react on contact with a specific contaminant or contaminant group, producing a colored or stained substrate medium. The length of the stain or colored sorbent is proportional to the concentration of contaminants in the air volume sampled. The concentration is typically read from a calibrated scale printed on the side of the tube. A gas sampling tube system is illustrated in Figure 9.2.

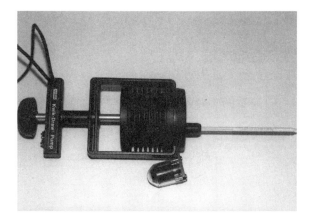

Figure 9.2   Gas sampling tube system.

Gas sampling tubes are commercially available for a large number (100+) of gases and concentration ranges. Their accuracy and precision varies. They are designed in most cases to provide an accuracy within ±25% of the true value with similar precision. Gas sampling tube systems are attractive for conducting sampling in indoor and other environments. They are relatively inexpensive, simple to use, and provide quasi-real-time sampling results. Unfortunately, they have significant limitations. These include, in many cases, relatively poor accuracy and precision, lack of specificity, high LODs, and limited shelf-life (1+ year). In addition the length of the color-stained substrate tends to be indeterminate (no sharp demarcation) and concentrations may be difficult to read. These limitations vary for the contaminants being sampled. Despite these limitations, gas sampling tube use is common in IAQ investigations. Best results have been reported for measurements of CO and CO_2.

b. Passive sampling.   Passive sampling techniques are widely used in measuring indoor contaminants. They are particularly useful for determining concentrations integrated or averaged over a period of hours, a week, or even months. In most cases, they are based on the principle of collecting contaminants by diffusion onto or into a sorbent medium. In passive samplers used for gas-phase substances, the sampling rate, and therefore volume of air that comes into contact with the collecting medium, can be determined from the cross-sectional area of the sampling face, the length/depth through the sampler that gases must travel to be collected on absorbing surfaces, and the diffusion constant of the gas being collected. The Palmes diffusion tube (Figure 9.3) was the first passive device to be developed and used. Palmes tubes are still commonly used to measure NO_2 levels in buildings. They have a sampling rate of 55 ml/hr, and concentrations are usually integrated over a period of 7 days.

Since the development of passive samplers in the 1970s, a number of new devices using the same sampling principles have been developed. These

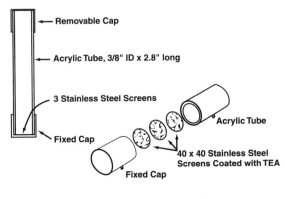

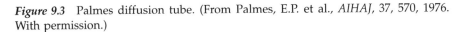

Expanded View of Sampler Bottom

Figure 9.3 Palmes diffusion tube. (From Palmes, E.P. et al., *AIHAJ*, 37, 570, 1976. With permission.)

include small badges that provide relatively high sampling rates over a period of 8 hours (Figure 9.4), and charcoal canisters used for radon measurements over a period of 2 to 7 days. Track-etch detectors are passive samplers that record alpha particle tracks on plastic film over a period of months. Unlike most passive samplers, track-etch detectors do not use the principle of diffusion directly.

Passive samplers have the advantage of low cost, simplicity of use, and reliability. They are designed to have an acceptable accuracy/precision (e.g., ±25%). Their accuracy/precision is diminished by changes in face velocity (due to changes in room airflow), which may occur during sampling. Passive samplers are not particularly useful in conducting problem building investigations because of the long sampling times required (generally 8 hours or more). They have, however, been widely used in screening measurements of HCHO, NO_2, and radon in indoor environments.

Figure 9.4 Badge-type passive samplers.

   c. Integrated sampling.   Unlike real-time or quasi-real-time sampling, where contaminant concentrations are measured (and recorded) as they occur, passive and many dynamic sampling procedures yield results that represent the integrated average of concentrations over the sampling period. Therefore the nature of contaminant fluctuations and peak concentrations are unknown. Short averaging times such as an hour (used with dynamic sampling for HCHO), when repeated over time, can provide a reasonable indication of the relative degree of fluctuation with time. Averaging or integration times reflect the combination of flow rates and time needed to achieve a minimum LOD value as well as times selected by investigators to reflect the nature of health concerns involved (e.g., acute symptoms vs. long-term cancer risk).

4. Sampling considerations

Once sampling objectives are defined and measurement methods selected, other concerns must be considered, depending on the nature of sampling conducted. These include the determination of whether sampling is to be conducted once or many times, sampling location, time and duration, and number of samples.

   a. One-time sampling.   Sampling conducted in many IAQ/IE investigations, as well as routine screening for radon and other contaminants, is usually done on a one-time basis; i.e., a single sample is collected or samples are collected in multiple locations at the same relative time. Such sampling is conducted once in order to reduce costs (with the assumption that one-time testing is sufficient). Such tests provide information only on contaminant concentrations present at the time of sampling. Unfortunately, significant time-dependent changes occur in the concentration of many contaminants. These may be episodic (associated with discrete, at times unpredictable events, e.g., CO); diurnal (e.g., increase of CO_2 levels over the course of a day in poorly ventilated buildings); seasonal (seasonally dependent HCHO levels in residences that vary by a factor of four have been reported); or associated with time-dependent contaminant decay (e.g., VOC levels in new buildings). Results of one-time sampling should always be interpreted within the context of known patterns of change in specific contaminant levels in indoor environments.

   b. Location.   The selection of sampling locations should reflect sampling objectives. In problem building investigations, these may include areas (1) with high complaint rates, (2) with potential sources, (3) where ventilation is expected to be poor, and (4) where complaints are relatively few (a control area). In some cases it may be desirable to collect a sample outdoors. Priority should be given to areas where complaints are most prevalent and where highest contaminant concentrations may be expected based on observations made in a walk-through inspection. Sampling in control areas may be conducted to identify differences in contaminant types and levels in order to determine whether complaints are consistent with exposures.

The number of sampling locations may be increased if electronic direct-read instruments are available. This is particularly true for CO and CO_2. Sampling for CO_2 should be conducted in return, supply, and outdoor air to determine the relative percentage of outdoor air being provided to a space, zone, floor, or building.

Sampling locations will, in many cases, be determined by the nature of the contaminant and potential exposures. In residential environments, USEPA recommends that sampling for radon be conducted in the lowest livable space. In some houses this would be the basement; in two-story houses on a slab or crawlspace, the first floor. Because of the effects of wind on building pressures, radon testing should always be conducted near the center of the dwelling.

Sampling locations in dwellings should be selected to represent potential sources as well as exposures. These would include rooms with obvious mold infestation or odor, and rooms occupied by individuals with health problems.

c. Time. When air sampling is conducted is an important consideration in problem building investigations, investigations in dwellings, and for screening measurements such as radon.

As indicated previously, concentrations of many airborne contaminants vary with time, and knowledge of such variation is important in conducting sampling and interpreting results. Time is described in the context of a single day when an investigation is being conducted; a specific hour or day when elevated concentrations may be expected; or a season when high, moderate, or low values may occur.

In poorly ventilated buildings, CO_2 levels reach their peak in early to mid-afternoon. If longitudinal (real-time sampling conducted over a period of hours) sampling is conducted, it should include those hours when peak CO_2 levels can be expected. Sampling CO_2 only at the beginning of building occupancy would be misleading.

In problem environments where exposure to CO is expected, high CO levels are likely to occur episodically. Multiple measurements may be necessary over the course of a single day, as well as measurements on other days if elevated levels are not initially observed. Elevated CO levels are more likely to occur during the heating season, especially when heating systems are active.

Formaldehyde levels vary over the course of the year and usually reach maximum values when temperature, relative humidity, and ventilation conditions are optimal. Such optimum conditions occur in the spring and fall, and in air-conditioned dwellings in summer.

With the exception of radon, the winter season in cold climates is a poor time to measure contaminant levels in residential buildings. Because of relatively high infiltration air flows, and lower thermostat settings and relative humidity (in the case of HCHO), many airborne contaminants in residential buildings may be at their lowest values in the winter. This includes, in many cases, airborne mold as well. However, because of induced convective soil gas air flows, radon levels in residential buildings may reach a peak during cold winter weather. Radon sampling may be desirable under such circumstances if one's objective is to maximize the probability that elevated radon levels will be measured, or undesirable if the objective of sampling is to obtain low test results for a real-estate transaction (as some would be wont to do).

The selection of a sampling time for contaminant measurements should be made with some knowledge of the range of variation that may occur with time. Assuming resources are limited, it is, in many cases, desirable to measure contaminant concentrations in the upper end of their range under normal building operating/environmental conditions. This is of particular importance when IAQ guidelines are being used to determine whether exposures are within acceptable limits or not.

d. Duration. Sampling duration is the amount of time required or selected to collect a sample. It reflects limitations of sampling and analytical methods used and potential exposures. With real-time or quasi-real-time sampling devices, sampling duration may be limited to the time necessary to collect the sample or stabilize instrument readings. With other instruments and analytical methods, the sampling duration may be determined by the time required to collect sufficient contaminant to meet the LOD. When the NIOSH chromotropic acid method is used to sample HCHO, a sampling duration of 1.5 hr is necessary to measure 0.02 ppmv at the optimum sampling rate of 1 L/min, and a sampling duration of 1 hr is needed to accurately measure a concentration of 0.05 ppmv. Using the DNPH–HPLC method, a sampling duration of 0.5 hr would be sufficient to meet a similar LOD value.

Relatively short sampling durations (0.25 to 2 hr) are often used to determine contaminant concentrations which may be responsible for acute symptoms. Sampling durations of a week to months may reflect the need to evaluate the potential for chronic effects, particularly for diseases such as cancer.

e. Sample number. A sufficient number of samples should be collected to ensure measured values are reasonably reflective of exposure conditions. Though there is no magic formula for determining the number of samples to be collected, it is advisable to collect multiple samples (which includes different locations) to avoid difficulties associated with bad samples or sample loss. As the number of samples increases, so does the time and cost of their collection and analysis. Investigators must use their professional judgment to determine the number of samples required to obtain meaningful results within resource constraints.

The number of samples may be determined by sampling objectives. In the case of radon, most homeowners choose to determine the average concentration near the center of the building by using one passive sampler. In many cases this may be sufficient. However, in the case of results near or above guideline values, repeated or long-term sampling is desirable. If the sampling objective is to determine the efficacy of a mitigation measure, several samples should be collected before and after the implementation of mitigation measures.

D. Sampling bulk materials/surface contaminants

Environmental sampling to identify/quantify contaminants on or in materials/surfaces is commonly conducted in buildings for such substances as asbestos, lead, pesticides, PCBs, dust allergens, and biological organisms such as mold. Many of the same principles/considerations described above for air sampling apply to material/surface sampling. These include performance considerations, resource limitations, and sampling considerations such as one-time sampling, sampling location, and number of samples to be collected.

1. Bulk sampling

Material sampling and analysis commonly described as bulk sampling is used to determine the presence and identity of asbestos fibers in building materials. USEPA requires the use of bulk sampling in the conduct of asbestos inspections in buildings for purposes of asbestos management and for major building demolition/renovation activities which would disturb asbestos-containing materials (ACM). In the former case, results of bulk sampling (in conjunction with the assessment of ACM condition and potential for disturbance) are used as indicators of potential human exposure to asbestos fibers in buildings.

Bulk sampling is used to identify lead-based paint in/on buildings and a variety of outdoor surfaces. Bulk sampling for lead is conducted using nondestructive *in situ* techniques such as X-ray fluorescence and paint chip and soil analysis using atomic absorption, inductively coupled plasma, and anodic stripping voltammetry.

Bulk sampling is also conducted on mold-infested materials to identify mold types as well as their relative abundance. Samples are dispersed in a liquid medium and plated out on nutrient agar plates.

2. Surface sampling

Surface sampling is conducted by vacuum or wipe methods to ascertain the degree of contamination by lead, pesticides, plasticizers, PCBs, and allergens. It is commonly used to determine the efficacy of remediation measures.

Surface sampling is commonly conducted for lead in house and other building dusts. Because of the very serious health hazards associated with lead exposure, significant efforts have been made to evaluate and use surface sampling techniques. These include vacuum and surface wipe sampling. Because of its longer history of use and observed significant associations with blood lead levels, most lead sampling of household dust is conducted using wipe sampling. Wipe sampling is recommended by USEPA and the

Department of Housing and Urban Development (HUD) for conducting risk assessments and clearance sampling in lead-based paint and lead-contaminated dust abatements. Guidance values include 100 µg Pb/ft^2 on floor surfaces, 400 µg Pb/ft^2 on window sills, and 800 µg Pb/ft^2 in window wells. USEPA has proposed changing the guidance value for floor dust to 50 µg Pb/ft^2. Samples are collected in a systematic fashion using a wet wipe on one square foot of floor surface and a measured area (equivalent to 1 ft^2) of window sills and wells. Wet wipe samples can be analyzed in a laboratory using atomic absorption analysis, or inductively coupled plasma or anodic stripping voltammetry. It is good practice for lead analyses to be conducted by a USEPA-approved laboratory. Wipe sampling appears to work relatively well on hard surfaces, with less reliable results on carpeting.

House dust containing lead can also be sampled using vacuum methods. Vacuum sampling has the potential for providing more accurate results on fabric surfaces such as carpet. Vacuum methods evaluated have included personal sampling pumps with a specially constructed surface vacuum attachment and a large, heavy-duty, specially designed USEPA vacuum cleaner. Concentrations in vacuum samples can be expressed on a unit surface area or on a mass basis. Mass concentrations are expressed as the amount of lead in dust per unit mass of dust collected (µg/g). Such concentration expression does not assess the absolute mass of lead that may be present in house dust.

Surface sampling for pesticides, plasticizers such as phthalic acid esters, and PCBs is conducted using vacuum sampling procedures on horizontal surfaces and by wipe sampling on hard horizontal and vertical surfaces. Concentrations in the former case are expressed on a mass-to-mass basis (ng/g, µg/g) and in the latter case on a mass per unit area basis (µg/m^2). Post-fire PCB remediation effectiveness is typically determined from surface wipe samples.

Surface sampling for mold can be conducted by collecting house dust or sampling surfaces of mold-infested materials with transparent sticky tape. In the former case, house dust samples are washed and separated into different fractions and then plated out on agar media to determine the number of colonies that are produced on a unit dust weight basis. At the present time there are no consensus guideline values to indicate the significance of culturable/viable mold concentrations in surface dust samples relative to human exposure and health effects. It is relatively common in building investigations to collect transparent cellophane tape samples of the surfaces of mold-infested building materials to identify major genera present. Of particular concern are toxigenic fungi such as *Stachybotrys chartarum* and *Aspergillus versicolor*. Mold may also be sampled qualitatively by the use of swab sampling in which Q-tips or similar products are brushed across the surface of infested areas and transferred to selected mold culture media.

Surface sampling is commonly conducted for allergens. Such sampling is typically conducted by using vacuuming techniques to collect surface dust. Dust samples are collected in locations where high allergen levels may be expected. The sample may be collected from a defined surface area (e.g., 1 m²) or time period (e.g., 5 min) to provide a sufficient sampling mass to express concentrations on a unit mass basis (e.g., µg/g). Collected dust samples are analyzed using monoclonal or polyclonal antibody procedures. Surface sampling for dust mite, pet dander, and cockroach allergens is preferred when assessing potential health risks associated with airborne allergen exposure.

Surface sampling has also been conducted for mineral fibers such as asbestos, fiberglass, mineral wool, and ceramic fibers using vacuum or wipe sampling techniques. The efficacy of such techniques in quantifying mineral fiber levels in surface dusts has not been well documented.

E. Measuring common contaminants in indoor environments

A variety of gas, vapor, and particulate-phase contaminants are measured in indoor environments. These include contaminants suspected of contributing to IAQ/IE complaints, and contaminants such as radon, which may pose long-term cancer risks. Contaminants associated with IAQ/IE complaints include CO_2, CO, HCHO, total volatile organic compounds (TVOCs), specific VOCs, and mold. Less commonly, airborne particulate matter levels are measured.

1. Carbon dioxide

Because it is widely used as the principal indicator of ventilation adequacy, CO_2 is the most commonly measured indoor air contaminant. Carbon dioxide measurements are made using direct-read, real-, or quasi-real-time instruments. These include electronic instruments that can provide instantaneous or near-instantaneous values and, if desired, a continuous record, and gas sampling tube systems. The measurement of CO_2 levels in electronic instruments is based on electrochemistry. Two types of CO_2 instruments are commonly available. These are pump-operated instruments which, on drawing air through a sensing chamber, quickly respond to changes in CO_2 concentrations. Other instruments determine CO_2 levels by passive absorption of CO_2 into the sensing unit. Pump-driven devices may have two sampling ranges: 0 to 2000 and 0 to 5000 ppmv; passive samplers have a sampling range of 0 to 2000 ppmv. Since CO_2 levels above 1000 ppmv are generally recognized as exceeding guideline values for ventilation adequacy, there is, in theory, no need to measure CO_2 levels above 2000 ppmv. Passive sampling devices are relatively inexpensive (circa $500). Both dynamic and passive battery-operated CO_2 samplers are available. Battery life in continuous operation is usually no more than an hour. Dynamic samplers can be operated continuously using a line cord.

Dynamic CO_2 monitors are calibrated using canisters of standard concentrations of CO_2 and zero gas (nitrogen). Passive sampling devices are factory calibrated and do not lend themselves to laboratory calibration. Examples of dynamic and passive direct-read electronic instruments can be seen in Figures 9.1 and 9.5.

Figure 9.5 Direct-read passive CO_2 sampler.

CO_2 measurements can also be made using gas sampling tubes. The principal advantage of using gas sampling tubes is their relatively low cost. Cost, as well as time required to collect samples, increases significantly when there is a need to collect many samples in different locations.

2. Carbon monoxide

Carbon monoxide levels are often measured in problem building investigations and residential environments when there is reason to suspect that a potential CO exposure problem exists. Carbon monoxide levels are usually measured by real-time, direct-read electronic instruments or by gas sampling tubes.

A variety of electronic direct-read CO measuring devices are available. These instruments measure CO concentrations in ranges from a few ppmv to several thousand ppmv. They determine CO levels by chemically oxidizing CO to CO_2. As oxidation takes place, an electrical signal is produced which is proportional to the CO concentration in the sampled air stream. Both pump-driven and passive, hand-held sampling devices are available. The former produces instantaneous real-time values, whereas the latter is somewhat slower, requiring a minute or more to respond.

Real-time or quasi-real-time CO monitors have advantages similar to those used for CO_2. Pump-driven devices can be easily calibrated with standard gas mixtures, providing relatively good accuracy. Electrochemical cells have a limited lifetime and must be periodically replaced.

Gas sampling tubes are commonly used in investigations where significant CO exposures may have occurred. Their accuracy is relatively good. They are available in several concentration ranges, with ranges of 1 to 50 ppmv and 5 to 500 ppmv used in IAQ/IE investigations.

3. Formaldehyde

Formaldehyde measurements were widely made in mobile homes, urea–formaldehyde foam-insulated (UFFI) houses, and a variety of stick-built houses in the 1980s. Requests for problem home investigations for potential formaldehyde-related health problems have decreased significantly over the past decade. Such investigations are now made only occasionally. Formaldehyde measurements are often made in problem building investigations, because HCHO is one of the contaminants included in routine screening.

Historically, the NIOSH bubbler/chromotropic acid method (NIOSH method 3500) has been the most widely used HCHO sampling/analytical technique. In this procedure, a sampling pump draws air through 15 to 20 ml of a 1% sodium bisulfite solution. Formaldehyde is collected by forming sodium formaldehyde bisulfite, an addition product. A sampling rate of 1 L/min is typically used, with a sampling duration of 1 to several hours. Samples are analyzed colorimetrically. The sampling/analytical accuracy is reported to be 92 to 95%. The method has a long history of use and is considered to be very reliable.

The DNPH–HPLC method has recently become a relatively popular technique for measuring indoor air concentrations of HCHO. The method can be used to collect HCHO, other aldehydes, and ketones on 2,4-dinitrophenyl hydrazone (DNPH)-coated substrates including glass fiber filters, silica gel, and C_{18} cartridges. Collected aldehydes and ketones are converted to stable hydrazones which are analyzed by high-performance liquid chromatography (HPLC). The DNPH–HPLC method has several major advantages. These include specificity for different aldehydes, including HCHO, and high sensitivity (it can detect a concentration of 9 ppbv in a 20 L sample). Though DNPH sampling is highly accurate in laboratory environments, it is subject to significant negative interference in the presence of O_3.

Both chromotropic acid and DNPH analytical methods are employed with passive sampling devices. Passive samplers are commercially available that allow quantification of HCHO after exposure for 8 hr, 24 hr, and 7 days. Passive samplers employing a 7-day sampling duration have been widely used for sampling residences and in research studies. Results are integrated and provide no indication of peak values which may be responsible for acute symptoms. They are generally not suitable for problem building investigations. Sampling devices based on the chromotropic acid and DNPH–HPLC methods are both relatively specific for HCHO and have low LODs based on conditions of use. A passive sampler based on the MBTH method is used for HCHO measurements. It, however, measures total aldehydes, and because of its lack of specificity has limited usefulness in IAQ investigations. It does have the advantage of high sensitivity with a sampling time of 2 to 3 hours. Passive samplers are relatively attractive as HCHO sampling devices because vendors provide both samplers and analyses at relatively low cost ($40 to 60 per sample).

Formaldehyde can also be measured by automated devices that provide quasi-real-time results. These are based on HCHO absorption in solution and colorimetric analysis using the pararosaniline method. This method is specific and highly sensitive, with an LOD significantly lower than the

NIOSH bubbler–chromotropic acid method. Depending on the range selected, it can measure HCHO levels as low as 0.002 ppmv and as high as 10 ppmv. Analytical instruments are expensive and relatively laborious to calibrate, set up, and clean.

4. Volatile organic compounds

Different approaches are used to sample or monitor VOCs, depending on aspects of their measurement and expression desired. In most buildings, levels of individual VOCs are very low (in the low ppbv range) and so cannot be monitored by direct-read electronic devices. An exception is methane, which is associated with sewer gas or heating and cooking gas leaks. Methane levels can be measured using portable flame ionization detectors (FIDs) (Figure 9.6).

VOCs can be sampled and expressed as concentrations of individual substances present or as TVOC concentrations. Similar sampling procedures are used in both cases. In practice, TVOCs are collected on solid sorbent media using dynamic (and occasionally passive) sampling or by using evacuated metallic canisters. In the former case, air is drawn through a glass sampling tube which contains one or more sorbents such as charcoal, tenax, XAD-2, Poropak Q, etc. Sorbent media vary widely in their ability to capture and retain specific VOCs; as a consequence, all have collection limitations. Multisorbent samplers have been developed for VOC sampling in indoor environments. They have the advantage of collecting VOCs over a broad range of volatilities with relatively high accuracy and precision using low sampling volumes.

VOCs can also be collected using evacuated metallic canisters. Canisters have several advantages over sorption methods. These include avoidance of chemical reactions on the sorbent and lower recoveries due to breakthrough or incomplete desorption from sampling media. As a result, a wide range

Figure 9.6 Portable flame ionization detector.

of VOCs can be detected and measured. Major disadvantages are the relatively small portion (1 ml) of the 1 to 10 L sample that is analyzed, and potential contamination of the sample from pumps, tubes, and fittings used with the technique.

Both sorbent tubes and samples collected in evacuated canisters are analyzed using gas chromatography, often in conjunction with a mass spectrometer. The mass spectrometer allows for relatively reliable identification of individual VOC species. Because of low concentrations (ppbv, $\mu g/m^3$), it is desirable to thermally desorb VOCs collected on sorbent tubes before analysis. Thermal desorption has several advantages over solvent (CS_2) desorption. These include fewer analytical operations, shorter operating time, recovery of the entire undiluted sample, and increased sensitivity to low VOC levels. In the evacuated canister method, a collected sample is injected directly into a gas chromatograph for analysis.

Total volatile organic compounds are usually quantified by using a gas chromatograph with a flame ionization detector (GC–FID), with toluene or hexane as a standard for determining concentrations. Concentrations of TVOCs, using a mixture of toluene, hexane, or cyclohexane and helium as a calibration gas, can be calculated based on measured FID response and the following equations:

$$f = c_s/V_s/C_s \tag{9.2}$$

where f = response factor (counts/μg)
$\quad$ c_s = FID count for a known volume of standard
$\quad$ V_s = volume of standard (m^3)
$\quad$ C_s = concentration of the cyclohexane/helium standard ($\mu g/m^3$)

The TVOC concentration is then calculated as

$$T = c/f/V \tag{9.3}$$

where c = FID count for air sample
$\quad$ V = volume of air sampled (m^3)
$\quad$ T = TVOC concentration, ($\mu g/m^3$)

The determination of TVOC concentration takes less effort than identifying and quantifying individual VOCs, with a shorter turnaround time for air sampling results. Total VOC concentration can also be compared to guideline values that have been proposed for use in assessing whether elevated VOC levels may be responsible for health complaints in a building.

The determination of TVOC concentrations alone may be less informative than a full quantitative analysis of individual VOCs. Such a full analysis is useful as a "fingerprinting" tool in identifying potential VOC sources such as photocopiers, etc.

Figure 9.7 Radon sampling devices.

5. Radon

Radon is sampled in indoor environments using both passive integrated techniques and direct-read quasi-real-time instruments. The former are used in screening measurements; the latter in evaluating sources and determining the effectiveness of abatement measures. Techniques are available to determine radon concentrations in air, water, and soil.

Most commonly, radon is sampled using charcoal canisters (Figure 9.7, right). The opened sampler is exposed for 2 to 7 days. Since radon diffuses to charcoal at a known rate, the sample volume can be calculated. Concentrations are determined by measurement of gamma ray emissions from radon decay products (RDPs) using gamma ray detectors. Concentrations are calculated based on sample volume, radon half-life, and radon/RDP equilibrium ratios. Radon concentrations collected on small charcoal samplers can also be determined using scintillation counters.

Short-term (2 to 7 days) radon concentrations can also be measured with passive devices which allow radon to diffuse into them where, on decay, they produce positive ions, which are attracted to and deposit on a specially charged plate (electret) (Figure 9.7, left). The change (decrease) in voltage can be directly related to the radon concentration and other sources of gamma ray energy. Electret devices can also be used for longer-term sampling (up to 3 months).

As indicated in Chapter 3, short-term measurements are not adequate to determine long-term exposures which may pose a cancer risk. For such determinations it is common to measure radon using a track-etch detector (Figure 9.7, center). The track-etch detector is a simple device that consists of a plastic strip or film affixed to a small cylinder. The radon concentration is determined by counting the microscopic tracks produced on the film by alpha-particles. Track-etch detectors are typically exposed for 3 to 6 months.

A variety of quasi-real-time portable instruments are available for sampling indoor radon concentrations. They use a variety of detection principles including ionization. Measurements are typically reported as hourly averages.

6. Airborne particles

A variety of sampling and instrumental methods are available for measuring particulate matter in the inhalable size range (≤ 10 μm). These include gravimetric methods, wherein airborne particles are collected on filters, with concentrations reported in micrograms or milligrams per cubic meter ($\mu g/m^3$, mg/m^3). Respirable particles can be measured with real-time or quasi-real-time instruments using optical techniques. Some instruments measure particles in the 0.1 to 10 μm diameter range by forward light scattering; respirable particles (≤ 3.0 μm diameter) are often measured by piezoelectric resonance. In the latter devices, nonrespirable particles are removed by an impactor or cyclone. Respirable particles are then electrostatically deposited on a quartz crystal sensor. The difference in oscillating frequency between sensing and reference crystals is determined and displayed as concentration in $\mu g/m^3$.

F. Sampling biological aerosols

Biological aerosols comprise the fraction of airborne particles that are of biological origin and remain suspended in air for some time. Bioaerosols vary in their composition depending on sources present. They may differ significantly between indoor and outdoor environments and from one indoor environment to another. Of interest in indoor environments are bioaerosols that consist of a variety of viable and nonviable microorganisms, reproductive propagules, and microbial fragments. Mold, bacteria, and actinomycetes are of special concern. There are two major bioaerosol sampling methodologies used in conducting air sampling: culturable/viable and total mold spore/particle sampling.

1. Culturable and viable sampling

Culturable/viable sampling is used to determine airborne concentrations of mold and bacteria (including actinomycetes) that are viable (alive) and can grow on the culture media used. In such sampling, air is drawn at high velocity through a multiholed orifice plate where inertial forces cause the impaction of airborne particles (including viable organisms) onto a nutrient agar surface. Various media are used depending on sampling objectives. Collection media are incubated at room or other selected temperatures, and colonies are allowed to grow. As colonies mature they may be identified to genus or even to species. Counts may be made based on genus/species types or on a total colony count basis. Concentrations are expressed as colony-forming units per cubic meter (CFU/m^3).

A variety of culturable/viable sampling devices and techniques are available for use (Table 9.1). These devices differ in the principle of collection, sampling rate, and collection time used. They are similar in that a solid culture medium is used to "grow out" culturable/viable particles for identification and enumeration. Samplers differ in their apparent collection efficiencies. Overall collection efficiency determines the suitability of a device for a given sampling application. It is determined by both the efficiency of

Table 9.1 Devices/Techniques Used for Viable/Culturable Bioaerosol Sampling

Sampler type	Operating principle	Sampling rate (L/min)	Recommended sampling time (min)
Slit or slit-to-agar impactor	Impaction onto solid culture medium on rotating surface	30–700	1-60, Depending on model and sampling circumstances
Sieve impactors			
Single-stage, portable impactor	Impaction onto solid culture medium in a "rodac plate"	90 or 180	0.5–5
Single-stage (N-6) impactor	Impaction onto solid culture medium in a 10 cm plate	28.3	1–30[a]
Two-stage impactor	As above	28.3	1–30[a]
Multiple-stage impactor	As above	28.3	1–30[a]
Centrifugal	Impaction onto solid culture medium in plastic strips	40±	0.5
Impingers			
All glass/AGI-30	Impingement into liquid; jet 30 mm above impaction surface	12.5	1–30
All glass/AGI-4	As above; jet 4 mm above impaction surface	12.5	1–30
Filters			
Cassette	Filtration	1–2	5–60

[a] Contemporary practice is 1 to 5 minutes.

Source: From Chatigny, M.A. et al., *Air Sampling Instruments for Evaluation of Atmospheric Contaminants*, ACGIH, Cincinnati, 1989, 10. With permission.

particle collection and the efficiency with which the viability of collected microorganisms is preserved.

Both multistage and single-stage impactors have been suggested as reference samplers to which results of other samplers are compared. Because of its widespread use in IAQ research and problem building investigations, the single-stage (Andersen N-6) sampler (Figure 9.8) has become the *de facto* reference sampler. It collects airborne particles onto a nutrient agar plate through an orifice plate with 400 tiny holes at a standard flow rate of 1 CFM (28.3 L/min); sampling durations of 1 to 2 minutes are commonly used.

Though the most widely used bioaerosol sampling devices, culturable/viable samplers have significant limitations. The most notable of these

Figure 9.8   Andersen single-stage bioaerosol sampler.

is that only viable particles can grow on collection media. In most cases, culturable/viable bioaerosol particles represent only a fraction of the total bioaerosol concentration in air. In the case of mold spores and particles, the number of total spores and particles averages 10+ times greater than the culturable/viable concentration. Differences can vary from several times to orders of magnitude higher. Apparent viability ratios vary from genus to genus, with viabilities ranging from <1% to >20%. In the case of bacteria, the ratio of culturable/viable to total particles may be 1% or less.

The viability of airborne bacteria and mold spores is determined by environmental stresses they are subject to during dispersal. These include desiccation and exposure to ultraviolet light. Though microbial particles may not be viable, they may nevertheless have considerable biological and medical significance. In the case of mold spores and hyphal fragments, nonviable and viable particles have similar antigenic properties.

Viable airborne microbial particles differ in their culturability on the various media formulations used in sampling. Apparent collection efficiencies for different organisms vary from one medium to another. The use of malt extract agar (MEA) and tryptocase–soy peptone agar have been recommended for general purpose use for mold and bacteria (including thermophilic actinomycetes), respectively. Malt extract agar appears to provide good results for a wide variety of fungal species, particularly those with high water activities. There is increasing use of diclorvan–glycerol agar (DG-18) for sampling xerophilic mold genera and species such as *Aspergillus,* and cellulose-amended agar for sampling *Stachybotrys chartarum*. For xerophiles, higher colony counts have generally been reported on DG-18 as compared to MEA.

2. Total spore and particle sampling

Total spore/particle sampling methods and devices are less widely used than culturable/viable methods described above, and their use has been, for the most part, limited to mold. This reflects the historical early development

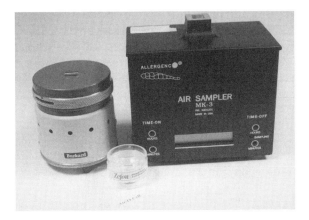

Figure 9.9   Total mold spore/particle sampling devices.

and use of culturable/viable sampling methods and their amenability to colony and organism identification.

In commercially available total spore and particle sampling devices, air is drawn through a tapered slit at relatively high velocities. Particles are inertially collected on a silicon grease- or adhesive-covered microscope slide. After staining and slide preparation, mold spores and hyphal fragments are enumerated at 1000× (oil immersion) magnification. Mold types can often be identified to genera (but not to species, as is the case with culturable/viable methods). Concentrations are usually based on a count of 5% of the deposition area and expressed as spores or structures per cubic meter (S/m³).

There are three impaction-type total mold spore/particle sampling devices commercially available. All three instruments draw particles through a tapered slitted orifice onto a greased impaction surface. Two are instruments with integrated pumps; the third, a sampling cassette. In the last case, particles are impacted on a greased microscope cover slip which is mounted on a slide and counted at 400 to 600× magnification. Its performance in providing accurate counts appears to be considerably less than devices with integrated pumps. The Burkard spore sampler, cassette sampler, and Allergenco spore sampler are pictured in Figure 9.9.

Not pictured here is a relatively new commercially available particle sampling device (button sampler) which can be used as a total mold spore sampler. It collects particles, mold spores, bacterial cells, etc. on a filter surface after passing through a porous cover which is designed to collect particles with maximum efficiency. Sample counts are made from the surface of light-transmitting filter surfaces. The button sampler is reported to have higher collection efficiencies for mold spores (such as *Cladosporium cladosporoides*) which have small aerodynamic diameters (≤2 μm). The nominal cutoff diameter for impaction samplers such as the Burkard and Allergenco is reported to be 2.5 μm.

Total mold spore and particle sampling devices have a significant advantage over culturable/viable samplers in one regard: because all mold particles are potentially allergenic, these devices are likely to better quantify health exposure risks. Because of the advantages and limitations of each, it is desirable to conduct concurrent sampling with culturable/viable and total mold spore/particle sampling devices to better interpret sampling results.

3. *Biological aerosol sampling considerations*

Though similar principles apply to all air sampling activities, there are sampling considerations unique to biological aerosols.

a. Limitations. The primary objective of biological aerosol sampling is to measure airborne concentrations of biological contaminants such as mold, bacteria, or actinomycetes in problem environments to determine whether measured levels are excessive or within acceptable limits. Unfortunately, there is no consensus as to what excessive or acceptable levels are. This reflects, in part, the fact that fluctuations in concentrations of an order of magnitude may occur in an environment over a period of days to months. In such a context, measured values, particularly at the low end of the range, are not meaningful and are even misleading. As a consequence, expert committees (such as the ACGIH Bioaerosols Committee) do not recommend the use of biological aerosol sampling results as a measure of exposure. Rather, sampling is recommended for the purpose of identifying sources in conjunction with building inspections. Nevertheless, biological aerosol sampling is widely used in problem building investigations and in evaluating the efficacy of remediation measures.

b. Area vs. aggressive sampling. There are two basic approaches to sampling airborne biological contaminants, area and aggressive sampling. Area sampling may be conducted to determine the concentrations and types of organisms that may be present in indoor spaces. It may also be conducted to determine the efficacy of remedial measures. Aggressive sampling is used to identify or confirm sources of mold/bacteria/actinomycete infestation by sampling after disturbance. Aggressive sampling may be used to evaluate the efficacy of remedial measures used in controlling toxigenic fungi.

c. Culturable/viable sampling.
i. Media selection. As indicated previously, media selection is a very important consideration in sampling biological aerosols. Media should reflect sampling objectives. It is important to understand the limitations and advantages of each medium used in biological aerosol sampling.

ii. Location and number. As with other airborne contaminants, sampling locations should reflect specific objectives. Samples should be collected from complaint and noncomplaint areas. Since outdoor air commonly has elevated biological contaminant levels that reflect outdoor sources, such samples should be collected to provide a reference for determining whether certain genera and species are being amplified in the indoor environment. Duplicate samples are recommended for each culture medium used at each location.

iii. Volume and duration.   As can be seen in Table 9.1, culturable/viable sampling devices vary in flow rates used and sampling duration recommended. Sampling rate is an important determinant of sampling duration, which may be influenced by biological aerosol concentration and the potential drying of culture media. Excessive collection of mold spores and particles may result in multiple mold particle deposition at the same collection site, overgrowth of colonies, colonies too numerous to count, and, even with multiple-hole correction procedures, a potential underestimation of airborne levels. It is often necessary to adjust sampling duration to assure a minimum level of detection in some cases and prevent media overload in others. Low sample volumes increase data variability; higher volumes obtained with longer sampling durations result in decreased sample counts due to the apparent desiccating effects of high velocity air flows on media and viable particles. Sampling durations of 0.5 to 2 minutes at a flow rate of 28.3 L/min are considered optimal.

iv. Disinfection.   Unique to biological aerosol sampling is the need to disinfect the sampler prior to field use. The sampler is subject to mold contamination on aerosolized media that deposits on sampler surfaces. Disinfection is commonly accomplished by wet wiping sampler surfaces (particularly interior surfaces) with a 70% ethanol solution.

v. Calibration.   As with other instruments, culturable/viable sampling devices need periodic calibration. Calibration must be conducted with devices that can measure high flow rates. A variety of calibration approaches have been recommended including pitot tubes, vane anemometers, and rotometers calibrated against primary or secondary standards. Wet test meters of sufficient volume work well under laboratory conditions. A wet test meter and an orifice cover connected to plastic tubing used to calibrate an N-6 Andersen sampler is illustrated in Figure 9.10.

vi. Handling and analysis.   After collection, samples can be incubated, enumerated, and individual organisms identified to specific taxa by the investigator, or as is commonly the case, sent to a laboratory that provides such services.

Mold sample plates are typically incubated at room temperature for 5 to 7 days before counts are made and taxa identified. Colony counts obtained from sieve plate impactor samplers (e.g., Andersen sampler) must be

Figure 9.10 Calibration of N-6 Andersen sampler with a wet test meter.

adjusted to account for multiple single site impactions when counts are relatively high (>50/plate); statistical charts have been developed for this positive-hole correction methodology. Selected positive-hole correction values are indicated in Table 9.2.

Sample handling for bacteria begins as soon as samples are collected. It is desirable to transport bacteria samples to the laboratory immediately, as delayed temperature control provides a selective advantage to organisms that grow at ambient temperatures.

d. Total mold spore/particle sampling. With the exception of media/culturing concerns and the need for disinfection, sampling considerations described for culturable/viable sampling above apply to total mold spore/particle sampling as well. These include flow rates, duration, and sample analysis.

i. Collection. Flow rates recommended for total mold spore sampling are in the range of 10 to 15 L/min. It is common to use a sampling duration of 10 minutes. In very dusty environments, particle overloading occurs and accurate counts of mold spores and particles cannot be made easily. In such environments, a sampling duration ≤ 5 min is recommended.

ii. Preparation. Samples can be prepared and counted microscopically within an hour after sampling. They do not, however, require immediate preparation and can be stored indefinitely for future microscopic evaluation and enumeration. Examination of slide samples at 1000× magnification is recommended because lower magnifications (circa 400 to 500×) result in significantly lower counts. Both spores and hyphal fragments are counted on a fraction of the collection surface (5%) using horizontal traverses. The count is converted to a concentration based on volume of air sampled, expressed as spores or structures per cubic meter (S/m^3).

Table 9.2 Selected Positive Hole Correction Values

Measured concentration (count/plate)	Corrected concentration (count/plate)
40	42
50	53
60	65
70	77
80	89
90	102
100	115
120	143
140	172
160	204
180	239
200	277
220	319
240	367
260	420
280	422
300	555
320	644
340	759
360	921
380	1198
400	2628

Source: Data extracted from Macher, J., *Am. Ind. Hyg. Assoc. J.,* 50, 561, 1989.

iii. Enumeration.   Enumeration of mold spores/particles on sample slides requires the use of a suitable collection medium (e.g., Dow Corning high vacuum grease, Dow Corning adhesive 280A, Dow Corning Silicone spray 316), a compatible mounting medium (aniline-blue-amended Calberla's solution adjusted to pH 4.2), a microscope with magnification capability of 1000×, a movable stage, and an ocular micrometer in the eye piece. It also requires technical skill to identify mold spores and particles against a heterogeneous mass of deposited particulate matter.

iv. Calibration.   Total mold spore samplers require frequent calibration since they are subject to significant flow rate changes. In the Burkard sampler, voltage has to be periodically adjusted to maintain the target flow rate near 10 L/min. Calibrations can be conducted using bubble-type calibration devices (Figure 9.11) or wet test meters connected to the top of the sampler using a specially designed rubber cork orifice.

Total mold spore/particle samplers have a major advantage over culturable/viable methods in that concentrations can be determined within an hour or so after collection. This makes them particularly suitable for sam-

Figure 9.11   Calibration of a total mold spore/particle sample device.

pling during toxigenic mold abatements and for clearance sampling when such abatements are completed.

    e. Biological aerosol sampling data interpretation.   As with gas and particulate-phase contaminants, the results of biological aerosol sampling must be amenable to meaningful interpretation. There is, however, little scientific evidence of a dose–response relationship between exposures and health effects. As a consequence, it is difficult to develop and establish guideline values for exposures to biological contaminants.

    Within certain limits, biological aerosol sampling results can be interpreted to indicate the potential nature of exposures. Based on genera identified and their relative abundance, samples can be evaluated to determine whether airborne mold spores are primarily associated with indoor or outdoor sources. They can also be used to determine the presence and relative abundance of mold taxa of concern such as species of *Aspergillus* and *Penicillium* and toxigenic species such as *S. chartarum* and *A. versicolor*.

    It is difficult to make firm conclusions about mold exposures based on one-time samples because of order-of-magnitude variations in airborne mold levels observed over time in building environments. This is particularly true when one-time sampling results are low. Comparisons to test results reported in the scientific literature for different building environments can be made to determine how individual results compare to those previously reported. Concentrations of airborne mold spores >500 CFU/m^3 or 5000 S/m^3 are not common in mechanically ventilated buildings operated under closure (windows closed) conditions. Concentrations above this are suggestive of an indoor infestation problem. Similarly, airborne mold concentrations >1000 CFU/m^3 or 10,000 S/m^3 are not common in residences under closure conditions. Higher concentrations and mold spora dominated by *Aspergillus* or *Penicillium* species are indicative of an indoor infestation problem with the potential for causing health effects as a result of exposures.

Table 9.3 Methods Used to Characterize Contaminants/Contaminant Emissions
Associated with Indoor Sources

Laboratory Studies

Extraction and Direct Analysis
  Provides information on material composition
  Does not provide information on actual emissions or rates
Static Headspace
  Provides qualitative and quantitative emissions data
  Does not provide data on emission rates
Static Equilibrium Tests
  Provides data on HCHO emission potentials from wood products

Dynamic Chamber Studies

Small Chambers
  Provides emission composition/rate data for controlled environmental conditions
  Chamber size limits use to small samples
Large Chambers
  Provides emission composition/rate data under controlled environmental
    conditions; used to evaluate emissions from large products
  Used to simulate real-world conditions

Full-Scale Studies

Test Houses
  Provides emission composition/rate data under semi-controlled real-world
    conditions; useful for validating chamber emission test results using IAQ models

Source: From Tichenor, B.A., Ed., *Characterizing Sources of Indoor Air Pollution and Related Sink Effects,* ASTM STP 1287, American Society of Testing Materials, West Conshohocken, PA, 1996, 10. With permission.

II. Source emissions characterization

A variety of techniques and systems are used to characterize emissions from indoor sources, emission rates, changes in emissions over time, and potential indoor concentrations. These are summarized in Table 9.3.

A. Laboratory methods

Several techniques are used to determine extractable compounds from source materials and to infer potential emissions. These may be directed to a single compound (e.g., HCHO) or multiple compounds (VOCs).

Extraction techniques typically attempt to identify and quantify substances present which are not intentionally a part of the product material (e.g., free HCHO associated with urea–formaldehyde resins, free styrene associated with styrene butadiene latex, etc.). Such free HCHO and styrene are extracted by solvents such as water, sodium bisulfite, or toluene in the first case, or an organic solvent in the second case. Extraction by methylene chloride or other solvents followed by gas chromatography/mass spectrometry has been used to identify volatile and semivolatile compounds in solid sources.

Solvent extraction and subsequent analytical quantification was widely used in the 1980s in attempts to characterize the potential potency of HCHO-emitting sources. Except for the perforator method (extraction with toluene), such techniques were not standardized nor were attempts made to relate extractable free HCHO concentrations to source emission potentials.

Wet products such as paints, varnishes, and other coatings can be sampled and analyzed directly to determine their composition as well as concentrations of potentially volatile or semivolatile substances. Concentrations are usually expressed on a weight-to-volume basis (with reference to the contents of either solids or liquids present or both). A standard method for the determination of VOCs from liquid or semiliquid compounds has been published by the American Society for Testing and Materials.

1. Static headspace analysis

Emissions from indoor sources can be better determined by using headspace analysis techniques, wherein samples of source materials are placed in a small airtight chamber made of or lined with inert materials (commonly glass). After a period of equilibration at room or elevated temperature (to increase emissions), samples are drawn from the chamber with a gas syringe and injected into a gas chromatograph linked to a mass spectrometer for analysis. In some cases, the chamber is purged with an inert gas, and contaminants are collected on a sorbent prior to analysis (purge and trap method). Headspace analyses continue to be widely used as a screening tool to identify potential contaminants of concern and their relative emission potential.

2. Static equilibrium chambers

Wood-product manufacturers use static equilibrium chambers and testing techniques on a day-to-day basis to evaluate emissions of HCHO from product batches. Samples of UF-bonded wood products are placed in a glass desiccator where, upon reaching equilibrium, the concentration of HCHO in an absorbing solution in a standard collecting vessel (usually a glass petri dish) is measured after a period of sample collection of 2 or 24 hours. Equilibrium chamber concentrations have been shown to be directly related to those obtained in large chambers. A simple linear regression equation is used to convert equilibrium chamber values to large chamber values.

3. Dynamic chamber testing

a. Small chambers. Emission rate data from source materials are determined using dynamic flow-through chamber testing. Usually, such determinations are made using small chambers, with volumes ranging from <1 L to 1.5 m³. Less commonly, large (15 to 30 m³) chambers are used. In either case, testing is conducted under carefully controlled conditions (temperature, humidity, airflow rate) so that emission rate data can be applied to typical indoor environment conditions. Testing is conducted by placing a sample of a source material in the chamber and measuring the concentration of individual substances at the chamber outlet or some well-mixed location. An exception to this is the widely used field and laboratory emission cell (FLEC) developed by Danish scientists. The stainless steel top of the FLEC chamber is placed on the surface of a source which becomes the bottom of the chamber. The size of the source sample is usually determined by expected loading factors under conditions of real-world use. A loading factor is described as the surface area of a source divided by the air volume of the space in which it is used. Loading factors are typically expressed as m^2/m^3 or, in the case of HCHO, ft^2/ft^3.

Concentration data are collected over a sufficient time interval to describe the history of emission rates. Emission rates may be determined for different chamber air exchange conditions or ventilation rates. They are often determined under very high ventilation rates to ensure that contaminant levels are above their LOD values.

b. Large chambers. Small chambers, while very useful in determining emission characteristics of many source materials, have obvious limitations. They cannot be used to test large assemblages such as furniture, work stations, office equipment, etc. Large (15 to 30 m^3) chambers are used to overcome these limitations but are much more expensive to construct, house, and operate. Nevertheless, they are the chamber of choice in emission and concentration studies, and testing when sources need to meet (1) performance criteria such as those required by the state of Washington for government office buildings or (2) compliance with the Department of Housing and Urban Development (HUD) HCHO product emission standards for particle board and hardwood plywood paneling used in the construction of manufactured housing.

B. Emission rates and rate modeling

Emission rates are expressed as mg/m^2 (hr) or µg/m^2 (hr). They are calculated from concentration data, the loading rate (source surface area divided by chamber volume), and air exchange rate using the following equation:

$$ER = C(N/L) \tag{9.4}$$

where ER = emission rate, mg/m^2(hr)
     C = chamber concentration (mg/m^3, µg/m^3)
     N = chamber air exchange rate, (hr^{-1})
     L = loading rate, m^2/m^3

Emission rates decrease with time. These time-dependent changes in ER are often modeled using a first-order decay equation.

$$ER = ER_0 e^{-kt} \tag{9.5}$$

where ER_0 = initial emission rate, $mg/m^2(hr)$
  e = natural log base
  k = first order decay constant (hr^{-1})
  t = time (hr)

When early emission rates are much faster than long-term emission rates, they may be better described by a double exponential model:

$$ER = ER_{01} e^{-k1t} + ER_{02} e^{-k2t} \tag{9.6}$$

where ER_{01} = the initial emission rate associated with evaporation, or in the case of UF-bonded wood products, free HCHO and HCHO released by hydrolysis
  ER_{02} = the emission rate associated with diffusion, or in the case of HCHO, UF resin hydrolysis
  k_1 = decay rate for initially rapid emissions (hr^{-1}, day^{-1})
  k_2 = decay rate for longer term emissions (hr^{-1}, day^{-1})
  t = time (hr, day)

Decay constants (k_1, k_2) in the double exponential model must be empirically derived.

As can be seen in Figure 9.12, the first-order exponential decay model predicts actual decreases in emission rates for HCHO emissions from particle board in a large dynamic chamber very well for the first couple of weeks, but poorly thereafter; the double exponential model, on the other hand, provides a much better fit for measured HCHO emission/concentration data over a 6-month period.

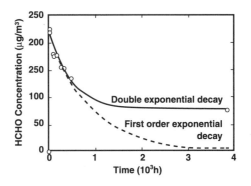

Figure 9.12 Emission decay rates of formaldehyde from particle board predicted by first order decay and double-exponential models. (From Brown, S.K., *Indoor Air*, 9, 209, 1999. With permission.)

Source emission models have also been developed that are based on mass transfer processes. Emission rates for wet sources (paint coatings, etc.) can be described by the simplified equation:

$$ER = K_g (C_s - C) \tag{9.7}$$

where K_g = mass transfer coefficient (m/hr)
   C_s = vapor concentration (mg/m³) just above the emitting surface
   C = chamber/room concentration (mg/m³)

The vapor concentration (C_s) is the vapor pressure expressed as a concentration in equilibrium with the source. It is linearly associated with the mass of source VOCs. The mass transfer coefficient (K_g) is determined from the vapor diffusivity, air velocity above the source, and source geometry. The calculation of the emission rate (ER) over time requires the use of a modeling equation of the form

$$ER = [C_v K_g/(r_1 - r_2)][(r_2 + N)e^{-r1t} - (r_2 + N)e^{-r2t}] \tag{9.8}$$

where coefficients r_1 and r_2 must be solved using another equation.

Emissions from source materials depend on a variety of factors. These include source moisture content and age, air exchange or ventilation rate, temperature, humidity (in some cases), loading rate, and sink effects. Wet sources, such as freshly applied coatings, have higher emission rates than dry products. Older sources have lower emission rates than new sources. Source materials subject to higher ventilation or air exchange conditions have higher emission rates and more rapid emission decay. Emission rates increase with temperature and, in the case of HCHO, this increase is exponential. Formaldehyde emissions increase linearly with increases in relative humidity. Humidity also appears to affect emissions of some VOCs.

The loading rate is a major determinant of emission rates, as can be seen in Equation 9.4. Since emission rate is inversely related to loading rate, high loading rates result in high chamber or space concentrations, which result in reduced source emissions.

The emission characteristics of many sources depend on sink effects. Sinks are indoor surfaces and materials that sorb or desorb contaminants generated by source materials. These sink effects result in lower initial air concentrations from which emission rate data are determined and, as a result of desorption, increase the emission decay period. In many cases, sinks are a source of contaminant emissions even after the primary emission source has been removed.

C. Full-scale studies

Full-scale or whole-house studies are occasionally used to validate chamber data. They provide the opportunity to evaluate the effect of variable air exchange rates, operation of heating and cooling systems, occupant activities, and interactions of source emissions with sinks. Such studies can be conducted on new unoccupied or existing lived-in houses. More commonly, new unoccupied houses are used for such evaluations. Full-scale studies are used to validate IAQ models.

D. IAQ modeling

The ultimate goal of source emissions assessment is to predict the impact of emissions of individual contaminants on IAQ. Such predictions require the development and use of single-room, whole-house, and large-building models. Because of the complexity involved, model development has been limited to single-room and residential house conditions.

Though models have been developed by USEPA and tested under real-world conditions, their performance at this time is relatively poor. This is likely due, in part, to the sink phenomena whose effects on air concentrations are not well understood. Indoor air quality modeling can best be described as being in its infancy, requiring considerable continuing research and development efforts.

Readings

American Conference of Governmental Industrial Hygienists, Bioaerosols Committee, *Guidelines for the Assessment of Bioaerosols in the Indoor Environment*, ACGIH, Cincinnati, 1989.

American Society for Testing and Materials, *Standard Guide for Determination of VOC Emissions in Environmental Chambers from Materials and Products*, ASTM, Philadelphia, 1990.

Cox, C.S. and Wathes, C.M., *Bioaerosols Handbook*, CRC Press, Boca Raton, 1995.

Dillon, H.K., Heinsohn, P.A., and Miller, J.D., Eds., *Field Guide for the Determination of Biological Contaminants in Environmental Samples*, American Industrial Hygiene Association, Fairfax, VA., 1996.

Hodgson, A.T., A review and a limited comparison of methods for measuring total volatile organic compounds in indoor air, *Indoor Air*, 5, 247, 1995.

Maroni, M., Siefert, B., and Lindvall, T., *Indoor Air Quality: A Comprehensive Reference Book*, Elsevier, Amsterdam, 1995, chaps. 23, 24, 26, 27.

McCarthy, J.E., Bearg, D.W., and Spengler, J.D., Assessment of indoor air quality, in *Indoor Air Pollution — A Public Health Perspective*, Samet, J.M. and Spengler, J.D., Eds., Johns Hopkins University Press, Baltimore, 1991, 82.

Nagda, N.L. and Harper, J.P., Eds., *Design and Protocol for Monitoring Indoor Air Quality*, ASTM STP, American Society for Testing and Materials, Philadelphia, 1989.

Spengler, J.D., Samet, J.M., and McCarthy, J.F., Eds., *Indoor Air Quality Handbook*, McGraw-Hill Publishers, New York, 2000, chaps. 51, 52, 56, 58.

Thorsen, M.A. and Molhave, L., Elements of a standard protocol for measurements in the indoor atmospheric environment, *Atmos. Environ.*, 21, 1411, 1987.

Tichenor, B.A., Ed., *Characterizing Sources of Indoor Air Pollution and Related Sink Effects*, ASTM STP 1287, American Society of Testing and Materials, Philadelphia, 1996.

USEPA, *Compendium of Methods for the Determination of Air Pollutants in Indoor Air,* EPA 600/S4-90-010, USEPA, Washington, D.C., 1990.

USEPA, *Indoor Radon and Radon Decay Product Measurement Device Protocols,* EPA 402-R-92-004, USEPA, Washington, D.C., 1992.

USEPA, *Protocols for Radon and Radon Decay Product Measurements in Homes,* EPA 402-R-93-003, USEPA, Washington, D.C., 1993.

Wallace, L.A., Volatile organic compounds, in *Indoor Air Pollution — A Public Health Perspective,* Samet, J.M. and Spengler, J.D., Eds., Johns Hopkins University Press, Baltimore, 1991, 253.

Wolkoff, P., Some guides for measurement of volatile organic compounds indoors, *Environ. Tech.,* 11, 339, 1990.

Yocum, J.E. and McCarthy, S.M., *Measuring Indoor Air Quality — A Practical Guide,* John Wiley & Sons, New York, 1991.

Questions

1. Describe why environmental sampling is conducted in building investigations.
2. Describe what the terms *sample* and *sampling* actually mean in the context of measuring indoor contaminants.
3. What are the advantages and disadvantages of routine screening measurements?
4. What is bulk sampling? When is it appropriate?
5. How do accuracy and precision differ?
6. Distinguish between sensitivity and specificity.
7. Characterize dynamic and passive sampling. What are the advantages and disadvantages of each?
8. Distinguish between real-time and integrated sampling. Describe advantages of each.
9. Under what circumstances would one use gas sampling tubes in IAQ/IE investigations? What are their limitations?
10. What is the significance of the term LOD in conducting indoor environment measurements?
11. What is one-time sampling? What are its limitations?
12. What is calibration? Why is frequent calibration desirable?
13. Describe the use of field and media blanks in conducting field measurements.
14. When conducting IAQ/IE measurements in buildings, what factors should be considered when selecting sampling locations?
15. In conducting a measurement of an indoor contaminant, what is the significance of time?
16. Identify factors that determine sampling duration.
17. How many samples should one collect during the conduct of a building investigation?
18. When would it be desirable to conduct surface sampling?
19. Describe several surface sampling techniques. What are their respective advantages?
20. What sampling approach would be most informative in determining whether *Stachybotrys chartarum* was a problem in a building?

21. Describe how you would sample the following gas-phase substances in a building:
    a. CO
    b. CO_2
    c. VOCs
    d. HCHO
22. What methods can be used to measure RSP levels in a building?
23. Describe the advantages and disadvantages of culturable/viable and total spore/particle sampling methods.
24. What is the significance of each of the following when conducting culturable/viable mold sampling:
    a. Media used
    b. Sampler disinfection
    c. Sampling duration
    d. Colony counting/identification

chapter ten

Source control

As indicated in previous chapters, a variety of contaminants are generated within, drawn in as a result of infiltration and ventilation, or passively transported into indoor environments. Because of potential health risks and other factors such as comfort and odor, it may be desirable to control either airborne or surface contaminants or both. There are two primary approaches to controlling indoor contaminants. These include controlling the contaminant at the source (source control) or controlling contaminants once they are produced or become airborne (contaminant control). Contaminant control measures, ventilation, and air cleaning are discussed in detail in Chapters 11 and 12.

It is easier, in theory, to control a contaminant at its source before it becomes airborne or causes significant indoor contamination by other mechanisms. Source control includes a variety of principles and applications based on individual contaminants and the nature of contamination problems. These include (1) measures that prevent or exclude, in some way, the use of contaminant-producing materials, furnishings, equipment, etc., in indoor environments; (2) elements of building design, operation, and maintenance that prevent or minimize contamination; (3) treatment or modification of sources directly or indirectly to reduce contaminant production and/or release; (4) removal of the source and replacement with materials with low or no contaminant production; (5) measures that prevent the infestation of indoor environments by biological organisms; and (6) removal of surface contaminants using cleaning measures.

Though source control is often described as an indoor contaminant control measure in a generic sense, its application depends on both sources and contaminants they produce. In most applications, source control is case specific.

Source control is typically used to prevent contamination problems from occurring in the construction, furnishing, and use of built environments; to prevent contamination associated with building renovation and abatement activities; and to reduce contaminant levels when screening measurements or a building investigation reveal that a contaminant-related problem exists. In the latter case, source control measures are implemented to mitigate an existing problem.

I. Prevention

It is more desirable to prevent indoor contamination problems than to mitigate them once they have occurred. Such problems occur as a result of (1) manufacturer inattention to, or denial of, potential health and safety problems involving the normal use of products; (2) consumer choices relative to the use of products; (3) decisions made by facilities personnel and homeowners; and (4) building design, construction, and operation and maintenance practices.

A. Manufacturing safe products and product improvement

Many products are used to construct, furnish, and equip indoor environments. Most manufacturers do not knowingly (at least initially) produce products that will pose minor or significant health risks to those who use them. When health risks do occur, they are inadvertent or unintended. Some products can be anticipated to pose potential health risks based on known toxicities and exposure potentials associated with hazardous or toxic components. Unfortunately, manufacturers did not address the potential health risks associated with asbestos, lead, and formaldehyde (HCHO). As health risks from exposures to such indoor contaminants became known, manufacturers often chose to deny that their products were harmful.

In the case of asbestos-containing building materials (ACMs) and lead-based paints (LBPs), exposures in new buildings and houses were reduced by regulatory prohibitions on the use of asbestos in building materials and lead in paint. In the case of HCHO, exposures in new housing were reduced as a consequence of regulatory limits on particle board and hardwood plywood use in new mobile homes and by voluntary industry efforts to reduce HCHO emissions from products.

Significant reductions in contaminant exposures can be achieved by the development of low-emission or no-emission products. Such improvements have been achieved for HCHO-emitting urea–formaldehyde-bonded wood products by changing manufacturing processes. These included changes in resin formulation and production variables, the addition of HCHO-scavenging compounds, attention to quality control, and use of various post-production steps.

Significant reductions in HCHO emissions from wood products were achieved by changing the molar HCHO-to-urea ratio (F:U) from 1.5:1 (commonly used in the 1970s and early 1980s) to 1.05:1. Additional reductions were achieved by changing process variables to decrease wood moisture levels and increase press temperature and time, and by adding HCHO-scavenging agents such as urea, ammonium compounds, and sulfites. The resultant decline in HCHO emissions was approximately 90+%.

Significant reductions in total volatile organic compound (TVOC) and 4-PC emissions from carpeting, and TVOC emissions from carpet adhesives, were made voluntarily by U.S. manufacturers in response to USEPA's carpet initiative.

As indicated in Chapter 7, health concerns associated with carbonless copy paper (CCP) have been reported for over two decades. Manufacturers, in response to scientific studies that implicated individual problem chemicals, changed product formulations. As a consequence, CCP products no longer contain Michler's hydrol of paratoluene sulfonate, phenyl novalac, and contain only very limited quantities of HCHO.

Other product improvements have included the voluntary elimination of mercury biocides in latex-based paints intended for use indoors and efforts by the wood-preservatives industry to limit pentachloraphenol use to outdoor wood products. Product improvements initiated by regulatory action have included the banning of asbestos for use in building products and limits on the lead content in paint.

B. Consumer avoidance

Exposure to indoor contaminants can be avoided by consumers of products and in the purchase or lease of building environments. Such avoidance can be total or selective. Consumers, however, must know what products and indoor environments to avoid and what alternative products and environments are available.

1. Products

For simplicity's sake, it would be desirable to have products and materials evaluated, rated, and labeled relative to their potential to cause health and indoor environment problems. However, because of the uncertainties inherent in such an undertaking, it would be very difficult to compile a list of problem and nonproblem products. In the U.S., such a process would, in many cases, not be able to sustain the legal challenges it would engender.

Nevertheless, sufficient information is available for informed consumer choices on a limited number of products. This is particularly the case with combustion appliances such as (1) unvented space heaters, (2) gas stoves and ovens, and (3) non-airtight wood burners. In each case, these products produce contaminants which, on short- or long-term exposure, may pose significant health risks. At greatest potential risk would be small children and individuals with a family history of respiratory disease such as asthma.

Unvented gas fireplaces and the burning of candles or incense pose relatively new exposure concerns. Though little scientific information is available on these potential indoor contamination and exposure problems, they have the potential for producing significant indoor emissions and exposures when used on a regular basis. As a consequence, regular candle burning should be avoided to prevent carbon deposition and soiling of interior building surfaces.

Total or partial avoidance can be applied to the use of UF-bonded wood products. Despite significant reductions in emissions from particle board, hardwood plywood, and medium-density fiber board (MDF), use of such products in high volume has the potential for producing elevated health-effecting HCHO levels. Such elevated HCHO exposures are likely to occur in new mobile homes (approximately 50% of new mobile homes are constructed with particle board decking).

Avoidance can also be applied to the use of CCP products. Avoidance would be appropriate for individuals who have a propensity for developing rashes and allergy symptoms. Total avoidance would be desirable when clerical workers develop severe symptoms on exposure. Less sensitive individuals may choose to minimize (but not totally avoid) contact with CCP materials.

Other products consumers may wish to avoid or use only under certain conditions include cool mist humidifiers, biocidal materials, fiberglass ductboard, and certain air cleaning devices. Significant microbial growth in water reservoirs of cool mist humidifiers, with subsequent aerosolization, commonly occurs when units are not cleaned regularly. Sonic humidification devices pose another indoor contamination problem, i.e., deposition of mineral dusts on interior building surfaces when water with a moderate-to-high mineral content is used. Such mineral deposits can be avoided only by using distilled water. Most consumers do not, however, use distilled water as recommended. Cool mist humidifiers should, in most cases, be avoided. Steam humidification in residences poses fewer risks of indoor contamination.

Fiberglass ductboard is widely used to form supply air trunklines and return air ducts in residences. Some products may produce irritating odors in response to heating system activation. Because such duct material is porous, it has the potential for collecting organic dust that can serve as a medium for microbial growth and subsequent building contamination. These potential problems may be avoided or reduced by the use of galvanized steel ductwork.

Biocides and pesticides are often used indoors. Their use should be limited to circumstances where major infestation problems must be controlled or can be controlled without significant biocidal exposures. Major infestations may include cockroaches, fleas, or ants. To minimize indoor contamination with cockroach-controlling pesticides, generalized or broadcast spraying should be avoided. It is more desirable and appropriate to use crack and crevice spray applications or the use of poison baits. Poison baits can also be used for ant control. Biocides are often used to treat interior residential duct surfaces after duct cleaning. The use of biocides for such applications is not known to have any useful purpose, and because commonly used biocides such as glutaraldehyde are potent irritants, their use should be avoided.

Air cleaning products that generate ozone (O_3) either deliberately or incidentally (electronic air cleaners) pose a risk to the health of exposed individuals as well as to building materials when used continuously. They also contribute to indoor chemical reactions. Complete avoidance of deliberately generated O_3, and limited use of equipment which incidentally generates it, would be appropriate.

2. Buildings

A residence in its totality may be considered a product. It is a product about which millions of Americans each year make major purchase and leasing decisions. Factors that primarily determine purchasing and leasing decisions include cost, size, appearance, and location. Environmental factors are less commonly considered. These include radon levels in certain eastern states and the potential for lead exposure to infants and young children. Increasingly, decisions are being made to avoid or accept risks associated with elevated radon levels and LBP based on radon test results and lead inspections or risk assessments, respectively.

Most potential purchasers/leasers of residences are unaware that many residential building units available in the marketplace are contaminated with common indoor allergens such as dust mite fecal matter, mold, and pet danders. Such exposure risks can be ascertained by professionals and, in many cases, lay individuals before purchase. Mold infestation and high dust mite populations are associated with moisture problems, and houses with mold infestation often have a characteristic musty odor. Individuals with a family history of asthma or allergy may choose to avoid such exposures by purchasing or leasing a newer dwelling on a dry site. Residential units with a significant history of indoor pets may not be a good choice for atopic individuals. Complete avoidance or remediation would be desirable.

C. Designing and constructing "healthy buildings"

Significant indoor contamination problems can, in theory, be prevented or avoided in new buildings by using appropriate design and construction practices. Designing and constructing buildings, particularly large nonresidential buildings, to achieve a "healthy building" environment is a significant undertaking fraught with many uncertainties. Nevertheless a number of "healthy" or "green" buildings have been designed and constructed using various design principles.

1. Identifying low-emission/low-toxicity products

A major imperative in designing and constructing truly "healthy buildings" is the identification and selection of low-emission/low-toxicity construction materials and furnishings. Such designs and construction must depend on a very limited source characterization database that has been developed in the U.S. and Northern Europe in the past decade. The use of product emission data is problematic since there is considerable uncertainty whether data

Table 10.1   Recommendations for Manufacturers/Suppliers of Products
                    Emitting Air Contaminants

1. Conduct testing of emission rates from:
    Coatings such as paints, varnishes, waxes
    Vinyl and fabric floor/wall coverings
    Adhesives
    Furniture/furnishings with pressed wood/fabrics
    Ductwork materials
    Office equipment/supplies
    Building maintenance materials
2. Provide MSDSs for chemicals used/products manufactured
3. Provide emission testing data for:
    Major organic compounds emitted
    Three product ages
    Compounds that are toxic or irritating at air concentration ≤ 5 mg/m^3
    Office machines
4. Provide documentation of:
    Chamber testing conditions
    Product storage and handling procedures

Source: From Tucker, W.G., *Proc., 5th Interntl. Conf. Indoor Air Qual. & Climate,* Toronto, 3, 251, 1990.

from small chambers can be reliably extrapolated to larger, more complex environments. Additional concerns include the relative significance of individual chemical species compared to TVOC (total volatile organic compounds) concentrations. In addition, toxicological and health effects information that would facilitate the use of product emission characterization data is lacking.

USEPA research scientists and engineers have suggested policy initiatives that would establish both a significant database for use by building designers and give manufacturers an incentive to voluntarily improve products. Under such an initiative, manufacturers and suppliers of building products, furnishings, and office equipment would be expected to comply with the needs described in Table 10.1. Based on data from manufacturers, indoor contaminant concentrations could be modeled and potential occupant exposures and health risks evaluated.

A low emission characterization of various building materials, furnishings, and office equipment based on an engineering assessment and the TVOC theory of exposure and mucous membrane irritation is summarized in Table 10.2. Under this classification, the maximum acceptable TVOC concentration from any one source would be 0.5 mg/m^3. The policy initiative, though not formally proposed by USEPA, provides a theoretical framework for the characterization of source emissions and potential human exposures associated with TVOC emissions from building products and furnishings. It would, in theory, provide an extensive database from which building designers could select low-emission and low-toxicity construction materials.

Table 10.2 Low TVOCs Recommended Emission Limits
for Building Materials and Furnishings

Material/product	Maximum acceptable emission rate[a] mg/h/m²
Flooring materials	0.6
Floor coatings	0.6
Wall coverings	0.4
Wall coatings	0.4
Movable partitions	0.4
Office furniture	2.5 mg/h/workstation

[a] Assumptions: air exchange = 0.5 ACH; maximum increment from one source = 0.5 mg/m³.

Source: From Tucker, W.G., *Proc., 5th Interntl. Conf. Indoor Air Qual. & Climate*, Toronto, 3, 251, 1990.

In the state of Washington's *Healthy Buildings* program, manufacturers of materials, furnishings, and finishes must provide emission testing information with their bids to ensure compliance with IAQ specifications, as well as emission profile data which details how product emissions change over time. The designer/builder of a Washington state office building must develop and implement an indoor source control plan and assure that maximum allowable air concentrations are not exceeded.

a. Target products. Since hundreds of products are used in most building projects, it would be difficult to evaluate all products to assure that they do not produce harmful emissions. It is therefore desirable to identify and evaluate those products (target products) that are more likely to pose significant indoor contamination and exposure problems. Such problems are likely to be related to product emission characteristics and the quantity and nature of materials used. In identifying target products, consideration is given to the overall building design, anticipated use of the space, material and products to be selected, and quantities and applications anticipated for each major product.

In selecting target products, emphasis is given to those materials which have large surface areas such as textiles, fabrics, and insulation materials. Materials considered to be significant contaminant sources because of their surface area include floor coverings, ceiling tiles, horizontal office workstation surfaces, and workstation partitions. Using floor area as a reference, the relative surface area for different floor coverings may vary from a fraction to 100%; ceiling tiles that serve as a decorative ceiling surface and base for return air plenums, 200%; workstation furniture, 15 to 35%; and interior workstation partitions, 200 to 300%.

Emissions testing of target products is essential to determine types of compounds emitted, emission rates, and changes in emissions due to environmental conditions. The burden of testing should, in theory, be borne by manufacturers and suppliers. At the present time, product testing is only conducted by a limited number of large manufacturers.

b. Emission labeling.   Another approach to the emission characterization problem is the labeling concept being pursued by the Danish Ministry of Housing. Danish scientists have made considerable progress in developing the scientific background for a product labeling system based on product impact on indoor air quality (IAQ). The Danish program combines emission testing over time, modeling, and health evaluations. The principal objective is to determine the time required (months) to attain an acceptable indoor air concentration for odor or mucous membrane irritation. The time value is then used to rank various products evaluated for their potential impact on IAQ and the health of those exposed.

2. Other design concerns

Identification and selection of low-emission products is a key element in designing a "healthy building." At present, building designers must select products based on very limited information and intuitive judgments as to what products may be acceptable. Design factors include site/planning/design, a variety of architectural considerations, and ventilation/climate control. Potential outdoor sources can be evaluated prior to site acquisition and project planning to avoid ambient air pollution problems nearby. Other design features include (1) placing motor vehicle access to garages, loading docks, and pedestrian drop-off points away from air intakes and building entries; (2) locating air intakes upwind of building exhausts and outdoor pollutant sources; (3) designing and specifying rooftop exhaust systems to minimize entrainment in the building wake and building re-entry; (4) locating pollutant generation activities such as printing, food preparation, tobacco smoking, etc., in areas where airflow can be easily controlled to avoid cross-contamination with adjacent spaces or to recirculating air; and (5) selecting HVAC (heating, ventilation, and air conditioning) system equipment that minimizes the deposition of dust on duct surfaces (avoid using porous insulation inside ductwork), provides improved drainage from condensate drip pans of fan coil units, and specifies steam rather than cool mist humidification. Steam humidification systems should be designed to operate without the use of volatile or semivolatile, and potentially irritating, boiler additives.

D. Building operation and maintenance

Significant, or even incidental, indoor contamination problems can be prevented in both residential and nonresidential buildings by implementing good building operation and maintenance (O&M) practices. In most cases these do not require special knowledge. They do, however, require that homeowners and facilities personnel operate and maintain building systems and environments to the standard for which they have been designed.

The operation and maintenance of a single-family residential building should in theory be a relatively simple task since, in most cases, it is owner occupied. Residential building O&M becomes more complex when occupancy is based on leasing contracts and in multifamily buildings.

1. Residential buildings

Residential indoor environmental contamination problems often result from inadequate or improper maintenance. Typically these include heating system operation, water damage, and building moisture problems. In older residences, it may include lead dust contamination associated with deteriorated LBP.

Heating systems based on the use of combustion appliances must be properly installed and maintained to prevent flue gas spillage during normal operation and when appliances deteriorate with age. Maintenance concerns include perforated heat exchangers and flue pipes, disconnected or partially disconnected flue pipes, obstructed chimneys, inadequate draft, etc. The likelihood of flue gas spillage increases with building system age since the likelihood of deterioration and malfunctioning increases. Maintenance of vented combustion appliances helps to assure that significant flue gas spillage will not occur. Good maintenance is also necessary to prevent mold infestation problems in residences. Such concerns are described in a contaminant-specific section of this chapter.

In residences built before 1978, it is good practice to maintain all painted surfaces in good condition and minimize contamination of exterior ground surfaces by LBP removed in preparing exterior surfaces for repainting. In building rehabilitation involving LBP, paint should not be removed by sanding or high-temperature paint removing devices.

2. Nonresidential buildings

Good O&M practices are also important in preventing indoor environmental problems in nonresidential buildings. They do, however, differ in scope (nonresidential buildings are more demanding in their O&M requirements, and building systems are often more difficult to maintain). In most cases, nonresidential buildings are operated and maintained by full-time facilities staff. Depending on individual circumstances, facilities staff may be poorly, minimally, or well-qualified to operate and maintain buildings and their mechanical systems. Because of staff and budget limitations, many nonresidential, nonindustrial buildings are poorly operated and maintained. In small or poorly funded school systems, custodians may be responsible for operating mechanical systems despite the fact that they are not adequately trained to do so. School corporations operating under significant budget restrictions often drastically cut maintenance budgets and defer important maintenance projects.

Air quality and comfort in mechanically ventilated buildings depend on the proper operation of mechanical ventilation and exhaust systems. This requires that HVAC systems be operated during periods of occupancy to provide a minimum of 20 CFM (9.5 L/sec) per person outdoor air to office building spaces and 15 CFM (7.14 L/sec) per person outdoor air in school buildings. Achievement of these requirements necessitates that mechanical systems be adequately maintained to provide desired air flows throughout the system, that adequate control be maintained over ventilation system air flows, and that facilities staff are sufficiently trained in the operation of HVAC systems.

HVAC system operation requires maintaining balanced air flows by (1) using correct damper settings, (2) ensuring that filters are changed frequently to maintain desired system flow rates, (3) ensuring that all equipment is operating properly, and (4) keeping condensate drip pans open and relatively clean of microbial growth. Outdoor air flows should be sufficient to balance or more than compensate for building exhaust to minimize re-entry and entrainment problems.

As indicated in Chapter 7, surface dust is apparently a major risk factor for SBS-type symptoms. As such, surface dust can only be reduced and maintained at acceptable levels by scrupulously cleaning horizontal building surfaces. Such cleaning by service personnel is inadequate in most buildings.

In a survey study of school teachers conducted by the author in Indiana, three factors were observed to be significantly and independently associated with SBS-type symptoms reported. These were inadequate ventilation, mold infestation, and surface dustiness or inadequate cleaning. Each of these school SBS-type symptom risk factors is directly associated with improper or inadequate building O&M practices.

A variety of other O&M practices can be implemented in buildings by facilities management to minimize contamination problems and occupant complaints. These include (1) scheduling renovation activities (such as painting) on days when the building is unoccupied (e.g., summer in the case of schools) or using high ventilation rates when renovation activities are conducted during occupancy; (2) wet shampooing carpeting under well-ventilated building conditions; (3) limiting the use of insecticidal applications for cockroach control to the crack and crevice method or poison baits, or, more appropriately, employing integrated pest management; (4) waxing floors after hours under ventilated conditions; and (5) using low-volatility/low-toxicity boiler additives in steam humidification.

Specific building O&M practices designed to manage potential asbestos and lead hazards in place, as well as biological contaminants, are described later in this chapter.

II. Mitigation measures

Source control measures have been described in the context of preventing or avoiding indoor contamination problems. Though highly desirable, source avoidance or prevention principles are, in many cases, not employed. As a consequence, source control measures must often be implemented to reduce contaminant exposures and resolve health, comfort, and odor complaints. A variety of measures may be employed on a generic basis or applied to specific contaminant problems. These include source removal and replacement, source treatment and modification, and climate control.

A. Source removal and replacement

In theory, source removal and replacement should be the most effective measure in reducing contaminant concentrations and human exposure. Sources, however, must be correctly identified and low-/no-emission replacements must be available for source removal to be effective. It is important that removal and replacement costs are reasonable, particularly when an indoor environment (IE) problem has low to moderate seriousness.

In some cases, materials that emit significant levels of HCHO (such as particle board flooring, MDF or particle board workstation surfaces, pressed wood cabinets) can be removed and replaced with materials that either do not emit HCHO and or have low emission levels. Alternative materials include softwood plywood and oriented-strand board.

Source removal associated with HCHO emissions from pressed wood products has historically posed significant challenges. Among these are interaction effects associated with the presence of multiple HCHO sources. In multiple source environments, it is common for a source with high emission potential (dominant source) to suppress emissions from other sources. Removal and replacement of lower-emission-potential sources (which may be present at relatively high loading rates) without simultaneously removing the dominant source typically fails to reduce HCHO levels to the degree expected when emission potentials of individual sources are considered singly. Other challenges involve mobile homes where interior and exterior walls are fastened to particle board floor decking. In such cases, removal and replacement is not practical. In office environments where pressed wood-based furniture and workstations are the major source of HCHO, removal and replacement with other materials would, in many cases, be considered excessively costly.

In the case of office equipment, removal and replacement of wet process photocopiers which have significant VOC emissions may be a desirable mitigation measure. Removal of badly soiled carpeting and its replacement with floor tile has been reported to result in a significant reduction in occupant symptoms in Swedish schools.

Source removal has seen its most significant use in efforts to control or prevent exposures to asbestos and lead. A further discussion of these is included in the contaminant-specific source control section of this chapter.

B. Source treatment and modification

Contaminant emissions from a limited number of sources can be reduced by (1) the application of materials and coatings that serve as diffusion barriers or contaminant scavengers, (2) exposure to gas-phase substances, and (3) heat treatment.

1. Diffusion barriers and surface coatings

Several source treatment techniques for HCHO emissions from pressed wood products have been evaluated and received limited use in reducing residential HCHO levels. European investigators have evaluated the efficacy of surface treatments and barriers on HCHO emissions from particle board under laboratory chamber conditions. Significant reductions in emissions were observed for vinyl floor covering, vinyl wallpaper, HCHO-scavenging paints, polyethylene foil, and short-cycle melamine–formaldehyde paper. Effectiveness was, however, observed to decrease with time. Several U.S. studies evaluated the effectiveness of potential HCHO barriers placed on particle board underlayment in controlled whole-house conditions. Vinyl linoleum and 6-mil polyethylene sheeting were observed to be effective HCHO barriers, resulting in a reduction of emission rates and air concentrations on the order of 80 to 90%. The effectiveness of several finish coatings and HCHO-scavenging paints applied to particle board underlayment under whole-house conditions were evaluated in other studies. Treatment effectiveness was highest for two applications of nitrocellulose-based brushing lacquer (70%), followed by alkyd resin varnish (53%), and polyurethane (43%). Variable performances were observed when three specially formulated HCHO sealant coatings were evaluated. The most effective product was observed to reduce whole-house HCHO levels by 78 to 87% on a long-term basis with but one application. Another was less effective (57 to 67%), and a third was ineffective with one application but reduced HCHO concentrations by 65% after two applications.

Based on these studies, it appears that source treatment in the form of surface coatings and physical barriers can reduce indoor HCHO levels associated with particle board underlayment. Such measures have not been evaluated for more complex HCHO-emitting sources such as kitchen and bath cabinetry, workstations, desks, furniture, tables, storage cabinets, etc. These materials are, in many cases, covered with paper overlays, vinyl coatings, plastic laminates, or clear finish coatings. They also contain raw surfaces including numerous joints which are not accessible to treatment. Wood furniture and cabinetry are often coated with acid-cured finishes which are significant sources of HCHO. It is unlikely that source treatment would be effective in controlling HCHO emissions from such complex sources in buildings.

2. Ammonia fumigation

During the late 1970s and early 1980s, Weyerhauser Corporation, a major wood-products manufacturer, developed and evaluated a technique for reducing HCHO levels in mobile homes based on the principle of using gaseous ammonia (NH_3) to react with HCHO and methylol end groups in U-F bonded wood products. On reaction with NH_3, free HCHO in the resin (as well as that which was airborne) formed hexamethylene tetramine. Methylol end groups on the U-F polymer were stabilized and were less susceptible to hydrolysis.

In conducting an NH_3 fumigation, a mobile home or other residential environment is exposed to high NH_3 concentrations produced by a strong ammonium hydroxide solution (28 to 30% NH_3) placed in the closed building at 80°F (26°C) for a period of 24 to 48 hours. Ammonia fumigations in real-world evaluations have been observed to be effective in building environments with high HCHO concentrations (≥ 0.20 ppmv), resulting in short-term reductions of HCHO in the range of 58 to 75%, and longer-term reductions on the order of 40 to 70%. They have been ineffective in building environments with relatively low HCHO concentrations (≤ 0.10 ppmv). Such concentrations appear to be the practical lower limit for HCHO reduction.

Ammonia fumigations were used by mobile home manufacturers and others to mitigate health and odor complaints in new mobile homes. Their primary advantage was that significant reductions in HCHO levels in buildings with large quantities of HCHO-emitting products could be achieved at low cost and with minimum disturbance. Major disadvantages included (1) the potential to cause stress corrosion and cracking of brass connections on critical appliances such as gas stoves, water heaters, and furnaces, as well as electrical connections; (2) the tendency to darken light oak cabinets; (3) the inability to reduce HCHO levels below 0.10 ppmv on a permanent basis; and (4) the persistence of irritating NH_3 odors up to several weeks after treatment.

3. Building bakeout

Building bakeout is a source treatment procedure developed in the early to mid-1980s to reduce VOC concentrations in new office buildings. It is based on the principle that the normal decrease of VOC levels with time can be accelerated by elevating building temperature for several days. Elevated temperature causes increased VOC emissions because of increased vapor pressures of residual solvents and monomers in finish coatings, caulks, adhesives, etc. As a result of elevated temperature and increased ventilation rate, bakeouts attempt to increase emission rates and reduce product emission potential.

The effectiveness of a building bakeout depends on several factors. These include the maintenance of building temperature in the range of 86 to 95°F (30 to 35°C), a bakeout duration of at least several days, an optimum ventilation rate, and initially high VOC concentrations. It is desirable to optimize temperature, duration, and ventilation to achieve maximum VOC reduction. Achieving optimal conditions is, however, difficult.

Complicating the task of conducting bakeouts are problems in attaining the desired temperature range for a sufficient duration. Constraints include the inability of some HVAC systems to reach desired temperatures without supplemental heating, concerns that elevated temperatures may damage building materials, and difficulties in providing sufficient ventilation to flush emitted VOCs (so they are not readsorbed by building materials) from the building environment.

The use of outdoor ventilation air makes it more difficult to attain and maintain optimal bakeout temperatures. In many instances, users of the bakeout procedure have had to minimize the use of ventilation during the bakeout period to achieve these temperatures, reasoning that temperature was more important in reducing VOC concentrations. Because of the large thermal masses involved, 2 to 3 days are usually required to reach desired temperatures.

Typically, the achievement of desired bakeout temperatures and their maintenance for 48 hours is constrained by the availability of the building for a sufficient duration. The time period available to conduct a bakeout is limited by ongoing construction activities and the desire of building owners to occupy spaces as soon as possible. In multiple-story building projects, construction activity varies, with some floors completed and occupied while others are in various stages of construction. Building bakeouts in such cases must be conducted on a floor-by-floor basis. Construction activities significantly complicate the process, making it more difficult to standardize bakeout conditions.

The efficacy of building bakeouts has been evaluated by a number of investigators. Results have been mixed. Most investigators report reductions in VOC levels in the range of 30 to 75%; some claim bakeouts were ineffective. In general, bakeouts appear to be most effective when initial VOC levels are high and optimal temperatures and duration are achieved.

C. *Climate control*

Climate control as a source control measure is based on the assumption that source emissions, and therefore air concentrations of target contaminants, can be reduced by controlling environmental conditions such as temperature and relative humidity. The effectiveness of climate control on reducing HCHO levels in a mobile home can be seen in Table 4.5. At the coolest temperature and lowest relative humidity (20°C, 30% RH), the HCHO concentration was 20% of that observed at the highest temperature and humidity conditions (30°C, 70% RH).

In theory, climate control can be used to control indoor HCHO and VOC levels; to the author's knowledge, it has never been deliberately used for this purpose. Nevertheless, lower temperatures employed by homeowners during the wintertime in colder climates has had the indirect effect of reducing indoor contaminant levels and exposures during that time. The relationship between IE conditions and SBS-type symptom prevalence rates was discussed in Chapter 7. Increased symptom reporting rates at high building temperatures is consistent with increased emissions and exposures to HCHO and VOCs at such temperatures.

III. *Contaminant-specific source control measures*

Source control is, in most cases, the preferred approach to reducing potential exposures to asbestos and lead in buildings and mitigating building-related health problems associated with biological contaminants. Contaminant-specific source control methods are described in detail for these three contamination and potential contamination problems in the following sections.

A. Asbestos

Source control is the only acceptable mitigation option for asbestos in buildings. This reflects its hazardous nature and the need to prevent asbestos fibers from becoming airborne. A variety of source control methods are used to minimize asbestos fiber exposures to service workers and building occupants. These include O&M practices, repair, enclosure, encapsulation, and removal.

1. Building operation and maintenance

Building O&M practices are considered to be an interim control approach (during the period between when friable ACM is identified and assessed in a building and long-term abatement efforts are implemented). In many cases, O&M is used to minimize exposures to service workers and building occupants until ACM must be removed as a result of USEPA regulatory requirements for removal during renovation or demolition.

Detailed guides that describe O&M practices for building ACM have been developed for schools and for managers of other buildings, and a guidance document for service and maintenance personnel has been developed and made available by USEPA.

Central to O&M programs is a requirement that service and maintenance personnel know how to recognize ACM and know where it is located so it is not inadvertently disturbed by their activities. Asbestos-disturbing activities to be avoided include: (1) improperly removing ACM during plumbing repairs; (2) changing light fixtures in ceilings with asbestos-containing acoustical plaster; (3) causing physical abrasion by moving construction equipment and furniture; (4) removing potentially contaminated ceiling tiles below ACM; (5) disturbing ACM while installing or repairing HVAC system ductwork, automatic fire sprinkler units, or electrical conduit or computer system wiring; (6) drilling holes in ACM; (7) hanging or attaching materials to ACM; and (8) resuspending asbestos fiber-contaminated dust under ACM by dry sweeping and dusting.

Custodians in buildings with surfacing ACM are advised to conduct all cleaning with damp cloths and mops to minimize resuspension of asbestos fibers. Maintenance and custodial activities in which building employees contact but do not disturb ACM are defined by the Occupational Safety and Health Administration (OSHA) as Class IV asbestos work. In cleaning small quantities of asbestos-containing dust from horizontal surfaces, custodial workers are subject to OSHA work practice requirements. They must be trained in accordance with OSHA's awareness training program, use wet methods and HEPA vacuums, and promptly clean up and properly dispose of ACM debris.

When service workers conduct routine maintenance and service activities involving thermal system insulation (TSI) and surfacing materials (SM) that are known or presumed to contain asbestos, their activities are also regulated by OSHA. These activities are described as Class III asbestos work. In such activities, service workers must wear, at a minimum, a half-mask NIOSH-approved respirator, and the building owner must have a respiratory protection program in place. Required work practices include isolating the work area; using plastic drop cloths where appropriate; and carefully cleaning up on work task completion using wet techniques, HEPA vacuuming, and proper waste disposal.

2. *Repair*

Asbestos-containing materials that have experienced limited damage can be easily repaired. Thermal system insulation applied to steam and hot water lines, as well as to boiler surfaces and steam condensate tanks, is commonly repaired. Such repairs may involve patching damaged surfaces with wettable cloth materials similar to those normally used to enclose ACM-containing TSI. It may also include removing short pieces of TSI using glove bag procedures. Repairs are regulated by OSHA as Class I asbestos work. Because of the significant engineering and work practice requirements, repairs are best conducted by abatement contractors using trained asbestos workers and supervisors.

3. *Enclosures*

Physical barriers can be used to enclose friable ACM to minimize its disturbance and asbestos fiber release. The use of enclosures, as recommended by USEPA, involves the construction of a nearly airtight barrier around ACM. Enclosures are a useful asbestos control measure when ACM is located in a small area or when the total area of ACM enclosed is small. They are inappropriate where water damage is likely and where damage or entry to the enclosure may occur.

Enclosures have the potential advantage of minimizing disturbance and associated fiber release from some ACM applications at relatively low cost (compared to removal). However, ACM remains and will eventually have to be removed on building demolition or renovation. Enclosures require periodic inspection for potential damage.

In theory, any physical barrier that reduces the potential for asbestos fiber dispersal into building spaces can be described as an enclosure. Because suspended ceilings are not airtight and are subject to entry by building service personnel, use of a suspended ceiling as an enclosure is not recommended. When a suspended ceiling serves as a base for a return air plenum, it may, in some measure, reduce the potential for fiber dispersal from that area into building spaces below. However, some released asbestos fibers may become entrained in the low-velocity flows that are characteristic of return air plenums. Removal of ceiling tile under damaged, friable ACM may result in a significant fiber release episode into the building space below.

4. Encapsulation

Encapsulation is widely used to reduce fiber release from asbestos-containing SM such as spray-applied acoustical plaster and similar materials. Encapsulants are designed to bind asbestos fibers to ACM and/or assure adherence of ACM to building substrates. Bridging encapsulants are applied to asbestos-containing SM to form a strong membrane that prevents fiber release. Penetrating encapsulants are designed to enter ACM to bind fibers to the ACM matrix and substrate.

Use of these two types of encapsulants depends on ACM characteristics. A bridging encapsulant is recommended for moderately friable ACM found on cementitious materials and complex surfaces such as pipes, ducts, and beams. Penetrating encapsulants are recommended for moderately friable ACM that is <1" (2.54 cm) thick. They are not recommended for painted or previously encapsulated surfaces.

Bridging encapsulants are widely used on acoustical plaster. The most common encapsulant is latex-based paint. Bridging encapsulants have several disadvantages: (1) they do not enhance adhesion to substrate surfaces; (2) they often significantly increase the weight of ACM; and (3) they make it more difficult and costly to remove ACM when it becomes necessary when the building is to be renovated or demolished.

Encapsulation is an appropriate asbestos control measure when ACM (1) retains its bound integrity, (2) is granular or cementitious, (3) is unlikely to become damaged, and (4) is highly-to-moderately accessible. If applied to ACM in poor condition, the material may delaminate, resulting in the fall of ACM onto horizontal surfaces with subsequent fiber release.

Since the source remains, encapsulation is not considered a permanent asbestos control method by USEPA. As a consequence, encapsulated ACM will require periodic inspection and possible reapplication of the encapsulant.

The cost of encapsulation varies. Under AHERA (see Chapter 13) school regulations, USEPA requires engineering controls and work practices similar to those used for asbestos removal. As a consequence, application costs are relatively high (approximately 50% of the cost of removal). Under OSHA asbestos construction industry rules, encapsulation is treated much like a repair and is classified as Class III asbestos work. Under this work classification, engineering controls and work practices may be significantly less demanding than for USEPA-regulated school abatement activities; therefore, encapsulation costs are significantly lower.

5. Removal

Removal is widely used as an asbestos abatement measure in buildings. Tens of thousands of asbestos abatement removal projects are conducted in the U.S. each year. In most cases, removal is conducted to comply with USEPA regulations that require removal of friable or potentially friable ACM when buildings are being renovated or demolished. Less often, ACM removal is prescribed to reduce potential exposures to building occupants or service workers. Removal is a permanent asbestos abatement measure and eliminates the need for O&M programs for ACM abated. Removal must be conducted in enclosures maintained under negative pressure to prevent contamination of adjoining building spaces. Removal must be conducted using wet techniques, HEPA filtration, and personal protective equipment.

B. Lead

Source control is the only acceptable means of minimizing potential lead exposures. Lead source control measures include repairing and repainting damaged materials, maintenance, in-place management, and a variety of lead abatement practices.

1. Repairs, repainting, and maintenance

Lead hazards in residential environments are typically associated with deteriorating or deteriorated lead-based paint (LBP). As a consequence, potential exposure risks may be reduced by repairing and repainting damaged materials that contain LBP. This may include rehanging doors and windows that are subject to friction and abrasion.

2. In-place management

An in-place lead management program is necessary when lead hazards have been identified on properties where young children may be exposed and interim control measures have been implemented. It should be designed to keep lead-based paint in good condition, minimize the production of dust from friction and impact surfaces, and control lead dust and paint chips during routine cleaning and maintenance activities. An in-place lead management program in multifamily dwellings or other leased housing should include: (1) lead hazard awareness training for maintenance workers; (2) work practice procedures to minimize lead dust production; (3) identification of buildings and sources producing, or having the potential to produce, lead hazards; (4) ongoing monitoring of lead painted surfaces and other identified lead hazards; (5) maintenance of records associated with lead activities; and (6) a designated lead program manager. In many respects, in-place management is similar to asbestos O&M programs.

3. Lead abatement

Lead abatement includes control options designed to "permanently" mitigate LBP-related hazards. These include removal and replacement of materials containing LBP, enclosure of LBP-coated materials, encapsulation, and LBP removal. They also include removal or covering of lead-contaminated soil. In the context of lead abatement, the term "permanent" is considered to be 20 years.

a. Building component removal and replacement. Replacement of building components coated with LBP is a very effective and widely used control measure. Replaced components commonly include windows and associated trim, sills, aprons, doors, baseboards, porch components, etc. Replacement is appropriate when there are no historic preservation or aesthetic concerns associated with rehabilitation activities. Replacement can produce moderate to significant quantities of lead-contaminated dust. As a consequence, care must be exercised in removing building components to minimize lead dust production and dispersal. Component removal and replacement must always be followed by specialized cleaning activities.

Replacement is one of the easiest and quickest ways to abate LBP. It truly is a permanent abatement measure. It can be combined with other rehabilitation activities to upgrade the quality of the building environment. Replacement can, however, be expensive, both in the costs of new materials and labor.

b. Enclosure. Enclosure is also commonly used in LBP abatement. An enclosure is a sealed, dust-tight barrier. It is designed to isolate LBP from building occupants. Materials used as enclosures include wood paneling, plywood, fiber board, wood underlayment, drywall, vinyl tile, and aluminum. Enclosures are affixed to floors, interior walls, and exterior wood siding. They are designed not to be easily breached, particularly by small children.

Since LBP remains, enclosures must be monitored periodically to assure they are in good condition. Because they can be disturbed during renovation activities, future owners must be informed of their presence.

The use of aluminum or vinyl materials to cover exterior wood siding with LBP is a low-cost, long-term solution to controlling the production of lead paint chips and dust around the perimeter of residential buildings. It reduces maintenance requirements and, in most cases, enhances the appearance of building structures.

c. Encapsulation. Encapsulation is recommended for materials that are in good condition. It involves the application of specially formulated coating materials over painted surfaces to form a strong, flexible, and durable surface that cannot be easily broken or chipped. Encapsulants are designed to prevent the production of paint chips by impact or friction and exposures that may be associated with children chewing encapsulated surfaces. Encapsulants, as a consequence, should be relatively nontoxic.

Encapsulants vary in their quality, reliability, and compatibility with painted surfaces. Because of incompatibility problems, encapsulants must be patch-tested on painted surfaces before they are permanently applied.

Encapsulation is often less expensive than other lead abatement measures. Though it reduces lead exposure hazards, lead nevertheless remains. Subsequent repair and renovation activities may disturb encapsulated surfaces. In some cases, encapsulants fail and separate from the covered LBP. The use of encapsulation as a lead abatement method, as a consequence, must be coupled with an ongoing monitoring program to assess damage or deterioration.

d. Paint removal. LBP can be abated directly by using techniques that remove it from painted substrates. Paint removal can be conducted both offsite and onsite. Onsite removal has the potential for producing large quantities of lead-containing dust that poses exposure risks to abatement workers and to occupants of abated structures if it is not followed by proper cleaning. Therefore, paint removal requires the application of very stringent work practices and cleaning measures.

Paint removal is employed in structures of historical or aesthetic significance to preserve the building's unique character. Some components can be removed and subsequently stripped offsite by furniture stripping companies. Such stripping services have become less available and more expensive because of waste disposal requirements and costs. Nevertheless, offsite paint stripping is the most desirable approach to removing paint from some building components.

Onsite paint stripping often requires engineering controls and work practices that are similar to those of asbestos. Paint removal must be conducted to minimize both the contamination of the abated building and abatement worker lead exposures. Paint removal may be accomplished by the use of wet scraping or wet planing, electric heat guns, hand tools with local exhaust systems, chemical stripping, and vacuum and water blasting.

i. Wet scraping and planing.   Wet scraping is used to remove loose paint. A painted surface is repeatedly misted with water during scraping to minimize lead dust production. Because wet scraping does not remove all paint on a surface, it is used to prepare a surface for painting or as an initial step in paint removal. Wet planing is similar to wet scraping. It is used to remove LBP from friction and impact surfaces like the edges of a door or window.

ii. Electric heat guns.   Heat guns are used to thermally soften LBP, which is then removed by scraping. Since they produce highly dangerous (to workers) lead fumes which can cause significant building contamination, heat guns that operate at temperatures >1000°F are not recommended. Low temperature heat guns work well in removing LBP but have significant disadvantages, e.g., the potential to cause burns and fires. Since heat guns involve dry scraping, significant quantities of lead dust are produced.

iii. Local exhaust hand tools.   Hand tools with local exhausts and HEPA filters have been developed for use in LBP removal. These include HEPA sanders, needle guns, saws, and drills. HEPA sanders and needle guns are commonly used in LBP removal. In both devices, a HEPA vacuum system is used to catch and filter lead dust as it is created. In needle guns, metal rods are contained inside a shroud. When activated, the metal rods break and loosen surface paint. They are best used on metal and masonry surfaces since they have a tendency to damage wood.

iv. Chemical stripping.   Chemical strippers (either volatile solvents or caustic pastes) are used to soften paint so that it can be removed by scraping.

The use of some chemical strippers may result in exposure to volatile vapors which may be quite toxic (e.g., methylene chloride). As a consequence, the use of such strippers is undesirable. Chemical stripping usually leaves some lead on painted substrates.

v. Vacuum and water blasting.   Blasting methods are used to remove paint on exterior surfaces such as building cladding. Both vacuum and water blasting methods produce significant waste and are very expensive. Vacuum blasting subjects a surface to bombardment with abrasive materials that are captured along with lead dust by means of a HEPA vacuum. It can be used on a variety of surfaces but works best on flat surfaces. Both methods can damage the treated surface.

C. Biological contaminants

A variety of source control techniques are used to prevent and mitigate building-related health problems associated with biological contaminants. This is true for Legionnaires' disease; hypersensitivity pneumonitis/humidifier fever and similar exposure problems; hospital-acquired infections; asthma and allergy symptoms associated with exposure to dust mites, cockroach, pet, and mold allergens; and other microbially produced contaminants such as mycotoxins, endotoxins, and microbial VOCs.

1. Legionnaires' disease

Prevention of outbreaks of Legionnaires' disease and mitigation of exposure problems once they have become manifest requires an understanding of environmental conditions that result in proliferation of the disease-producing organism, *Legionella pneumophila*, its dispersal, and techniques that are effective in its control.

Outbreaks of Legionnaires' disease can be avoided. Warm water environments (>86° <140°F, >30° <60°C) favor the growth of *L. pneumophila*. Outbreaks or sporadic cases of Legionnaires' disease have been reported for mechanical draft cooling towers, evaporative condensers, whirlpool spas, hot tubs, and hospital hot water heating systems. Such systems are at risk of serving as reservoirs for significant *L. pneumophila* population growth. It is therefore incumbent on facilities management to review its use of such systems and O&M practices.

a. Cooling towers.   One approach to reducing the potential occurrence of Legionnaires' disease is to abandon the use of mechanical draft cooling towers and evaporative condensers. Many hospitals in the U.S. no longer use mechanical draft cooling towers because of risks to immunocompromised patients. Another approach is to upgrade older cooling towers by installing more efficient mist eliminators. Newer systems have significantly lower emissions (17 to 20%) of aerosolized water than older cooling towers. Hospital guidelines in the United Kingdom recommend replacement of wet towers with dry towers when the former reach the end of their useful life, and avoidance of wet towers in new applications.

b. Cooling tower operation and maintenance. The key to avoiding Legionnaires' disease infections associated with wet cooling towers is for building management to have a preventive O&M program in place. Elements in such a program include periodic sampling of cooling tower water to determine *L. pneumophila* levels present and biocidal treatments when deemed necessary. Guidelines for the operation and maintenance of cooling towers have been developed and published by the World Health Organization (WHO).

i. Biocidal treatment of cooling tower waters. A number of biocides have been evaluated for use in controlling *L. pneumophila* populations in cooling tower water. Though many have been found to be effective under laboratory conditions, they are often not effective under actual cooling tower operation. As a result, it would be prudent for building management to follow the recommendations of the Centers for Disease Control (CDC). Populations of *L. pneumophila* in cooling tower water can be significantly reduced in routine maintenance activities and following disease outbreaks. The CDC-recommended procedure includes the establishment and maintenance of a neutral pH with careful addition of acid, addition of hypochlorite sufficient to maintain a free residual chlorine content of 100 to 250 ppm (w/v) for a minimum of 48 hours, flushing and refilling the system, and, if necessary, maintaining a free chlorine residual of 2 to 5 ppm until *L. pneumophila* populations have been reduced to acceptable levels.

c. Potable water systems. *Legionella pneumophila* survives low-level chlorination in water treatment systems. As a consequence, hot water systems in hospitals, hotels, nursing homes, and possibly residences may be subject to significant *L. pneumophila* population development. Since such waters are used for whirlpool spas and hot tubs, they pose potential exposure risks as well.

Exposures from potable water systems can be reduced by the implementation of good system O&M practices. These would include periodic collection of samples to assess the presence of *L. pneumophila* and population levels. Risk factors for proliferation of *L. pneumophila* include elevated water temperature ($>86°$ $<140°F$, $>30°$ $<60°C$), older hot water tanks, vertical configuration, and elevated magnesium and calcium levels. Water with temperatures $>140°F$ ($60°C$) are typically free of *L. pneumophila*.

Several methods have been evaluated for use in preventing and controlling the growth of *L. pneumophila* in hot water systems. These include thermal eradication and hyperchlorination. The "heat and flush" method (thermal eradication) is the most widely used *L. pneumophila* control method in hospitals. The method requires heating water to $158°F$ ($70°C$), followed by flushing all faucets and shower heads with heated water. This procedure only provides short-term control. However, if temperature is maintained at 140°F (60°C), the "heat and flush" method will be effective for 2 to 3 years.

Hyperchlorination is also widely used to control *L. pneumophila* in potable water systems. Its primary advantage is that it provides a residual chlorine concentration throughout the entire potable water system. It is a relatively costly procedure, and operators experience some difficulty in maintaining a stable chlorine concentration.

Once building (including cruise ship) management begins a program of disinfecting hot water systems, it must make a long-term commitment to maintain the effort. The effectiveness of efforts to prevent the growth of disease-producing populations of *L. pneumophila* requires continuing vigilance and implementation of appropriate eradication techniques.

2. *Hypersensitivity pneumonitis and humidifier fever*

Outbreaks or sporadic cases of hypersensitivity pneumonitis and humidifier fever-type illness can be prevented by the implementation of O&M practices. The most common apparent source of microbial contamination associated with hypersensitivity pneumonitis is the condensate drip pan associated with cooling fan coil units in HVAC systems. Routine inspection and maintenance can reduce the likelihood that drainage is impaired and significant microbial growth allowed to occur.

Drain pans should be sloped a minimum of 0.25" (6.25 mm) per 1.2' (30 cm) toward the drainage point, which should be flush with the lowest surface in the pan. Water should flow to a trapped water system. Microbial slimes or biofilms in drain pans and other wet surfaces should be physically removed by periodic cleaning.

Humidifier fever is reported in Europe, where cool mist humidifiers are used in association with HVAC systems. These outbreaks of humidifier fever can be eliminated by the cessation of humidification, replacement of cool mist humidifiers with steam humidification, or regular cleaning of cool mist humidifiers to maintain low bacteria levels. Cleaning and maintenance of water reservoirs in humidifier fever episodes has been reported to be effective in mitigating symptoms and health complaints. This includes removal of mineral scale and disinfecting with sodium hypochlorite. Water reservoirs must be kept free of scale and cleaned with biocides when not in use. Use of scale and rust inhibitors or biocides on a continuous basis is not recommended because they may be aerosolized and produce an indoor contamination problem. In industrial operations, exposure to bacterial endotoxins is a major occupational health problem. Biocidal treatment of metal cutting fluids and product wash waters is typically used with limited success.

3. *Asthma and allergic rhinitis*

As indicated in Chapter 5, both asthma and allergic rhinitis are caused by exposure to allergens associated with a variety of organisms. Most commonly these include dust mites, cockroaches, pet danders, and mold. Control measures used to reduce exposures are case specific; i.e., unique control measures are used to control individual allergen sources.

a. Dust mites. The probability that an individual will develop allergy or asthma from exposures to dust mite antigens (allergens) depends on his/her genetic predisposition and dust mite populations that produce significant levels of mite antigens. Dust mites must have a suitable environment to live and multiply. Primary environmental requirements are a source of human skin scales and elevated relative humidity. These are facilitated by high building moisture conditions, carpeted flooring, and bedding. A number of control measures have been evaluated to control exposures to dust mite antigens.

i. Dust control.   Many allergists advise their patients to use dust control measures. These include enclosing mattress tops and sides with a plastic cover; thoroughly vacuuming (with a HEPA vacuum) mattresses, pillows, and the base of the bed; daily damp dusting the plastic mattress cover; and weekly changing and washing of pillow cases, sheets, and underblankets. These practices are designed to reduce dust mite populations, allergen production, allergen exposure, and allergen-transporting dust. The plastic sheeting prevents allergen-containing fine dust particles from being embedded in the mattress and facilitates vacuuming. Other dust control measures include frequent vacuuming of textile floors and soft furnishings.

Though physicians recommend the use of dust control measures, studies evaluating their effectiveness in reducing dust mite allergens and asthmatic and allergic symptoms have shown mixed results, varying from completely ineffective to highly effective. The reason for such differences is likely due to differences in patient compliance with dust control measures. Good compliance typically occurs with procedures that are easy (bedding changes) and poor compliance with those that are more demanding (frequent vacuuming). Vacuuming must be done with a HEPA vacuum, as conventional devices increase airborne dust concentrations and aggravate symptoms.

ii. Climate control.   Dust mites obtain their water requirements from the atmosphere. They require a minimum relative humidity (R.H.) of 70%. Humidity requirements to sustain dust mite population growth are achieved by a combination of elevated indoor humidity and higher humidity in microenvironments associated with textile floor coverings. Therefore, the control of dust mite populations, and, indirectly, mite allergen levels, can be achieved by various climate control measures. These include building on a dry site, dehumidifying to R.H. <50%, air conditioning during humid months, and removing carpeting from bedroom flooring of mite-sensitive individuals. These control measures have been shown to significantly reduce dust mite populations and allergen exposure potential.

iii. Biocides.   Dust mite populations can be controlled effectively by application of aqueous and powdered acaricides to mite-infested surfaces (e.g., bedding, carpeting, and upholstery). Mite acaricides include formulations of benzoic acid esters and other compounds. European studies have shown that they can be highly effective in controlling dust mite populations and reducing allergen levels. However, studies showing concomitant reductions in asthmatic/allergic symptoms have not been reported.

Benzoic acid esters such as benzyl benzoate are commonly used as acaricides in Europe. Health risks based on what is presently known appear to be small, as benzoates are rapidly metabolized to hippuric acid and excreted in the urine. Nevertheless, a major question remains as to the advisability of using biocides whose safety is relatively unknown in indoor environments, particularly when applied to bedding materials immediate to one's breathing zone. Benzoic acid esters have not been approved for use as acaricidal treatments by USEPA, though they are available from suppliers of various allergy-related products in the U.S.

iv. Denaturants and other compounds. Though acaricides can reduce dust mite populations and allergen levels on a long-term basis, they cannot reduce allergen levels that are already present. As a consequence, commercially available mite acaricides are formulated with denaturants that are designed to degrade protein allergens, as well as with other substances that prevent mite allergens from becoming airborne. The most commonly used denaturant is tannic acid. It has been shown to be effective in laboratory trials.

b. Cockroaches. The role of cockroach allergens in causing sensitization, allergy symptoms, and asthma has only been elucidated in the past decade. Few studies have been conducted to evaluate control techniques to reduce exposure to cockroach allergens. Therefore, the best approach would be to implement measures designed to control cockroach populations.

i. Pesticides. Cockroaches in buildings are primarily controlled by using pesticides. Prior to the year 2001, chlorpyrifos, in a product called Dursban, was the most widely used pesticide for cockroach control. It has since been replaced by other pesticides. Recommended application procedures included spraying cracks and crevices, with particular attention to kitchens and bathrooms because they are associated with the highest cockroach populations. Poisoned baits and integrated pest management procedures are also recommended.

Cockroach populations are very difficult to control because of their high reproduction potentials and favorable environmental conditions that are common in low-income housing. Though cockroaches are known to infest "relatively clean" houses, populations often become enormously high in the relatively unsanitary conditions found in poor housing. In many cases significant cockroach control on a sustained basis can only be achieved in community-based programs.

ii. Integrated pest management. Integrated pest management (IPM) is a systematic approach to controlling pests in individual environments. It is designed to reduce the potential for harm to people and the environment while controlling pests in a cost-effective manner. It consists of using available knowledge of the biology and behavior of pest species.

Integrated pest management employs a combination of tactics including sanitation, pest monitoring, habitat modification, and judicious use of pesticides when such use is deemed essential. Sanitation includes removing food from accessible areas, cleaning, and managing wastes (e.g., keeping refuse in tight containers). Habitat modification includes fixing leaking pipes, promptly sealing cracks, adding physical barriers to pest entry and movement (e.g., using screens), removing clutter, etc. Insect populations may be reduced by using traps (e.g., glue boards) or poison baits, or spraying judiciously (crack and crevice spraying).

When pesticides are used in an IPM program, they should be species-specific and attempt to minimize toxic exposures to humans and nontarget species. When pesticides are to be spray-applied, spraying should be conducted during unoccupied periods under ventilation conditions that can flush volatile components from the building before occupants return. Pesticides should only be applied in target locations, with minimum treatment of exposed surfaces. Care should be employed to minimize contamination of other parts of the building.

Application of IPM is more difficult in residential environments because homeowners/occupants do not, in most cases, have the requisite knowledge. Nevertheless, homeowners can seek advice from public health personnel on the use of IPM methods to control infestations of pest species such as cockroaches.

   c. Pet danders.   Allergens associated with cats and dogs are widely present in North American and European buildings where pet ownership is high and small pets are kept indoors. When an allergist diagnoses a patient with a high sensitivity to pet allergens, the patient is advised to remove pets from the home environment. However, in many cases pets are seen as members of the family, and removal decisions may be traumatic.

Removal of a pet that is contributing to allergic and asthmatic reactions in one or more members of a household should, in theory, result in reduction of allergy and asthmatic symptoms. However, removal is unlikely to be effective immediately since the allergen may persist in the building environment for months and even years after the pet is removed. Pet danders are common in building environments where pets have never been present. They are passively transported into buildings on pet owners' clothing. In theory, rigorous cleaning of horizontal surfaces after pet removal should result in significant reductions in allergen levels. The effectiveness of cleaning measures has not been reported in the scientific literature, though washing cats has been shown to have limited effectiveness in reducing allergen production.

   d. Mold.   Mold infestations in residential and nonresidential buildings are common. Such infestations can, in theory, be avoided by selection of appropriate building sites, provision of adequate site drainage, and attention to good building maintenance. Once a building (or some portion thereof) becomes infested, the problem may be mitigated easily or only with great difficulty. The prevention and mitigation of mold infestation problems require an understanding of factors that contribute to mold growth. These are described in detail in Chapter 6.

i. Avoidance.   Moisture, both in the liquid and vapor phase, is a key factor in causing mold infestation problems. As indicated in Chapter 6, many species require a minimum R.H. of 70% for spore germination but can grow at lower R.H. Avoidance of mold infestation problems requires that buildings and materials be kept relatively dry.

ii. Building sites.   Many mold infestation problems, particularly in residences, are associated with building sites that are poorly drained. Such sites are typically characterized by heavy clay soils and persistent or seasonal high water tables. Moisture problems on wet sites may be exacerbated by trees that produce heavy shade. Individuals and occupants who have a history of allergy or asthma should avoid living on a poorly drained site. Sites can be evaluated for drainage hazards using soil maps, or by professional soil scientists. Many existing homes on poorly drained sites (depending on their age) have developed mold infestation problems. These may be identified from a casual building inspection (e.g., strong mold odor) or a more rigorous professional inspection.

iii. Building management.   Mold infestation in many buildings occurs as a result of poor management (and sometimes construction) of the building and its environment. This is true for both residential and nonresidential buildings. In both instances, mold infestation problems are often a result of neglect or various occupant behaviors.

Problems in residential buildings. Mold problems in residences not associated with poor building sites include infested cool mist humidifiers; overuse of humidification systems; lack of dehumidification in basements; use of low nighttime and seasonal setback temperatures; poorly constructed brick or stone veneer; inadequate wall and ceiling insulation; poorly ventilated bathrooms; inattention to water spills and plumbing leaks; inadequate caulking of windows and doors; and a less than expeditious response when cleaning up, drying, and repairing materials associated with water intrusion through the building envelope.

Humidifiers and humidification. Use of cool mist humidifiers in a residential environment can result in significant localized mold exposure from contaminated reservoir water. Such contamination results when humidifiers are not cleaned routinely. Studies have shown that cool mist humidifiers require regular cleaning to maintain low microbial levels. Because of the significant cleaning that is often required, it is better not to use cool mist humidifiers. Use of humidification based on recommendations of allergists cannot be supported by medical science. It would be more desirable to use heated vaporizers if humidification is required to protect indoor materials and increase humidity levels during the wintertime in cold climates.

It is not uncommon for building occupants to overhumidify their residences. Such overhumidification can result in the formation of condensate on cold windows, with subsequent mold infestation on windowsills and frames. Overhumidification should be avoided.

Dehumidification. Elevated humidity is common in basements, even when a building site is relatively dry. Basement environments are influenced by the cool environmental conditions associated with the ground. Consequently, relative humidities are often high. Relative humidity can be reduced by heating basements to achieve acceptable humidity levels during winter in northern climates and dehumidifying during summer. Dehumidification units require a minimum temperature of approximately 65°F (18°C) to operate properly (to prevent coil freezing).

Water intrusion. A common, but hidden, problem in brick veneer and masonry-clad homes is penetration of rain water into building cavities through the mortar, particularly during windy, rainy weather. Such cladding should be constructed with weepholes that are open and functioning. If not, it should be provided with functioning weepholes on a retrofit basis.

Rain penetration into building cavities can also occur through cracks around windows and doors due to poor caulking practices and caulk failure. Good maintenance requires that windows and doors be well caulked at all times. It also requires that settling cracks in mortar be caulked early in their development to prevent water intrusion and mold infestation.

Miscellaneous condensation. Water condensation on poorly insulated walls and ceilings during winter weather can be prevented by use of insulation; mold growth in bathrooms due to high humidity can be avoided by installation and use of bathroom exhaust fans; mold growth associated with water spills in bathrooms and leaking plumbing can be avoided by quickly (within 24 to 48 hours) drying the wet materials and, in the case of plumbing, repairing leaks as soon as possible; borders around windows and door frames need to be well caulked at all times and routinely inspected and attended to; and all water leaks associated with roofs and other damaged materials need to be quickly repaired and all materials dried within 24 to 48 hours.

Problems in nonresidential buildings. Though nonresidential buildings experience mold infestation problems, they are usually somewhat different in their origin. Common causes of mold problems include roof leaks, plumbing failures, leaks from AHUs, improper operation of HVAC systems, and infestation of filters and duct liner materials.

Roofing leaks. Leaks are common on flat-roofed buildings. Significant maintenance attention and sealing roof surfaces are necessary to prevent water intrusion. Once leaks develop, they must be repaired quickly, and wetted materials dried, to prevent infestation.

Plumbing failures. Plumbing failures may include leaks at joints or sink fixtures, overflowing toilets, pipe breaks, sewage backups, etc. The key to preventing mold infestation is a rapid effort (within 24 to 48 hours) to repair plumbing problems and dry the wet materials. The more quickly wet materials are dried, the less likely they will become infested.

HVAC system problems. Leaks from AHUs are common. They, like other water problems, must be repaired as soon as possible, and wet ceiling tiles removed and dried or discarded.

One of the most common causes of significant mold infestation in schools is improper operation of cooling systems during summer months. In many cases, building spaces are cooled to temperatures of 70 to 74°F (21 to 23.5°C) during daytime hours and the HVAC system is subsequently shut down during nighttime, unoccupied hours. On high dewpoint days (75 to 77°F, 24 to 26°C), outdoor air infiltrates the building and condenses on cooler building surfaces which are below the dewpoint. Significant mold infestation generally occurs, particularly on older books. Such problems can be avoided by operating cooling systems continuously or maintaining cooling temperatures at higher settings (78 to 80°F, 25 to 26°C).

Infestation of HVAC system filters is often associated with excessive dust accumulation as a result of infrequent filter replacement. More frequent replacement is desirable. Duct liner materials are porous and collect organic dust, which has the potential for becoming infested, particularly near water sources such as fan coil units. Insulating materials are best placed on the exterior of ducts to avoid soiling and mold infestation.

Remediation measures. Once a mold infestation problem has become manifest, remediation measures are necessary to eliminate or reduce mold growth and reduce occupant exposures. The most commonly used remediation measures are removal and disinfection.

Removal can be viewed in two contexts. These are the removal of mold growth from an infested surface by application of cleaning measures and, in the case of wood, scraping. Such cleaning and scraping of mold-infested materials are commonly done on relatively hard surfaces such as window frames, window glass, flooring, drywall, and wall cavity materials. It is typically followed by other treatments to prevent reinfestation.

In the case of soft materials such as paper products, furnishings, carpeting, and heavily infested structural timbers and gypsum board, the materials themselves may be removed and replaced. Such removal is essential to mitigating problems associated with toxigenic fungi such as *Stachybotrys chartarum* and *Aspergillus versicolor*. In the case of heavy infestations by toxigenic fungi, removal is conducted to the minimum standards of asbestos removal, with a higher degree of worker supervision. Such abatements are conducted under negative pressure enclosures with full respiratory protection and personal protective equipment for workers. The abatement must meet clearance values using aggressive sampling for specific toxigenic fungi before the contractor can be released. Clearance values of zero spores of *S. chartarum* on a 100% count of total mold spore samples are recommended.

Biocidal treatments are commonly used in mold mitigation activities. In small infestations (such as a wall surface), cleaning is usually followed by a biocidal treatment, typically using a hypochlorite solution (1 cup per gallon recommended). Such treatments may be ineffective on heavily infested materials. In many cases, biocidal treatment is followed by repainting or painting with a mold-inhibiting paint.

Readings

Bayer, C.W., The effect of "building bake-out" on volatile organic compound emissions, in *Indoor Air Pollution — Radon, Bioaerosols, and VOCs*, Kay, J.G., Keller, G.E., and Miller, J.F., Eds., Lewis Publishers, Boca Raton, FL, 1991, 101.

Building Performance and Regulations Committee/Committee on the Environment, *Designing Healthy Buildings: Indoor Air Quality*, American Institute of Architects, Washington, D.C., 1996.

Coluccio, V.M., *Lead-based Paint Hazards: Assessment and Management*, Van Nostrand Reinhold, New York, 1994.

Federal–Provincial Committee on Environmental & Occupational Health, *Fungal Contamination in Public Buildings: A Guide to Recognition and Management*, Health Canada, Ottawa, 1995.

Flannigan, B. and Morey, P.R., *ISIAQ Guideline: TFI-1996 — Control of Moisture Problems Affecting Biological Indoor Air Quality*, ISIAQ Inc., Milan, Italy, 1996.

Girman, J.R., Volatile organic compounds and building bakeout, in *Problem Buildings: Building-associated Illness and the Sick Building Syndrome*, Cone, J.E. and Hodgson, M.J., Eds., *State of the Art Reviews: Occupational Medicine*, 4, No. 4, Hanley and Belfus, Inc., Philadelphia, 1989, 695.

Godish, T., *Sick Buildings — Definition, Diagnosis and Mitigation*, CRC Press, Boca Raton, FL., 1995, chap. 9.

Godish, T., 1989, *Indoor Air Pollution Control*, Lewis Publishers, Chelsea, MI., 1989, chaps. 2, 3, 4.

Hansen, W., Ed., *A Guide to Managing Air Quality in Healthcare Organizations*, Joint Commission on Accreditation of Healthcare Organizations, Oakbrook Terrace, IL, 1997.

HSE, The control of legionellosis including Legionnaires' disease, *Health & Safety Series Booklet HS(G)70*, Health & Safety Executive, Sudbury, U.K., 1994.

Illinois Dept. of Public Health, *Integrated Management of Structural Pests in Schools*, Springfield, 1994, www.idph.il.us/envhealth/entpestfshts.htm

Knudsin, R.B., Ed., *Architectural Design and Indoor Microbial Pollution*, Oxford University Press, New York, 1988.

Levin, H., Building materials and indoor air quality, in *Problem Buildings: Building-associated Illness and the Sick Building Syndrome*, Cone, J.E. and Hodgson, M.J., Eds., *State of the Art Reviews: Occupational Medicine*, 4, No. 4, Hanley and Belfus, Inc., Philadelphia, 1989, 667.

Maroni, M., Siefert, B., and Lindvall, T., *Indoor Air Quality. A Comprehensive Reference Book*, Elsevier, Amsterdam, 1995, chaps. 28, 30, 31.

Miller, R.P., Cooling towers and evaporative condensers, *Ann. Int. Med.*, 90, 667, 1979.

NIOSH/USEPA, *Building Air Quality Action Plan*, U.S. Government Printing Office, Washington, D.C., 1998.

Raw, G.J., Roys, M.S., and Whitehead, C., Sick building syndrome — cleanliness is next to healthiness, *Indoor Air*, 3, 237, 1993.

Spengler, J.D., Samet, J.M., and McCarthy, J.F., Eds., *Indoor Air Quality Handbook*, McGraw-Hill Publishers, New York, 2000, chaps. 1, 2, 59-63.

Tasaday, L., *Residential Lead Abatement*, McGraw-Hill Publishers, New York, 1995.

USEPA, *Evaluating and Controlling Lead-based Hazards: A Guide for USEPA's Lead-based Paint Hazard Standard*, (Public review draft), USEPA, Washington, D.C., 1998.

USEPA, *Radon Prevention in the Design and Construction of Schools and Other Large Buildings*, EPA 625-R-92-016, USEPA, Washington, D.C., 1994.

USEPA, *Radon Mitigation Standards*, EPA 401-R-93-078, USEPA, Washington, D.C., 1994.

USEPA, *Flood Cleanup: Avoiding Indoor Air Quality Problems*, EPA 402-F-93-005, USEPA, Washington, D.C., 1993.

USEPA, *Managing Asbestos in Place: Building Owner's Guide to Operations and Maintenance Programs for Asbestos-Containing Materials*, EPA 745K93013, USEPA, Washington, D.C., 1993.

USEPA, *Use and Care of Home Humidifiers*, EPA 402-F-91-101, USEPA, Washington, D.C., 1991.

USEPA, *What You Should Know About Combustion Appliances and Indoor Air Pollution*, EPA 400-F-91-100, USEPA, Washington, D.C., 1991.

USEPA, *Guidance in Controlling Asbestos-containing Materials in Buildings*, EPA 560/5-85-024, USEPA, Washington, D.C., 1985.

Questions

1. What source control practices can be implemented on a proactive basis?
2. Why is it better to control a contaminant at a source rather than to control it after it is released into the building?
3. What is a building bakeout? What factors determine its effectiveness?
4. How have wood products been improved to reduce formaldehyde emissions?
5. As a consumer, what products should you avoid in order to minimize exposure to toxic airborne contaminants?
6. In designing a low-emission/low-VOC building, how would you attempt to identify products that would have the greatest impact on indoor air quality?
7. What value is there in labeling products based on emissions?
8. In designing a building, how could one reduce the potential for re-entry of exhaust gases and entrainment of outdoor contaminants?
9. Steam humidification is often used in buildings. Describe advantages and disadvantages of its use.
10. How can building O&M programs be used to maintain a healthy indoor environment?
11. What limitations are there in removing sources that can contribute to indoor environment complaints?
12. Describe source treatment as a source control method.
13. What is an ammonia fumigation? What are its advantages and limitations?
14. How does indoor climate affect indoor contaminant levels?
15. Under what conditions would it be appropriate to use an O&M program to control asbestos exposures in buildings?
16. Describe the use of enclosures, encapsulants, and repairs in controlling potential asbestos exposure to service workers.

17. Your building is scheduled for an asbestos removal abatement project. How are such abatements conducted? What are the risks involved?
18. Describe abatement measures used to control lead exposures in housing.
19. What special measures should be used in removing lead-based paint from building surfaces?
20. An outbreak of Legionnaires' disease has occurred in a hospital. Describe mitigation measures that should be applied to minimize exposures to at-risk patients.
21. How can outbreaks of hypersensitivity pneumonitis be controlled in a building? Humidifier fever?
22. What control measures can be implemented to reduce dust mite allergen exposures to dust mite-sensitive asthmatic patients?
23. How does one control exposure to mold on a proactive basis? On a retroactive basis?
24. Under what circumstances would one use biocides to control exposures to common allergens?

chapter eleven

Ventilation

Ventilation has historically been applied and viewed as both a desirable and effective technique in improving thermal comfort and general air quality and comfort in buildings. It has been used to dilute and exhaust unwanted contaminants such as combustion by-products, lavatory and cooking odors, heat and moisture, etc. from residential and nonresidential buildings. Exhaust ventilation is widely used in industry to remove contaminants at their source to reduce worker exposure and health risks. General dilution ventilation induced and delivered by mechanical fans is used in office, commercial and retail, and institutional buildings to maintain acceptable air quality.

Ventilation is a physical process that involves the movement of air through spaces. It has two dimensions. When air flows into a space, ventilation is characterized by mixing, dilution, and partial replacement. When air flows (is removed) from a space (because it is under negative pressure), it is replaced by air from a nearby area. General dilution ventilation describes the first case; exhaust ventilation, the second.

Ventilation, because it involves the flow of air due to pressure differences, is a natural phenomenon. However, natural ventilation is often too variable or inadequate. As a consequence, it is necessary to use systems that deliver controlled, mechanically induced ventilation to provide for the needs of a variety of building spaces.

Ventilation in its dilution, displacement, and replacement aspects causes changes in chemical composition and environmental factors in the air environment of ventilated spaces. These changes may result in reduced overall contaminant levels or a decrease in concentrations of some substances, with an increase in others. The outcome depends on the nature of air in ventilated spaces as well as air used for ventilation. The same is true for environmental factors such as temperature and relative humidity.

The desired effect of ventilation, whether it is natural or mechanically induced, is to enhance or protect the quality of air in the space being ventilated. Ventilation causes an exchange of air within building spaces and between building interiors and the outside environment. Ventilation is used to dilute and remove contaminants, enhance thermal comfort, remove excess moisture, enhance air motion, improve general comfort, and in large buildings, maintain pressure differences between zones.

I. Natural ventilation

All buildings are subject to natural forces that result in air exchange with the ambient (outdoor) environment. Natural ventilation depends on the inflow of air as a result of (1) pressure differences when a building is under closure conditions and being heated or cooled, (2) pressure-driven flows when building windows and doors are open, or (3) the continual movement of air through a building as it enters through some openings and exits through others (Figure 11.1). In the last case, small buildings will experience relatively high air exchange rates. Maximum exchange rates will depend on wind speed, the position of open windows and doors relative to each other, and prevailing winds. Though limited scientific data are available, air exchange rates should be at their maximum when buildings are ventilated by opening windows and doors when outdoor conditions are favorable.

Residential buildings, which include both single- and multi-family structures, are increasingly being provided year-round climate control. As a consequence, the practice of opening windows and doors for ventilation purposes is decreasing. Such residences, and those in seasonally cold or warm climates, are maintained under closure conditions for extended periods of time (upwards of 9 months or so).

Under closure conditions, ventilation occurs as a result of infiltration and exfiltration processes. These processes involve pressure-driven flows

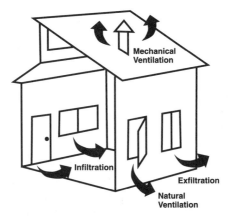

Figure 11.1 Ventilation air flow through a single-family house under open conditions. (From USEPA, *Introduction to Indoor Air Quality — A Self-paced Learning Module*, EPA/400/3-91/002, Washington, D.C., 1991.)

associated with temperature differences (between indoor and outdoor environments) and the speed of the wind.

Infiltration occurs as a result of the inflow of air through cracks and a variety of unintentional openings (leakage areas) in the building envelope. Infiltration only occurs through leakage areas where internal pressures are negative relative to those outdoors. In residential structures, infiltration typically occurs at the base and during windy conditions on the lee (downwind) side of buildings.

When infiltration occurs, it replaces and displaces air in the building interior. As a consequence, air must also flow outward through cracks and other leakage areas. In residential structures, such air outflows (exfiltration) typically occur through ceiling areas and upper wall locations where internal pressures are positive.

A. Stack effect

On calm days or during calm periods during the day, infiltration and exfiltration occur as a consequence of pressure differences associated with differences between the inside and outside temperature (ΔT). During the heating season in seasonally cool to cold climates, warm air rises and creates a positive pressure on ceilings and upper walls of small residential (and non-residential) buildings. This upward flow of warm air produces negative pressures, which are at a maximum at the base of the structure. These negative pressures cause an inflow (infiltration) of cool or cold air, with maximum inflows where negative pressures are (in absolute terms) the highest. Infiltration causes air to be drawn in from both the outdoor environment and from the ground (soil gas).

An idealized characterization of pressure conditions in a single-family dwelling on a cool day is diagrammed in Figure 11.2. As can be seen, an area of neutral pressure exists between negative and positive pressure environments. This is the neutral pressure level or neutral pressure plane (NPP). At

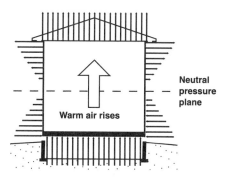

Figure 11.2 Generalized pressure conditions in a small house on a cool, calm day under closure conditions. (From Lstiburek, J. and Carmody, J., *Moisture Control Handbook*, Van Nostrand Reinhold [John Wiley & Sons], New York. With permission.)

the NPP, indoor and outdoor pressures are equal. In Figure 11.2, the NPP is located at approximately mid-level, suggesting that leakage areas are uniformly distributed over the building face. The location of the NPP depends on the distribution of leakage sites. In older single-family dwellings, the NPP is often above mid-height, and in the case of houses with flue exhaust of combustion by-products, the NPP may be above the ceiling level during exhaust operation. The construction of residential dwellings has included a number of energy-conserving measures, most notably tighter building envelopes. Since leakage areas have been reduced, pressure characteristics have changed, resulting in reduced infiltration and exfiltration. In such houses, the NPP would be expected to occur at mid-height (absent the active operation of combustion exhaust systems).

In tall buildings, the NPP may vary from 30 to 70% of the building height. The inflow and outflow of air discussed above is called the stack effect because airflows are similar to those which occur in a smokestack. The magnitude of the stack effect increases significantly with building height. The change in pressure with height in a large building has been reported to be approximately 0.001″ H_2O (0.25 pascals) per story. Stack effect flows upward are particularly noticeable in elevator and other service shafts and in open stairwells. Each story, if constructed in an airtight way, can behave more or less independently, i.e., have its own stack effect.

The influence of stack effect on building air exchange rates is, for the most part, proportional to ΔT, the difference between indoor and outdoor temperatures. This relationship can be seen in model predictions graphed in Figure 11.3 for $\Delta T = 0°F$ and $40°F$. As ΔT increases, air exchange increases, with maximum values on cold days. Minimum air exchange occurs when

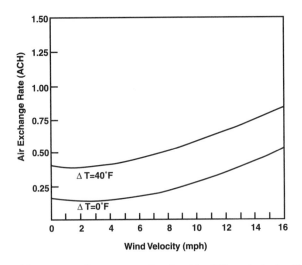

Figure 11.3 Building air exchange associated with different stack effect and wind speed conditions.

indoor and outdoor temperatures are the same or little different from each other. Such conditions exist for brief periods (hours) during diurnal changes in outdoor temperatures, and for more extended periods on mild overcast days, and mild days in the fall and autumn in temperate climatic regions. They commonly occur in coastal regions where maritime climates produce outdoor temperature conditions which in many cases are in the same range as those indoors.

B. Wind

Infiltration and exfiltration are also significantly influenced by wind. The effect of wind on pressures both inside and outside of buildings is a relatively complex phenomenon.

As wind approaches a building it decelerates, creating significant positive pressure on the windward face. As wind is deflected, its flow separates at a building's sides as well as its top or roof. This produces negative pressures around the sides of the building, the roof, and the leeward side. The effect of these pressure differences is to cause an inflow of air on the wind side and an outflow on all exterior surfaces which are negative relative to indoor pressures. The effects of wind on pressure conditions on a rectangular building oriented perpendicular to the wind is illustrated in Figure 11.4. The distribution of positive and negative pressures on a building depends on wind speed, building geometry and size, and the incident angle of the wind. The magnitude and distribution of infiltration air is influenced by the type of building cladding; tightness of the building envelope; and barriers to air

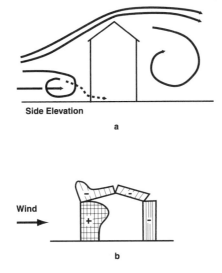

Side Elevation

a

Wind

b

Figure 11.4 Effect of wind on pressure conditions in/on a rectangular building oriented perpendicular to the wind. (From Allen, C., *Technical Note AIC 13*, Air Infiltration Centre, Berkshire, U.K., 1984. With permission.)

movement such as trees, shrubbery, and other buildings. Such barriers induce turbulence that reduces wind speed and alters wind direction.

The effect of wind on building infiltration is a squared function of wind velocity or speed (mph, m/s). The modeled effect of wind speed on house air exchange rates can be seen in Figure 11.3 for a ΔT of 0°F (0°C) and 40°F (22°C). Though the effect of wind speed on air exchange rates is exponential, significant increases in infiltration-induced ventilation are only seen at high wind speeds (>8 mph, 3.56 m/s). As can been seen in Figure 11.3, the combined effect of stack effect and wind speed on building air exchange rates (ventilation) can be significant.

Figure 11.3 describes the effect of indoor/outdoor temperature differences and wind speed on a single house. Because of differences in building tightness, distribution of leakage areas, and orientation to the wind, air exchange rates in other houses under similar stack effect and wind speed conditions are likely to be different (though the general form of the relationship will be much the same). These curves are based on the following linear model:

$$I = A + B(\Delta T) + C(v^2) \tag{11.1}$$

where I = infiltration rate (ACH)
    A = intercept coefficient, ACH (ΔT = 0, v = 0)
    B = temperature coefficient
    C = wind velocity coefficient
    ΔT = indoor/outdoor temperature difference, (°F)
    v = wind speed (mph, m/sec)

Both the temperature and wind speed coefficients are empirically derived and differ for each building. Coefficient differences among buildings are relatively small.

C. Infiltration and exfiltration air exchange rates

As indicated above, building air exchange rates associated with infiltration/exfiltration-induced airflows vary with indoor/outdoor temperature differences (which vary considerably themselves), wind speed, and tightness of the building envelope. They may also be influenced by pressure changes associated with the operation of vented combustion appliances, bathroom/lavatory fans, and leaky supply and return air ducts. Each of these can increase infiltration and air exchange rates above those associated with combined stack effect and wind infiltration and exfiltration values. Leaky supply/return air ducts may be responsible for upwards of 30+% of infiltration/exfiltration-related air exchange in residential buildings.

In response to energy concerns in the late 1970s and early 1980s, the U.S. Department of Energy supported several studies to evaluate infiltration and exfiltration rates in U.S. housing stock. The average air exchange rate for

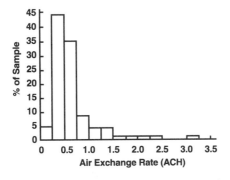

Figure 11.5 Infiltration rates measured in 312 North American houses in the early 1980s. (From Grimsrud, D.T. et al., LBL-9416, Lawrence Berkeley Laboratory, Berkeley, CA, 1983.)

more than 300 U.S. houses was measured on a one-time basis. As can be seen in Figure 11.5, approximately 80 to 85% of houses tested had a daily average air exchange rate of <1 air change per hour (ACH). Infiltration/exfiltration rates in low-income housing were on average significantly higher; approximately 40% had infiltration values >1 ACH.

On a population basis, these one-time measurements of infiltration/exfiltration-induced air exchange were likely to have demonstrated a reasonable estimate of ventilation conditions in housing stock existing at the time (early 1980s). Since then, construction practices have changed (tighter building envelopes are now the norm), and significant weatherization measures have been implemented to reduce energy losses in low-income housing. Weatherization measures using retrofit tightening of building envelopes in low-income housing have, however, only been moderately effective (on average, ≤25% reduction in building leakage and infiltration-associated air exchange).

It is highly probable that construction practices in the past several decades have significantly increased the stock of housing units in North America, northern Europe, and other developed regions and countries which have lower air exchange (and thus ventilation) rates than older houses. Decreasing natural ventilation rates have been a cause for concern among policy makers in various governmental agencies, utilities which have supported weatherization measures, public health groups, and research scientists. It was and is widely believed among environmental and public health professionals that decreasing natural ventilation rates associated with infiltration/exfiltration-reducing measures are likely to cause an increase in indoor contaminant levels and health risks associated with increased exposures.

D. Leakage characteristics

Air exchange rates in buildings associated with thermal and wind-induced pressure differences are affected to a significant degree by building leakage characteristics. Typical leakage areas are indicated in Figure 11.6 for a single-

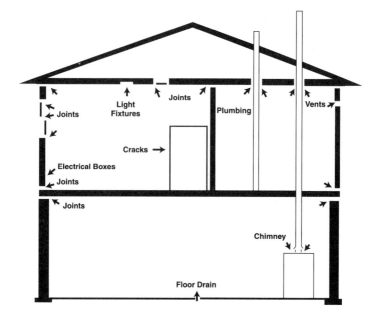

Figure 11.6   Air leakage sites on a single-family house with basement substructure.

story house on a basement. Major structure-related leakage areas include the sole plate where the building frame is fastened to the substructure, and cracks around windows, doors, exterior electrical boxes, light fixtures, plumbing vents, and various joints. Leakage also occurs through exhaust fans, supply/return ducts, and combustion appliance flues. Leakage is particularly pronounced when combustion appliances such as furnaces, hot water heaters, and fireplaces are in operation. Leakage can also occur through duct systems which provide heating and cooling when duct runs are in crawlspaces, attics, and garages.

The size of leakage areas and their distribution in a building determine the magnitude of infiltration and exfiltration air exchange when measured under similar environmental conditions. They also determine the nature of air flow patterns into and out of buildings.

Building leakage potentials are commonly assessed using fan-pressurization or blower-door techniques. By pressurizing buildings with a fan installed into a test door, the overall leakage potential of residential buildings can be determined to identify leakage areas that need to be caulked or sealed. Such leakage characterization is commonly conducted in weatherization programs which target low-income housing.

II. Measuring building air exchange rates

Building air exchange rates (ACH) associated with wind and thermally induced pressure flows, as well as those associated with mechanical ventilation, can be measured using tracer gas techniques. Tracer gases used in such measurements are characteristically unreactive, nontoxic, and easily measured at low concentrations. Sulfur hexafluoride and perfluorocarbons are commonly used to measure air exchange rates because they can be detected and quantitatively determined in the parts per billion range (ppbv). On occasion, nitrous oxide (N_2O) or carbon dioxide (CO_2) are used. Carbon dioxide has the advantage of being measured in real time on relatively inexpensive continuous monitors. It has the disadvantage of being produced by humans. Thus it cannot be used when building spaces are occupied. Silicon hexafluoride, perfluorocarbons, and N_2O are collected using one-time sampling techniques and require analysis on sophisticated, expensive instruments. Perfluorocarbon measurements are usually made using permeation tubes as sources and passive samplers as collectors. As such, air exchange measurements based on perfluorocarbons typically provide 7-day averages.

The concentration decay method (previously described in Chapter 8) is the most widely used air exchange measuring technique. It involves initial injection of a tracer gas into a space or building with the assumption that the tracer gas is well mixed in building/space air. The decrease, or decay as it is often described, of the tracer gas concentration is measured over time. From these measurements the air exchange rate I in ACH can be determined from the following exponential equations:

$$C_t = C_0\, e^{-(Q/V)\, t} \tag{11.2}$$

where C_t = tracer gas concentration at the end of the time interval ($\mu g/m^3$, ppmv, ppbv)
C_0 = tracer gas concentration at time $t = 0$ ($\mu g/m^3$, ppmv, ppbv)
V = volume of space (m^3)
Q = ventilation rate (m^3/hour)
e = natural log base
t = time (hours)

The ratio Q/V, considered in the context of hours, yields the air exchange rate I in ACH.

$$C_t = C_0\, e^{-It} \tag{11.3}$$

Equation 11.4, used to determine I, the air exchange rate, can be derived from Equation 11.2.

$$I = (\ln C_0/C_t)/t \tag{11.4}$$

Let us assume that the initial tracer gas concentration was 100 ppbv and at the end of 2 hours it was 25 ppbv. We could then calculate I as follows:

$$I = (\ln 100/25)/2 \tag{11.5}$$

$$I = 0.69 \text{ ACH}$$

In perfluorocarbon determinations of air exchange, perfluorocarbons are injected at a constant rate. The air exchange I is calculated from the ventilation rate (Q), which is the ratio of the rate of injection (F) to the measured concentration C.

$$Q = F/C \tag{11.6}$$

Since Q is expressed in m^3/sec or cubic feet per minute (CFM), it must be multiplied by appropriate time units and divided by the volume of the space to obtain ACH values.

III. Mechanical ventilation

Most large commercial, office, and institutional buildings constructed in developed countries over the past three decades are mechanically ventilated. Use of mechanical ventilation is often required in building codes and represents what can be described as good practice for building system designers and architects. Increasingly, buildings are being designed to provide year-round climate control. To ensure optimum operation of heating, ventilating, and air-conditioning (HVAC) systems, windows are sealed so they cannot be opened by occupants to provide ventilation. The availability of outdoor air for space ventilation depends on the design and operation of HVAC systems as well as air that enters by infiltration and exfiltration processes.

Mechanical ventilation is used in buildings to achieve and maintain a comfortable and healthy indoor environment. Two ventilation principles are used to accomplish this goal; general dilution and exhaust ventilation. Both principles are used in most buildings. General dilution ventilation is the dominant ventilation principle used to ventilate buildings. Local exhaust ventilation is used for special applications: removing lavatory and kitchen odors, combustion by-products, combustion appliance flue gases, etc.

A. General dilution ventilation

Ventilating buildings to provide a relatively comfortable, healthy, and odor-free environment is based on the premise that a continual supply of outdoor air can be introduced into building spaces. As ventilation air mixes with contaminated air, contaminant levels are reduced by dilution.

In general dilution theory, a doubling of the air volume available for dilution is expected, under episodic or constant conditions of contaminant generation, to reduce contaminant concentration by 50%. If the volume of ventilation air were to be doubled again, the original concentration would be reduced to 25% of its original value. By decreasing ventilation air required for dilution, contaminant concentrations would be expected to increase in an inverse manner to the dilution-induced decreases described above.

In the real world, the level of contaminant reduction associated with general dilution ventilation is often less than that described above. Contaminant concentrations are determined by how they are generated (episodic, constant, increasing), ventilation capacity (the HVAC system's ability to mechanically deliver outside air to dilute contaminants to acceptable levels), and the operation and maintenance of HVAC system equipment. Elevated contaminant levels may result from high source generation rates, inadequate design capacity, reduced outdoor air flows associated with energy-management practices, and poor operation and maintenance of HVAC systems.

Because of the large air volumes required to reduce elevated contaminant levels, general dilution ventilation is not the most efficient control measure. However, when contaminant sources are diffuse and cannot be easily controlled in other ways, it is the control method of choice. Such conditions are common in many large nonindustrial, nonresidential public access buildings.

1. HVAC systems

The ventilation function is incorporated into systems that provide climate control, such as heating and cooling (and sometimes humidification), in most buildings where general dilution ventilation is used to reduce contaminant levels and provide a comfortable and healthful indoor environment. These HVAC systems include one or more air-handling units (AHUs), supply ducts, diffusers, return air grilles, return air plenums, dampers, exhaust fans and exhaust outlets, intake grilles, mixing boxes, etc. A relatively simple HVAC system design that provides conditioned air to a single space is illustrated in Figure 11.7. In this case, the AHU is a small box suspended above ceiling level. It consists of a filter, thermal sensors, and heating and cooling coils. Air that enters the AHU can be 100% outdoor air, nearly 100% recirculated air, or various mixtures of recirculated (from the conditioned space) and outdoor air.

Outdoor and recirculated air percentages are varied to meet building operating needs. In middle latitudes in North America, such systems are operated with 100% outdoor air during limited periods of spring and autumn months when temperatures are favorable for using outdoor air for free-cooling. That is, outdoor air is used to cool building spaces in lieu of activating energy-consuming cooling units. During cold weather, HVAC systems may be operated on nearly 100% recirculated air to prevent freezing of system coils. In many instances, it is still common (particularly in school buildings) for facilities managers to operate HVAC systems on or near 100% recirculated air as an energy management strategy. In the latter case, HVAC systems are not being operated as designed, and building air quality is unlikely to be acceptable. Good operating practice would require that at a minimum 15 to 20% outdoor air ventilation be provided during normal building occupancy hours to maintain acceptable air quality. In older AHUs, dampers that regulate the percentage of recirculated and outdoor air have

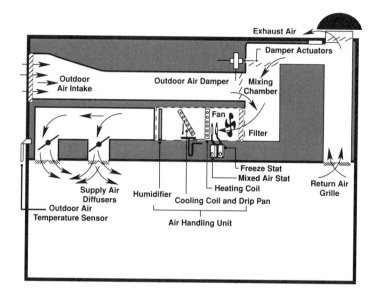

Figure 11.7 HVAC system providing conditioned and ventilation air to a single occupied space. (From USEPA, *Building Air Quality Manual*, EPA 402-F-91-102, Washington, D.C., 1991.)

to be manually adjusted. In most modern units, such adjustments are made automatically by use of thermal or CO_2 sensors, or timers.

a. Types of HVAC systems. Buildings are climate controlled and ventilated with a variety of HVAC systems. There are three basic system types used in buildings, each with its own variations. These include all-air, air–water, and all-water systems.

i. *All-air systems.* In all-air systems, air is conditioned as it passes over heating and cooling coils. It is then delivered to occupied spaces through a single duct or through individual hot and cold ducts to a mixing box (and then to occupied spaces). Air flow through these systems may be at a constant rate (constant-air-volume [CAV] systems), or air flow may be varied to individual spaces (variable-air-volume [VAV] systems). Simplified CAV and VAV system designs are illustrated in Figures 11.8 and 11.9.

In CAV systems, temperature is varied to meet heating and cooling needs. Supply air flows at a fixed and constant rate. In VAV systems, air is initially conditioned in an AHU. It is delivered to different zones or spaces through VAV boxes which regulate air flow and temperature in response to heating/cooling demands of individual zones. Both systems provide outdoor air for ventilation and are able to control the relative amount of outdoor and recirculated air that spaces are provided. Variable-air-volume systems were developed in response to energy management concerns and, when properly installed and operated, can reduce building energy consumption

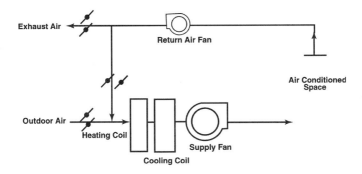

Figure 11.8 Simplified constant-volume HVAC system design. (From McNall, P.E. and Persily, A.K., in *Ann. ACGIH: Evaluating Office Environmental Problems*, 10, 77, ACGIH, Cincinnati, 1984. With permission.)

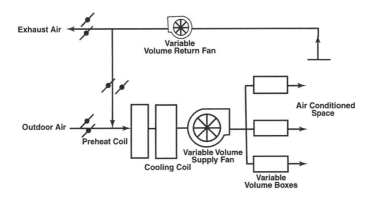

Figure 11.9 Simplified variable-volume HVAC system design. (From McNall, P.E. and Persily, A.K., in *Ann. ACGIH: Evaluating Office Environmental Problems*, 10, 77, ACGIH, Cincinnati, 1984. With permission.)

requirements (compared to CAV systems.) Variable-air-volume systems are more complicated than CAV systems and, as a consequence, are more difficult to operate effectively. Not surprisingly, VAV system operation in many buildings has been plagued with problems. One of the most common has been the complete closing of VAV valves and cessation of air flow through diffusers when a space's thermal requirements have been met. Under such conditions, one or more spaces may receive little or no ventilation air.

ii. Air–water systems.   Air–water systems differ from all-air ones in that air can be reheated or cooled by passing over a fan coil before it enters conditioned spaces. A terminal reheat system is illustrated in Figure 11.10.

iii. All-water systems.   In all-water, or hydronic systems, heating and cooling occurs in terminal reheat units located in each space. The terminal

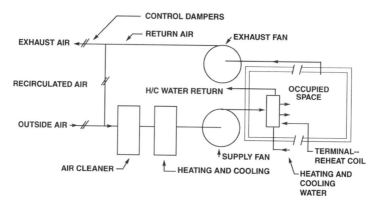

Figure 11.10 HVAC system with terminal reheat unit. (From Hughes, R.T. and O'Brien, D.M., *Ind. Hyg. Assn. J.*, 47, 207, 1986. With permission.)

reheat unit provides for air circulation in a space, but no ventilation. A separate duct system must be used to provide outdoor air for ventilation.

iv. Unit ventilators. In many school buildings constructed in the 1960s and 1970s, exterior classroom spaces are heated and ventilated through modular units installed beneath windows. These units (Figure 11.11) are typically provided with hot water that passes through a fin tube system. Air warmed by convection is delivered through the top of the unit ventilator (univent). Return air is drawn into the base through a filter. It is then mixed with outdoor air, which is drawn through intake grilles on the building exterior before it is heated and delivered. Some univent systems are equipped with chilled water lines for air conditioning. Unit ventilators may also be used to condition interior classrooms or larger spaces. In such cases, they are suspended from the ceiling and outside air must be ducted to them.

Unit ventilators are relatively simple mechanical systems. They are often poorly operated. Their operation and ability to adequately ventilate spaces is compromised by the force of numbers (many individual units in a building), and dampers (which regulate the percentage of outdoor and recirculated air), which typically require manual adjustment. Because of poor operation and maintenance, unit ventilators often do not adequately ventilate building spaces.

b. Ventilation standards and guidelines. It is widely accepted by building designers, owners, research scientists, and policy makers that large, nonresidential, nonindustrial buildings with significant occupant densities must be provided with adequate outdoor ventilation air to provide a comfortable and relatively odor-free building environment. It is important, therefore, that consensus standards and guidelines be available for use by building design professionals and by state and local governments, which set building codes. Such standards and guidelines are used to design HVAC

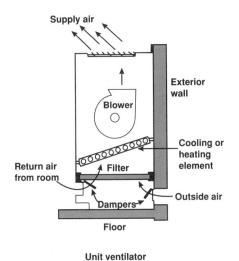

Supply air

Blower

Exterior wall

Cooling or heating element

Return air from room

Filter

Dampers

Outside air

Floor

Unit ventilator

Figure 11.11 Unit ventilation (univent) system. (From Spengler, J.D., *Environ. Health Perspect.*, 107, Suppl. 2, 313, 1999.)

systems with sufficient capacity to provide outdoor ventilation rates to attain and maintain acceptable air quality.

In North America, ventilation standards have, over the years, been developed by a consensus process by professional or standards-setting organizations. The American Society of Heating, Refrigerating, and Air-Conditioning Engineers (ASHRAE) has been the lead organization in the development of ventilation guidelines. ASHRAE guidelines are given further authoritative status by their acceptance and publication by the American National Standards Institute (ANSI).

i. Ventilation Rate Procedure. The primary approach to setting ventilation standards is to specify ventilation rates. Ventilation rates are expressed as volumetric air flows needed per building occupant, commonly cubic feet per minute (CFM) per person or liters per second (L/s) per person.

The Ventilation Rate Procedure is based on the pioneering experimental work of Yaglou in the 1930s, who determined the amount of outdoor air needed to maintain relatively human odor-free environments in buildings. As a consequence of Yaglou's work, a ventilation guideline of 10 CFM (4.76 L/s)/person was used by building designers in the period 1936–1973. In 1973, ASHRAE, taking energy conservation concerns into account, reduced its consensus standards to 5 CFM (2.34 L/s)/person for nonsmoking office and institutional environments. Higher ventilation rates were recommended for other environments such as hotel rooms, taverns, auditoriums, residential living areas, industrial environments, and where building occupants smoked. This standard was widely used until 1989, when it was revised.

The 1989 ASHRAE ventilation standard/guideline significantly increased the amount of outdoor air which was believed needed to provide office, commercial, and institutional buildings with sufficient outdoor ventilation air for a comfortable and healthful environment. The revised ASHRAE ventilation standard reflected the explosion in problem building complaints which occurred in the late 1970s and the early- to mid-1980s. It was apparent to many building and indoor air quality (IAQ) professionals that the 1973 ASHRAE standard of 5 CFM/person was not adequate and that increased ventilation was necessary. The recommended ventilation rates for general office environments became 20 CFM (9.5 L/s)/person, with 15 CFM (7.14 L/s)/person in school buildings.

In the Ventilation Rate Procedure, the minimum ventilation rate for building spaces is based on the maximum acceptable bioeffluent levels, using carbon dioxide (CO_2) emissions and building concentrations as reference points.

Ventilation guidelines are based on the use of the following mass balance equation:

$$Q = G/(C_i - C_a) \tag{11.7}$$

where Q = ventilation rate (L/s)
$\quad C_i$ = acceptable indoor CO_2 concentration
$\quad C_a$ = ambient CO_2 concentration
$\quad G$ = CO_2 generation rate (0.005 L/s)

Solving this equation when C_i = 0.10% and C_a = 0.0365%:

$$Q = 0.005/(0.0010 - 0.0003) = 7.14 \text{ L or } 15 \text{ CFM} \tag{11.8}$$

Assuming a metabolic CO_2 generation of 0.005 L/sec at design adult occupant capacity, a ventilation rate of 15 CFM/person would be required to assure that an indoor CO_2 guideline value of 1000 ppmv would not be exceeded. A ventilation rate of 20 CFM/person was recommended for office spaces by ASHRAE in 1989. Based on the use of the above equation or the graph in Figure 8.2, a ventilation rate of 20 CFM would, in theory, result in peak steady-state CO_2 levels of 800 ppmv.

In the Ventilation Rate Procedure, CO_2 concentrations are used as a surrogate for bioeffluents which cause human odor and are believed to contribute to general discomfort. Carbon dioxide itself is unlikely to cause discomfort or health effects at concentrations found in buildings, including those that are not well-ventilated. Carbon dioxide is used as an indicator because it is the bioeffluent with the highest rate of emission and is relatively easily measured. Research studies have shown a strong correlation between building CO_2 levels and human odor intensity.

The relationship between CO_2 levels and building occupancy can be seen in Figures 11.12a and 11.12b. The effect of different ventilation rates on

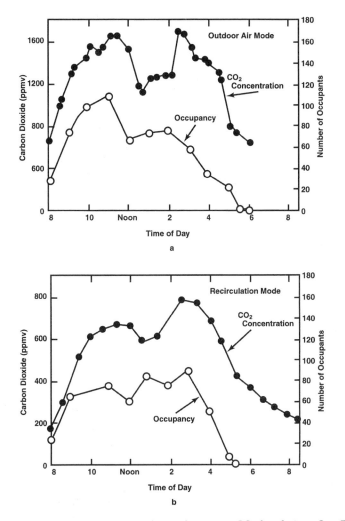

Figure 11.12 Effect of occupancy and ventilation on CO_2 levels in a San Francisco office building. (From Turiel, I., et al., *Atmos. Environ.*, 17, 51, 1983. With permission.)

CO_2 levels can also be seen when Figures 11.12a and 11.12b are compared. In Figure 11.12a, the HVAC system is being operated on 100% outside air; in Figure 11.12b, it is being operated using a mixture of recirculated and outdoor air.

ii. Indoor Air Quality Procedure.   In 1982, ASHRAE introduced what it described as the Indoor Air Quality (IAQ) Option. It was developed to provide building system designers with an alternative to the Ventilation Rate Procedure described above. It recognized that the Ventilation Rate Procedure focused primarily on bioeffluent control and was not designed to control contaminants generated in building spaces from construction materials, furnishings, and a variety of occupant activities. In the IAQ Option or Procedure, building designers could use any amount of outdoor air as long as guideline values of specific contaminants such as formaldehyde (HCHO), ozone (O_3), etc., were not exceeded. The IAQ Procedure was intended to (1) encourage innovative energy-conserving solutions for providing building ventilation and (2) give recognition to the fact that ventilation standards based on bioeffluents alone were not adequate to deal with the many contaminant problems that exist in indoor spaces.

Before the IAQ Procedure could be used by design engineers, it would be necessary to develop performance standards from knowledge of the health effects of all or targeted pollutants in a building. Acceptable levels would be set after the relative risks of exposures to specific contaminants had been determined. A prescriptive path would be developed to give building designers specific procedures to satisfy performance standards. This would include upper bounds on source strength and lower bounds on ventilation rates required. These bounds would be determined from air quality models designed to simulate contaminant emissions and transport, ventilation parameters, and occupant exposures.

The IAQ Procedure makes considerable sense from a scientific perspective because it focuses on the problem of controlling a variety of contaminants which may adversely affect air quality and pose health risks. Not surprisingly, the IAQ Procedure has rarely been used to design building ventilation systems. Unlike the Ventilation Rate Procedure, which is known for its simplicity and long history, the IAQ Procedure is characterized by knowledge requirements that are beyond those of the average design engineer and numerous uncertainties related to the health effects of targeted contaminants, acceptable levels of exposure, sources and their emissions, and how installed ventilation systems will actually perform to achieve IAQ Procedure objectives.

The IAQ Procedure requires development of guideline values for acceptable levels of specific contaminants. Manufacturers of products that emit contaminants may not accept the fact that their products are unsafe (i.e., they emit contaminants that exceed acceptable guideline values). A classic case of this has been HCHO. In 1981, ASHRAE published a guideline value for HCHO. After being threatened with a lawsuit by wood-products manufacturers and others, ASHRAE deleted HCHO from its guideline values in 1989 (and placed other guideline values in the appendix).

Despite new efforts to revive it (see Chapter 13), the IAQ Procedure has a limited future relative to its general use as a ventilation specification tool in designing buildings. Any appeal that it has is lost in the complexities of its use and the politics of prescribing guidelines for acceptable contaminant levels. Variations of this concept have received limited application. In Washington, new state office buildings must use materials that meet emission limits under specified ventilation conditions. A number of other large buildings in the U.S. have also been constructed using IAQ Procedure concepts.

The IAQ Procedure is, in reality, a program of source control and ventilation specification, design, and implementation.

    iii. Perceived Air Quality Procedure.   Ventilation requirements can also be determined using the Perceived Air Quality (PAQ) Procedure, developed by Danish engineer/scientist, Ole Fanger. It is based on the premise that sufficient outdoor air should be provided to reflect the pollution load of the building and the resultant perceived air quality. The pollution load is determined from the irritating/discomfort effects of building contaminants generated by humans, tobacco smoking, and emissions from building materials on the olfactory and common chemical senses. The two olfactory senses combine to assess perceived freshness or pleasantness or various degrees of stale, stuffy, or irritating air.
    Perceived Air Quality is expressed in units called decipols. One decipol is the equivalent of the PAQ in a building space with a pollution strength of one olf ventilated by 10 L/s (21 CFM) of clean air. Pollutant emissions from a "standard" person are equal to one olf. Using the standard person as a reference, other pollution sources are assessed relative to their effects on PAQ. Olf ratings for sources indicate the number of standard persons required to make air as unacceptable as their emissions. The effect of ventilation on perceived dissatisfaction (rated by a trained panel of judges) can be seen in Figure 11.13. These are expressed in L/s/olf (person). Equivalent values in CFM/person can be approximated by multiplying each L/s/olf value by a factor of 2. The curve in Figure 11.13 was derived from panel ratings of the acceptability of bioeffluents from more than 1000 sedentary men and women.
    The PAQ procedure requires quantification of olf values of pollution sources in building spaces. Olf values from different sources, including peo-

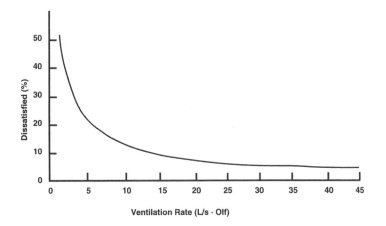

Figure 11.13 Relationship between ventilation rate and dissatisfaction with air quality. (From Fanger, P.O., *Proc. 5th Internatl. Conf. Indoor Air Qual. Climate*, 5, 353, 1990.)

ple (1 olf/individual), smoking (6 olfs/individual smoker), and building materials and furnishings are added together. The source strength is determined by adding the square meters of source materials present. Olf values are not available for many materials typically found in building spaces. As a consequence, olf values are estimated for different types of spaces such as offices, schools, assembly halls, etc.

In the PAQ procedure, a desired air quality is identified, perceived air quality is estimated to determine the olf load, and ventilation rates required to handle the total olf load are calculated.

In the PAQ Procedure, ventilation rates of 4, 7, and 16 L/s/person (8.4, 14.7, and 33.6 CFM/person) would be required to achieve minimum, standard, and high levels of PAQ. This would be equivalent to 30, 20, and 10% occupant dissatisfaction and 2.5, 1.4, and 0.6 decipols, respectively.

The PAQ Procedure differs from the IAQ Procedure in that it is designed to provide sufficient outdoor air to maintain human comfort. It does not address health concerns directly. It also differs from the IAQ Procedure in its level of development and evaluation (much more developed) and its relative simplicity. The PAQ Procedure, however, would be difficult for building designers to use because, in addition to being based on people, it is based on a variety of other sources whose contribution to PAQ cannot be known with certainty *a priori*, that is, before the building is constructed.

Information generated in the development of the PAQ Procedure is interesting in a number of ways. In Figure 11.13, the percentage of dissatisfied occupants decreases rapidly with increasing ventilation rates. The greatest improvement in occupant satisfaction is observed as ventilation increases up to 10 L/s/olf, with limited improvement in satisfaction scores as ventilation rates increase further. It appears that in the case of PAQ the optimum ventilation rate is approximately 10 L/s/olf, which is nearly equivalent to the 20 CFM/person ventilation rate specified in ASHRAE's 1989 standard. Figure 11.13 appears to confirm the wisdom of ASHRAE's recommendations on ventilation requirements for mechanically ventilated office buildings.

c. Flush-out ventilation. Conventional ventilation systems can be operated to speed up the decay rate of initially high material-, finishes-, and furnishings-based VOC concentrations in new or remodeled buildings and thereby improve indoor air quality. Such reductions can be achieved by what is described as "flush-out" ventilation. In flush-out ventilation, AHUs in a building, wing, or floor are operated to provide a continuous supply of 100% outdoor air for 3 to 6 months. Flush-out ventilation may begin during construction or when the building is available for occupancy.

"Flush-out" ventilation is based on laboratory and field studies that have demonstrated that VOC and HCHO levels decrease with time and that emission rates increase with increasing ventilation rates, resulting in an acceleration of their normal decay with time. "Flush-out" ventilation has been used for new building occupancy in Scandinavian countries and been recommended for use in Washington and California.

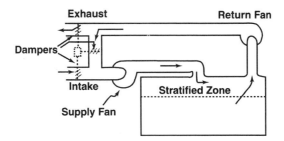

Figure 11.14 Diagrammatic representation of short-circuiting/stratification in a mechanically ventilated space. (From Janssen, J., in *Ann. ACGIH: Evaluating Office Environmental Problems*, 10, 77, ACGIH, Cincinnati, 1984. With permission.)

  d. Ventilation system problems. Ventilation systems designed to provide adequate outdoor air for occupant comfort, and a presumably healthy environment, experience a variety of problems that compromise their effectiveness, and in some cases contribute to or create IAQ/IE (indoor environment) problems. These may be related to design and operational factors.

  i. Ventilation efficiency. In buildings served by mechanical ventilation systems, conditioned air (which includes outdoor air) is delivered to individual spaces through diffusers located in the ceiling. Because it is economical to do so, return air outlets are located in the ceiling as well, sometimes within several yards (meters) of diffusers. This proximity, diffuser design, low air flow rates, and conditioned air that is relatively warm, may create an environment in which short-circuiting takes place. A diagrammatic representation of short-circuiting of building supply air can be seen in Figure 11.14; supply air is seen to stratify. Under such circumstances, mixing of supply air with air in occupied spaces may be limited. Stratification can significantly reduce ventilation effectiveness.
  Ventilation effectiveness may be determined by the design and layout of office spaces. Ventilation efficiencies have been reported to be significantly better in closed office spaces compared to open-plan spaces. This was presumed to have been due to the fact that supply air diffusers and return grilles were in closer proximity in open-plan office spaces. However, placement of supply air diffusers and return air grilles varies considerably in closed office spaces. In many cases, they are located within 6 feet (2 m) of each other; in other cases closed office spaces are served by a supply air diffuser, with return air moving through a grille located in the door. In the latter case, ventilation efficiencies would be expected to be higher than in open-plan spaces. Relative ventilation effectiveness values (as a decimal fraction, with 1 or 100% most desirable) for different spaces in an office building are illustrated in Figure 11.15. Ventilation effectiveness was determined from tracer gas studies.

  ii. System imbalances. System imbalances contribute to ventilation inadequacies in some building spaces relative to others. These imbalances

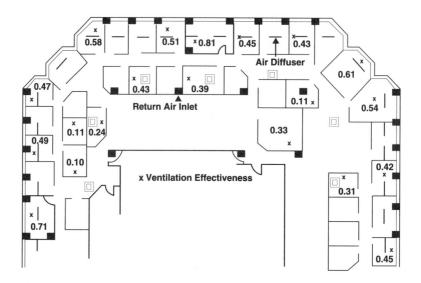

Figure 11.15 Ventilation effectiveness (decimal fraction) in different spaces in an office building. (From Farant, J.P. et al., *Proc IAQ '91: Healthy Buildings*, ASHRAE, Atlanta, 1991. With permission.)

may result when supply air dampers are improperly adjusted or maintained. They may cause high velocity flows in some areas and much lower flows in others. Occupants in high velocity flow areas may complain of annoyingly high air velocities, while in other areas occupants report a lack of air movement.

Imbalances may occur when spaces are provided with supply air but no return air, or return air without supply air. Such circumstances occur in buildings with moveable partitions and in those remodeled to provide closed-door offices without reconfiguring supply and return air outlets accordingly. It is not uncommon for enclosed spaces to have no return air flows through ceiling or door grilles.

   iii. Cross-contamination.   A common IAQ/IE problem in multipurpose buildings (e.g., hospitals, research laboratories, commercial buildings, and industrial buildings with adjoining office spaces) is contaminant movement from high-generation areas to other parts of the building. Cross-contamination occurs when pressure differences exist between building zones served by different AHUs. Cross-contamination is one of the most frequently encountered problems in problem-building investigations. It is commonly perceived by occupants as chemical odors or as an odor-out-of-place phenomenon.

Cross-contamination complaints arise when contaminants migrate from special-use areas where odors and even minor symptoms associated with solvents and other source materials are tolerated to areas where they are not.

An example of cross-contamination associated with shops in a strip mall is indicated in Figure 11.16. In this illustration, air flows into the restaurant

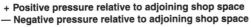

| AHU | | AHU | | AHU | | AHU | | AHU |

Exhaust Hood

☐

E   Restaurant   Clothing Store—1   Print Shop   Clothing Store—2   Business Supply   W

+ **Positive pressure relative to adjoining shop space**
— **Negative pressure relative to adjoining shop space**

Figure 11.16   Cross-contamination in adjoining strip mall business places.

from the adjoining clothing store (1) through wall penetrations associated with plumbing, electrical wiring, etc., because the restaurant is strongly negative relative to clothing store 1. Occupants of the restaurant may or may not find clothing store odors objectionable. However, the negative pressure in clothing store 1 and slight positive pressure in the print shop is likely to result in the movement of solvent vapors from the print shop into clothing store 1, where occupants are likely to find them objectionable. Since the east wall of clothing store 2 is positive relative to the adjoining print shop, vapors from the print shop are unlikely to move into clothing store 2 unless some factor causes the indicated pressure differences to change. Such changes do, in fact, take place when doors are opened and remain open for several minutes or more. In Figure 11.16, air can flow from clothing store 2 to the business supply store because the latter is negative relative to the former. The degree of cross-contamination and its objectionableness depends on the nature of contaminants and extent of their migration from one shop to another. Contaminant migration depends on the magnitude of pressure differences and the relative "leakiness" of wall structures separating each unit.

Assuming that the building (Figure 11.16) is under single ownership, it would be desirable to modify and operate ventilation systems in the print shop so that both east and west walls are negatively pressurized relative to adjoining spaces. The print shop should be equipped with an exhaust ventilation system to remove solvent vapors and maintain the space under negative pressure.

iv. Re-entry.   Re-entry of exhausted contaminants is also a common problem in buildings. This is the case when exhaust outlets and outdoor intakes are located relatively close together, with intakes downwind of exhaust outlets. Re-entry also occurs by infiltration when imbalances between intake air and exhaust air flows occur. When exhaust air flow exceeds air flow through intakes, the building or parts of it become depressurized (under strong negative pressure). Depressurized indoor spaces may cause the re-entry of contaminants which were initially removed through one or more exhaust systems. Buildings that are inadequately ventilated (because of energy management concerns) are often under negative pressure relative to the outdoor environment. As a consequence, building air may be at significant risk of becoming contaminated due to re-entry or entrainment (as described in the next section).

In addition to the location of exhaust and intake air systems and building depressurization, other factors contribute to re-entry. These include the design and height of exhaust vents, exhaust velocity, and air flow patterns over building roof surfaces where exhaust vents are located.

For both practical and architectural (aesthetic) reasons, exhaust vents on many large buildings are relatively short and, in many cases, flush with the building surface. Exhaust vents are often covered with rain shields, which significantly reduce the upward momentum of exhaust gases.

Exhaust gases released near the roof surface have limited upward momentum. As a consequence, they become entrained in what is described as the building wake. The building wake is a layer of air close to the roof surface that is only weakly coupled with the free-flowing air above it. It is produced as a result of airflow patterns over and around building roof surfaces. This phenomenon is illustrated in Figure 11.17. As contaminants enter the wake, their dilution is limited because of the relatively stable air that characterizes it. Contaminants within the wake move in the same direction as air flowing above it. Commonly this results in a downwash phenomenon on the lee side of buildings, with subsequent re-entry of contaminants through outdoor air intakes or by infiltration through building leakage sites.

Re-entry is less likely to occur, or less likely to cause significant indoor contamination, when exhaust vents are elevated above the roof surface and operated to obtain significant upward momentum. In such instances, contaminants may be released above the building wake or break through it as a result of upward momentum.

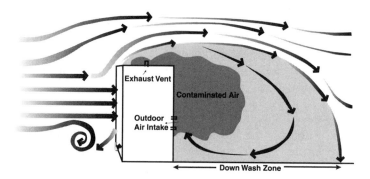

Figure 11.17   Building roof-top wake and re-entry phenomena.

v. Entrainment. In re-entry, exhaust gases, vapors, and particles may become entrained in outdoor air intake flows, resulting in varying degrees of indoor air contamination. Entrainment and subsequent indoor contamination may also occur from contaminants generated from nearby sources. They may move through outdoor air intakes, loading docks with open doors, and leakage sites by infiltration. Entrainment of contaminants generated external to buildings is a common occurrence. It is more notable when high-emission sources are located close to outdoor air intakes and when buildings or portions thereof are depressurized relative to the outdoor environment. Commonly reported entrainment problems include motor vehicle exhaust, vapors from asphalt sealers and asphalt or rubber compounds used in sealing roofing materials, particulate and vapor-phase substances associated with lawn or agricultural pesticide applications, and nearby commercial (restaurants, print shops) and industrial operations.

e. Ventilation innovations. Ventilation engineers have developed new systems designed to improve the effectiveness of conventional ventilation in controlling contaminants, and alternative systems that may be more effective than conventional general dilution ventilation systems. In the first case, demand-controlled systems have been developed to quickly respond to changing ventilation needs by increasing and decreasing volumetric outdoor air flow rates in response to changing indoor contaminant concentrations. In the second case, displacement ventilation systems have been designed to more effectively move contaminants generated in occupied spaces to exhaust outlets in ceilings.

i. Demand-controlled ventilation. Demand-controlled ventilation (DCV) systems differ from conventional ones in that outdoor air flow rates are varied in response to sensor-measured contaminant concentrations located in the return air stream of a room, ventilated area, floor, or whole building. These systems have the advantage of decreasing outdoor air flows to conserve energy when selected contaminants are below acceptable levels and of increasing outdoor air flows when contaminant levels begin to rise to unacceptable ranges.

A variety of sensors are available for use in DVC systems. These include CO_2, relative humidity, and mixed gas sensors. Though CO_2 is a relatively poor indicator of air quality other than human bioeffluents, CO_2 sensors are commonly used in DCV systems. This reflects their relatively low cost, reliability, simplicity of use, and the historical role of using ventilation rates based on CO_2 as the determinant of ventilation adequacy. The relationship between CO_2 levels and outdoor air flow rates can be seen in Figure 11.18.

ii. Displacement ventilation. In conventional ventilation systems, contaminant levels are reduced by mixing ventilation air with room air. Maximum contaminant reduction will result with "perfect" or complete mixing which, in many cases, does not occur.

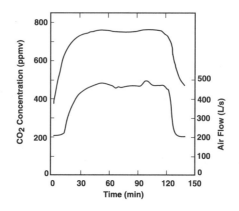

Figure 11.18 Demand-controlled increases in outdoor air flows associated with increases in indoor CO_2 levels. (From Strindehag, O., *Proc IAQ '91: Healthy Buildings*, ASHRAE, Atlanta, 1991. With permission.)

Displacement ventilation (DV) has been developed in an effort to ventilate spaces more effectively in industrial and office environments. Its use is much more widespread in Europe than in North America. It is based on the principle of moving contaminated air upward from floor or near-floor level to exhaust grilles in the ceiling in a piston-like fashion. Displacement ventilation attempts to move excess heat and contaminants toward the ceiling in order to improve air quality in the portion of the space occupied by humans.

In DV, relatively low flow rates (< 10 L/s) are created at floor level at temperatures of approximately 65°F (18°C). This air is warmed by sources in the space or by use of heaters. Upward convection air flows create, by piston action, stratified zones of contaminants and temperature, with the highest contaminant and temperature levels near the ceiling. Displacement ventilation appears to work best when supply air and convection flow rates are equal. The effect of DV on contaminants is illustrated in Figure 11.19. In

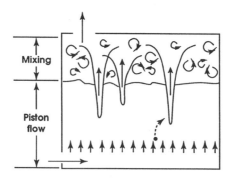

Figure 11.19 Effect of displacement ventilation on contaminant removal from a space. (From Koganei, M. et al., *Proc. 6th Internatl. Conf. Indoor Air Qual. Climate*, 5, 241, 1993.)

theory, DV is approximately 3 times more effective in removing contaminants from a space than general dilution ventilation at the same air flow rate.

f. General dilution ventilation and air quality. As indicated previously, ventilation, including both general dilution and exhaust ventilation, is widely used in many large, private, commercial, and institutional buildings to maintain acceptable air quality and comfort and control odors.

The relationship between building ventilation conditions and human bioeffluent levels determined from calculations of CO_2 emission rates and indoor measurements has been well established. It has been widely assumed that general dilution ventilation can be used to effectively control other contaminants as well and reduce complaints associated with "sick building syndrome" phenomena.

The use of mechanical ventilation (as opposed to natural ventilation) has been suggested to be a contributing factor to occupant complaints of illness symptoms which have been reported over the past three decades. The apparent relationship between "sick building" complaints and mechanical ventilation has been circumstantially linked to the "epidemic" of building-related health complaints which coincided with (1) efforts of building facilities managers in the late 1970s and early 1980s to reduce outdoor air ventilation flows to conserve energy and (2) the increasing construction of new buildings designed to be more energy efficient.

i. Ventilation and sick buildings. A number of cross-sectional epidemiological studies have been conducted to evaluate potential relationships between illness symptoms and building ventilation systems. Several studies have shown significantly higher complaint rates in mechanically ventilated, as compared to naturally ventilated, buildings. In other studies, different types of ventilation systems were compared to determine potential relationships with occupant health symptoms. These studies showed higher symptom prevalence rates associated with air-conditioned (with cooling) buildings (Table 11.1). The scientific evidence indicates that mechanical ventilation *per se* is not associated with increased symptom prevalence rates; rather, it suggests that mechanical ventilation with cooling is.

ii. Ventilation conditions and symptom prevalence. A number of cross-sectional epidemiological studies have been conducted to determine potential relationships between ventilation rates and building occupant symptom prevalence rates. No apparent relationships with ventilation rates were reported in some studies, whereas others have shown limited reductions in symptom reporting rates with increasing nominal or measured ventilation rates. Some evidence exists to suggest that increasing ventilation rates from minimum values results in significant reductions in symptom reporting rates and increases in occupant satisfaction with air quality. At increasingly higher ventilation rates, no significant improvements are observed. In general, stud-

Table 11.1 Relationship Between Building/Work-Related Symptoms and Ventilation System Type

Building ventilation type	Sickness index	% with symptoms									
		Dry eyes	Itchy eyes	Runny nose	Blocked nose	Dry throat	Lethargy	Headache	Flu	Difficulty breathing	Chest tightness
Natural (N = 442)	2.49	18	22	19	40	36	50	39	15	6	6
Mechanical (N = 944)	2.18	20	20	16	32	33	42	33	14	6	5
Local induction/fan coil (N = 508)	3.81	34	33	29	58	56	66	52	28	14	11
Central induction/fan coil (N = 1095)	3.70	31	32	28	57	54	68	47	29	12	14
"All air" (N = 1384)	3.12	31	29	21	45	46	56	43	25	9	8
Whole group (N = 4374)	3.10	27	28	23	47	46	57	43	23	9	9

Source: From Burge, S. et al., Ann. Occup. Hyg., 31, 493, Elsevier Science Ltd., Amsterdam, Netherlands, 1987. With permission.

ies in which ventilation rates are increased from the low to the high end of the range of 0 to 20 CFM (0 to 9.54 L/s)/person have shown significant reductions in symptom reporting rates, whereas studies at ventilation rates >20 CFM (9.54 L/s)/person have not. This indicates that ventilation rates <20 CFM/person are a risk factor for sick building-type symptoms.

The mixed results observed for studies designed to evaluate relationships between ventilation and building-related health complaints may be due to a number of confounding factors. These include: (1) differences in study designs, (2) problems encountered in controlling actual ventilation rates by mechanical system adjustments, (3) differences in ventilation efficiencies, (4) presence of sources with high emission strengths, (5) dynamic interactions between source emissions and ventilation, (6) relationships between contaminant concentrations and health effects, (7) ventilation systems serving as contaminant sources, and (8) multifactorial causes of building-related symptoms.

Only a few of these potential confounders are described here to suggest why few research studies have been able to establish a strong relationship between illness symptoms and building ventilation conditions. Most notable are relationships between contaminant concentrations, ventilation and health effects, ventilation systems as sources of contaminants, and the multifactorial nature of building-related health complaints.

It is often assumed that increasing ventilation by a factor of 2 will result in a reduction of contaminant levels by 50% and a concomitant 50% decrease in symptom prevalence. For many contaminant sources, emissions are coupled to ventilation rates as a consequence of vapor pressure phenomena. Increasing the ventilation rate increases the emission rate of emitting materials. The net result of this coupling effect is to reduce the effectiveness of ventilation in reducing contaminant concentrations in indoor air.

In addition to this nonlinear relationship between contaminant levels and ventilation, the effectiveness of ventilation is further diminished by the nonlinear relationship between irritation symptoms (typical SBS symptoms) and contaminant exposure. The relationship between symptoms and exposure concentrations is typically log-linear as evidenced by mouse bioassay studies of irritant responses (Figure 11.20). Given the log-linear response, reductions in contaminant levels of 90% or so would be required to achieve a 50% reduction in irritation/symptom response. Such reductions cannot be easily achieved by mechanical ventilation systems in buildings except when the initial ventilation condition is near zero. This is consistent with study results that show reductions in symptom reporting rates and occupant dissatisfaction with increasing ventilation rates up to 20 CFM/person. In Figure 11.13 it can be seen that increases in ventilation rates above 10 L/s/olf (or 10 L/s/person) apparently do not result in any significant improvement in occupant satisfaction with air quality. Thus, above a maximum ventilation rate (circa 20 to 30 CFM [10 to 15 L/s]/person), ventilation is relatively ineffective in reducing symptom reporting rates and increasing occupant satisfaction with air quality.

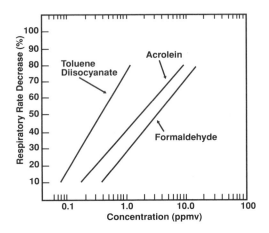

Figure 11.20 Log-linear relationship between irritant response and exposure concentrations in mouse bioassays. (From Alarie, Y., *Environ. Health Perspect.*, 42, 9, 1981.)

HVAC systems themselves may be a source of contaminants. This is notably the case with units that provide cooling. Condensate waters and wetting of nearby porous insulation have been widely reported to be sources of biological infestation and contaminant generation. Such infestation may result in exposures that cause allergenic responses in some building occupants and, in some cases, outbreaks of hypersensitivity pneumonitis. Duct liner materials may also serve as a source of particulate dusts on deterioration, and may emit a variety of gas-phase substances.

In addition to contaminant/human response relationships, other factors may confound the effectiveness of ventilation in reducing occupant symptoms and improving satisfaction with air quality. Significant evidence exists to indicate that health complaints in buildings are caused by a variety of factors, so that the prevalence of symptoms or a group of symptoms may be a cumulative result of responses to a variety of exposures. In cross-sectional epidemiological studies, it is common to identify multiple risk factors which appear to contribute to health complaints. These have included exposures to VOCs, floor dust components, handling carbonless copy paper, working with photocopy machines and photocopied paper, working with video display terminals, psychosocial factors, biological contaminants, etc. A number of these factors may not be air quality concerns (e.g., handling photocopied and carbonless copy paper, working with video display terminals, etc.), and thus would not be responsive to changes in ventilation rates. In other cases, exposure may be localized (e.g., resuspension of surface dust), and therefore not controllable by general dilution ventilation.

B. Local exhaust ventilation

When contaminants or potential contaminants are generated by high-emission sources such as specific equipment types, appliances, processes, or room environments, it is both desirable and practical to control them at the point of generation rather than allow them to disperse in building air where general dilution ventilation is less likely to be effective and economical.

1. Applications

Local exhaust ventilation is used for specialized purposes in large office, commercial, and institutional buildings, as well as in residential structures. Use of local exhaust ventilation is evidenced by the often numerous exhaust outlets or stacks on rooftops and sides of large buildings. Its use is also evidenced by the pressure imbalances that occur when building designers and facility operators fail to design and operate HVAC and local exhaust systems properly.

Specific applications of local exhaust ventilation in large buildings include the control of lavatory odors, odors and combustion by-products associated with cafeteria and restaurant emissions, ammonia from blueprint machines, solvent vapors from printing equipment and silk screening, paint and varnish vapors from school wood and automobile shops, dusts from wood-working shops, vapors from photography developing rooms, and water vapor from swimming pools and spas. Special applications in residential environments include exhaust fans in bathrooms to control water vapor generated in bathing, and in kitchens to control odors, cooking by-products (such as grease), and combustion by-products associated with gas cooking.

2. Performance

Local exhaust systems are designed to capture contaminants at or near their source of generation before they are dispersed into the surrounding environment. This is particularly the case with hood-based systems. It is less so when simple fans are used to exhaust contaminants from a localized area such as a bathroom, darkroom, etc.

Hood exhaust systems are used to capture combustion and cooking by-products in cafeterias, restaurants, and homes; solvent vapors in laboratories and shop rooms, etc. Such exhaust systems include one or more hoods, blower fans, ducts, and exhaust vents. Some exhaust systems may have low-efficiency particle filters that provide protection for the blower fan and minimize the accumulation of potentially hazardous deposits (e.g., cooking greases) in the blower and ducts.

The effectiveness of local exhaust ventilation systems depends on the configuration of the hood, its proximity to the contaminant generation source, air velocities at the face of the hood and volumetric flow rates, availability of makeup/replacement air, and a variety of environmental factors such as temperature of contaminated air and the presence or absence of turbulent air flows near the hood. Exhaust systems which are not properly designed and operated are unlikely to effectively remove contaminants generated from localized sources. This is particularly true with kitchen range hoods. Such hoods, in many cases, have insufficient fan power to capture

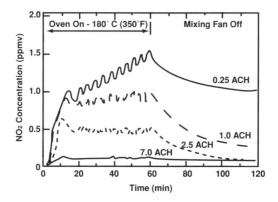

Figure 11.21 Effect of range hood volumetric air flow (air exchange) on NO_2 levels in a kitchen. (From Traynor, G.W. et al., *Atmos. Environ.*, 16, 2927, 1982. With permission.)

and remove combustion/cooking by-products. Contaminant (NO_2) removal efficiencies as a function of volume flow rates can be seen in Figure 11.21. In many cases, homeowners do not activate hood fans at all and if any contaminant removal occurs, it results from the buoyant upward flow of warm gases/vapors.

3. Special applications

a. Combustion exhausts. Special applications of local exhaust ventilation widely used in buildings are the exhaust systems that are integral parts of building space heating equipment such as boilers, furnaces, fireplaces, and wood-burning stoves. Such systems are equipped with ducts and chimneys that convey flue gas into the atmosphere. In the case of fireplaces, furnaces, and many boilers, flue gases are carried upward and outward by their thermal buoyancy and upward flow of building air through a natural draft hood. Mechanical fans are used to exhaust combustion gases when they may not have sufficient buoyancy. Mechanical exhaust is commonly used in oil-burning and all high-efficiency natural gas and propane-fueled furnaces. As indicated in Chapter 3, combustion gas exhaust systems are subject to flue gas spillage associated with equipment malfunction and weather conditions.

b. Soil-gas ventilation. Soil gas contaminated with radon is commonly removed in radon mitigation efforts by the use of sub-slab ventilation, a special application of local exhaust ventilation. By exhausting soil gases from beneath, and in the case of basements, around, building substructures, radon concentrations can often be reduced to acceptable levels.

In practice, a variety of approaches are used to exhaust soil gas to control indoor radon levels. These include the installation of systems to exhaust (1) air flowing into basement sumps, (2) soil gas beneath slab-on-grade and basement substructures, and (3) soil gas flowing upward through concrete

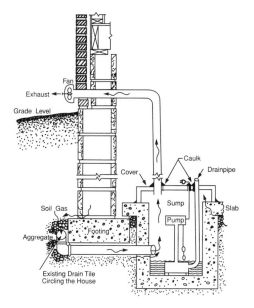

Figure 11.22 Sump/subslab ventilation used to control radon in a house. (From USEPA, *Radon Reduction Techniques for Detached Houses*, EPA/625/5-86/019, 1988.)

blocks. An application of a soil-gas ventilation system applied to a sump is illustrated in Figure 11.22. A soil-gas ventilation system coupled with sealing substructure cracks is the most effective mitigation method available to reduce high radon levels in residences. Typical radon reduction efficiencies are on the order of 90 to 95%.

In houses built on crawlspaces, radon reduction can be achieved by mechanically ventilating the air space between the ground and the floor. Typically, this is accomplished by using an exhaust fan. Such exhaust fans may also be used to control moisture levels in crawlspaces as well. It is desirable in such applications to use fans resistant to corrosion.

Readings

ASHRAE, *Ventilation for Acceptable Indoor Air Quality*, Standard 62-1989, American Society of Heating, Refrigerating and Air-conditioning Engineers, Atlanta, 1989.

Bearge, D.W., *Indoor Air Quality and HVAC Systems*, CRC Press/Lewis Publishers, Boca Raton, F., 1993.

Fanger, P.O., The comfort equation for indoor air quality, *ASHRAE J.*, October, 33, 1989.

Godish, T., *Sick Buildings: Definition, Diagnosis and Mitigation*, CRC Press/Lewis Publishers, Boca Raton, FL, 347, 1996, chap. 10.

Godish, T., Ventilation, in *Indoor Air Pollution Control*, Lewis Publishers, Chelsea, MI, 175, 1989, chap. 5.

Godish, T. and Spengler, J., Relationships between ventilation and indoor air quality: a review, *Indoor Air*, 6, 135, 1996.

Hunt, C.M., King, J.C., and Trechsel, H.R., Eds., *Building Air Change Rate and Infiltration Measurements, ASTM STP 719*, American Society of Testing Materials, Philadelphia, 1980.

Janssen, J.J., Ventilation for acceptable indoor air quality: ASHRAE Standard 62-1981, *Ann. ACGIH: Evaluating Office Environmental Problems*, 10, 59, 1984.

Maroni, M., Siefert, B., and Lindvall, T., *Indoor Air Quality. A Comprehensive Reference Book*, Elsevier, Amsterdam, 1995, chap. 29.

McNall, P.E. and Persily, A.K., Ventilation concepts for office buildings, *Ann. ACGIH: Evaluating Office Environmental Problems*, 10, 49, 1984.

Sandberg, M., What is ventilation efficiency?, *Building Environ.*, 16, 123, 1981.

Spengler, J.D., Samet, J.M., and McCarthy, J.F., Eds., *Indoor Air Quality Handbook*, McGraw-Hill Publishers, 2000, chapters 7, 8, 13.

USEPA, *Building Air Quality Manual*, EPA 402-F-91-102, USEPA, Washington, D.C., 1991.

Woods, J.E., Measurement of HVAC system performance, *Ann. ACGIH: Evaluating Office Environmental Problems*, 10, 77, 1984.

Questions

1. What is general dilution ventilation? Under what circumstances is it used?
2. What is natural ventilation?
3. Describe the forces responsible for infiltration-/exfiltration-induced air exchange in buildings.
4. Under what atmospheric conditions is building air exchange in a closed residence at its highest value? Its lowest?
5. Use Equations 11.2 to 11.4 to calculate the air exchange rate in a building when C_t = 400 ppmv CO_2, C_0 = 1500 ppmv CO_2, and t = 2 hours.
6. What factors contribute to ventilation system imbalances in large buildings?
7. Describe ventilation air flow through an air handling unit, into a conditioned space, and from and into outdoor air.
8. What is short circuiting? What factors contribute to it?
9. Describe differences between continuous and variable air volume HVAC systems and advantages and disadvantages of each.
10. How are ventilation guidelines and standards used to assure low odor and relatively good human comfort in mechanically ventilated buildings?
11. What is a unit ventilator? Under what conditions is it used?
12. Describe the Ventilation Rate Procedure; under what conditions is it used?
13. Describe Indoor Air Quality and Perceived Air Quality Procedures and their potential for being used to better ventilate buildings.
14. What is flushout ventilation? How is it used?
15. What is demand-controlled ventilation? How does it work?
16. What is ventilation efficiency? What factors affect it?
17. Why does cross-contamination occur in buildings?
18. What factors increase the risk of contaminant re-entry into buildings?
19. What is entrainment? How does it differ from re-entry and cross contamination?
20. How effective is general dilution ventilation in reducing SBS-type symptom complaints in buildings?
21. What factors confound the evaluation of the effectiveness of general dilution ventilation in mitigating SBS-type symptom complaints?

22. How is local exhaust ventilation used in a mechanically ventilated building?
23. Why is local exhaust ventilation in many cases more effective than general dilution ventilation in reducing contaminant exposures in buildings?
24. What factors determine the effectiveness of local exhaust systems?
25. How is local exhaust ventilation used to control radon levels in buildings?
26. In one mechanically ventilated building, CO_2 levels were observed to peak at 10 a.m. and again at 2 p.m. Explain the reason or reasons for such differences.

chapter twelve

Air cleaning

Air cleaning is commonly used in industrial operations to reduce emissions of particulate dusts as well as gaseous and vapor-phase contaminants to the ambient atmosphere. It is also used as a contaminant control measure in indoor environments. Air cleaning is used both generically and to control specific contaminant problems. It is often used by individuals in an attempt to reduce exposure to allergens and other suspected indoor air contaminants. It is rarely used as a mitigation measure to resolve problem building complaints.

In most applications, air cleaning is a process in which airborne contaminants are removed from a moving airstream by some type of physical, chemical, or physical–chemical process. Less commonly, air cleaning is attempted by introducing charged ions into indoor spaces, or using oxidizing substances such as ozone (O_3) or passive systems such as plants.

Though air cleaning systems may be used generically to improve overall cleanliness in indoor spaces and protect mechanical equipment in (1) heating, cooling, and ventilation (HVAC) systems in mechanically ventilated buildings and (2) residential heating and cooling systems, air cleaners cannot, as many lay individuals believe, be universally applied for the control of all airborne contaminants. Most air cleaners have been designed to collect airborne particles, and thus cannot control gas/vapor-phase contaminants. However, air cleaners can be designed to control specific gas/vapor-phase substances or classes of substances, or a combination of particle and gas/vapor-phase contaminants.

I. Airborne particles and dusts

Most indoor spaces are sufficiently contaminated by airborne particles to warrant at least a minimum level of air cleaning. Particles may be generated indoors from a variety of activities and sources or enter indoor spaces from the ambient atmosphere through HVAC systems or natural ventilation processes. Particles may include fabric lint and paper dust, tobacco smoke, organic dusts (see Chapter 5), mineral particles, pollen and mold spores, industrially and photochemically derived particles, etc.

Airborne particles can be removed from indoor and outdoor air used for ventilation by application of relatively simple physical principles adapted from industrial gas cleaning. Air flows through air cleaning devices used for indoor environments are orders of magnitude less than those used for stack gases and large industrial local exhaust ventilation systems. Particle/dust loading rates are significantly lower as well. Airborne dust concentrations in indoor air rarely exceed 200 µg/m³ and are usually <50 µg/m³. Airborne particles are typically collected and removed from ventilation air by fibrous media filters; less commonly by electrostatic precipitation.

A. Filtration

A wide variety of filter types and filtration systems are used to remove airborne particles from supply air systems and indoor spaces. Filter panels are inserted into air-handling units (AHUs) upstream of blower fans in HVAC systems and domestic heating and cooling systems. These panels contain a medium which varies from a simple metal grid to the more commonly used fibrous mats which are oriented perpendicular to the direction of air flow. Filter materials may include glass, cellulose, or polymeric fibers, which vary from <1 to 100 µm in diameter. Filter mats vary in density and depth, with porosities in the range of 70 to 99%. Because of differences in fiber diameters and mat densities, filters vary in their ability to capture airborne particles.

1. Collection processes

Large particles (such as lint) may be collected on filters by sieve action. Most smaller particles are collected as a result of a number of particle deposition processes. These include interception, impaction, diffusion, and electrostatic attraction. Deposition of particles on filter fibers by interception, impaction, and diffusion processes can be seen in Figure 12.1.

Interception occurs when particles follow a streamline within one radius of a filter fiber (Figure 12.1a). As a particle collides with the perimeter of a fiber, it loses velocity and is captured. Interception is important in collecting particles with diameters >0.1 µm, and is a particularly important collection mechanism in the range of minimum collection efficiency. Collection efficiency increases with increasing filter density and is independent of particle velocity.

Impaction occurs when particles of sufficient size and mass cannot flow along fluid streamlines around filter fibers. Due to their inertial mass, large particles collide with filter fibers and are collected (Figure 12.1b). Collection efficiency increases with increasing particle size, particle velocity, and Stokes number (ratio of particle stopping distance to fiber diameter). Impaction is

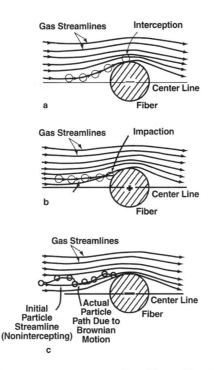

Figure 12.1 Particle deposition processes on filter fibers. (From Hinds, W.C., *Aerosol Technology — Properties, Behavior, and Measurement of Airborne Particles*, John Wiley & Sons, New York, 1982. With permission.)

an important particle collection mechanism for particles with aerodynamic diameters >1 μm.

Particles that are <1 μm in diameter behave like gases; i.e., they follow fluid streamlines and are subject to Brownian motion (random motion of molecules). As a consequence, they can diffuse to surfaces. Brownian motion/diffusion increases the probability that small particles (<1 μm) will move into an intercepting streamline and be deposited on a filter fiber (Figure 12.1c). Diffusion is the only collection mechanism for particles with diameters < 0.1 μm (Figure 12.2). Collection efficiency increases with decreasing particle size.

Particles can be collected on ordinary filter fibers by electrostatic processes in which particles which naturally carry a charge are attracted to fibers that carry an opposite charge. The relative role of electrostatic deposition (as compared to other deposition processes) in filtration has not been well-defined.

2. Filter performance

The efficiency or performance of a filter is determined by parameters involving particles, filters, and air flow. These include particle diameter, fiber diameter, filter packing density, filter depth, and air flow rate.

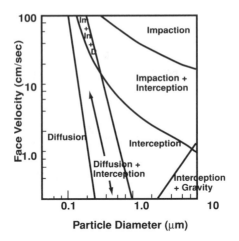

Particle Diameter (μm)

Figure 12.2 Relationship between particle size and deposition processes. (From Hinds, W.C., *Aerosol Technology — Properties, Behavior, and Measurement of Airborne Particles,* John Wiley & Sons, New York, 1982. With permission.)

As can be seen in Equation 12.1, the ability of a particle to penetrate a filter decreases exponentially with increasing filter thickness.

$$P = e^{-rt} \tag{12.1}$$

where P = penetration, %
      e = natural log base
      r = fractional capture per unit thickness
      t = filter thickness, mm

The value of r depends on particle diameter, packing density, fiber density, and face velocity of the airstream.

Not surprisingly, easy-to-collect particles are collected on or near the surface of a filter. As particles move through it, particle size distribution changes (based on aerodynamic diameters). Each filter type is characterized by a particle size, usually between 0.05 and 0.5 μm, at which fractional collection efficiencies are minimal. This generalized relationship can be seen in Figure 12.3 for particles moving through a filter at 1 and 10 cm/sec. As face velocity increases, a minimum value for collection efficiency is reached. Collection due to diffusion at this point is at a minimum. These minimum collection efficiencies are used to specify the performance of high-efficiency filters. High-efficiency particulate absolute (HEPA) filters are reported to be 99.97% efficient at 0.3 μm. This means that they have a minimum 99.97% collection efficiency for the most difficult-to-collect particle size.

Collection efficiency is also a function of fiber diameter (Figure 12.4). The highest minimum collection efficiency and overall collection efficiency occur at the smallest fiber diameter tested (0.5 μm).

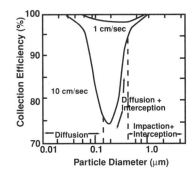

Figure 12.3 Relationship between particle size and collection efficiency. (From Hinds, W.C., *Aerosol Technology — Properties, Behavior, and Measurement of Airborne Particles*, John Wiley & Sons, New York, 1982. With permission.)

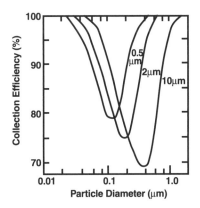

Figure 12.4 Relationship between filter fiber diameter and collection efficiency. (From Hinds, W.C., *Aerosol Technology — Properties, Behavior, and Measurement of Airborne Particles*, John Wiley & Sons, New York, 1982. With permission.)

As indicated in Equation 12.1, particle penetration decreases with increasing filter thickness. Collection efficiency increases as well. However, as filter thickness increases, resistance to air flow also increases, with a corresponding pressure drop and decreased air flow through the filter. Pressure drop is directly proportional to filter thickness and inversely related to fiber diameter. A significant pressure drop is undesirable because it reduces the air flow rate and volume of air treated.

3. Filter applications and types

Filters are used in a variety of air cleaning applications. Dust-stop filters used in HVAC systems and residential heating and cooling units are the major application. Higher-efficiency in-duct filters may be used to improve the physical cleanliness of indoor surfaces and reduce airborne allergen levels in residences. Free-standing modular air cleaners employing fibrous

Figure 12.5   Dry panel-type filter.

filters are designed and widely used to clean air in small spaces such as rooms in residences. Filtration systems which require high performance characteristics (HEPA) are used in clean rooms in the manufacture of pharmaceuticals and semiconductors, in nuclear power plants and other nuclear facilities, and in hospital operating rooms. They are also used in portable home air cleaning devices.

Commercially available filters include wire mesh units used in older HVAC systems and cooking range hoods, and the more commonly used fibrous media filters. In the latter case, filters are described as dry, viscous-media, or charged-media filters (see electronic air filters). They may be single-use or renewable panels.

a. Dry-type panel filters.   Dry-type panel filters (Figure 12.5) have high porosities and low dust spot efficiencies. As a consequence, their use is limited to dust-stop filters in HVAC systems and home heating and cooling units. They collect large particles such as lint by sieve action, impaction, and interception. They are typically used in systems with air velocities in the range of 200 to 700 ft/min (fpm) (60.9 to 213.3 m/min [mpm]). Newly exposed filters have low associated pressure drops (0.05 to 0.25" H_2O, 12.4 to 62 pascals). Typically, they are allowed to reach pressure drops of 0.50 to 0.75" H_2O (124 to 186 pascals) before being replaced. Collection efficiency increases significantly as filters become soiled. Replacement is required because of the associated increased resistance to air flow. Filter media used in dry-type panels include fiberglass, open cell foams, nonwoven textile cloths, and cellulose fibers.

b. Extended-surface dry-type filters.   Extended-surface media filters have been developed to overcome the air resistance/pressure drop problems associated with use of thick, high-density, high-efficiency filtration media. Filter surface area is extended by pleating the filter medium; this significantly increases the surface area available for particle collection (relative to the flat

Figure 12.6 Bag-type filter.

face area of the filter). Pleat depth may be several inches to upwards of 12 to 18″ (30.4 to 45.7 cm) or more. As the medium surface area is increased, pressure drop decreases to acceptable levels (despite the increase in medium density and/or thickness).

Extended-media filters vary in thickness, density, fiber size, media composition, pleats per nominal face area, and depth. As such, their performance varies from medium to high efficiency. Cellulose, glass fiber, wool felt, or a variety of synthetic fibers are used in extended-media filters. Fibers are typically oriented randomly to form a mat.

Extended-media filters are available in several designs. They may be constructed in the form of bags which are similar in concept, and somewhat similar in design, to bags used in industrial applications. A bag-type extended-media filter used in HVAC systems is pictured in Figure 12.6.

In typical pleated filters, the medium is held in place by a panel frame or box (Figure 12.7). The V-shaped pleats may have a depth of 2 to 36″ (5.1

Figure 12.7 Medium efficiency pleated-panel filter.

Figure 12.8   HEPA filter.

to 91.4 cm) or more, depending on the application. Increasing pleat number and filter depth are used to increase filtration efficiency, air volumes that can be cleaned, or both. The filter medium may be rigid enough to be self-supporting. In many high-efficiency filtration appliances, the filter medium is held in place by a combination of rigid corrugated metal spacers and a thick adhesive application between the filter medium and its surrounding panel, case, or box.

As indicated, extended-media filters provide higher performance than dry-panel filters. They are classified as medium, high, and very high efficiency. Medium-efficiency filters have dust spot efficiencies (described later) in the range of 40 to 60%. They consist of 5 to 10 μm diameter fibers that are in $1/4$ to $1/2''$ (6 to 12 mm)-thick mats. High-efficiency filters have dust spot efficiencies in the range of 80 to 90% or more, with fiber diameters in the range of 1 to 4 μm.

Very high particle cleaning efficiencies are achieved by HEPA filters (collection efficiencies of 99.97% or greater at a 0.3 μm DOP test [see DOP smoke penetration tests]). A HEPA filter is illustrated in Figure 12.8. Their small fiber diameters and high packing densities favor the collection of very small particles (circa 0.01 μm) by Brownian motion. Electrostatic forces cause small particle agglomeration and adherence to media fibers. In some applications, filter surfaces are coated to discharge static build-up so that they may be cleaned by a pneumatically operated pulse cleaning system and reused.

c. Viscous-media panel filters.   Filter fibers are often coated with viscous, low-volatility oils to enhance particle collection and retention. Viscous-medium panel filters have high porosities and low resistance to air flow. They also have very low collection efficiencies for particles commonly found in indoor air, but are very efficient in collecting fabric dusts and very large particles (>10 μm). Typical operating velocities through such filters are in the range of 300 to 600 fpm (91.4 to 182.8 mpm).

d. Renewable-media filters. Renewable-media filters were developed to provide a filter surface that is slowly being moved to accommodate particle collection and HVAC system operating needs. They consist of a slowly moving curtain/filter unit which advances in response to pressure drop or a timer. As the medium becomes excessively soiled, it moves to a takeup roll on the bottom of the filter system. When the roll has been completely soiled, it must be removed and replaced. Need for replacement is usually signaled by an alarm. Such filters have high arrestance (described later) efficiencies (60 to 90%) and low dust spot efficiencies (20 to 30%). They are available as both dry and viscous types. In viscous-media renewable filter systems, the filter may pass through a reservoir of a viscous medium where it sheds its dust load and is recoated.

4. Air flow resistance

Because of their nature, all air cleaning systems using the principle of filtration must be designed and operated with due consideration to the effects of the filter on air flow. Filters with high packing densities have both increased collection efficiencies and resistance to air flow. As a filter becomes soiled, its resistance to air flow increases; as resistance increases, air flow decreases.

As indicated previously, extended-media filters are designed to provide high collection efficiencies at acceptable air resistances described as pressure drop (ΔP). Pressure drops associated with high-efficiency filters and filter soiling must be considered in the design and operation of cleaner units and HVAC systems. Of major importance is the selection and use of system blowers that will develop sufficient static pressure to overcome resistance to air flow and maintain desired volumetric air flows. As a consequence, blower fans must be adequately sized in terms of volumetric flow rate and horsepower rating.

For illustrative purposes, let us use two different fans for a small freestanding residential air cleaner (Table 12.1). Though both fans have the capacity to move 125 CFM of air at zero static pressure (no air resistance), they differ in horsepower. Fan #1 has a higher horsepower rating than fan #2. As resistance to air flow increases to 0.5″ H_2O, air flows associated with both fans decrease. At 0.5″ H_2O ΔP, fan #2 can pull virtually no air. Fan

Table 12.1 Air Flow Rates Under Different
Static Pressures for Two Fans

Static pressure (″ H_2O)	Air flow rate (CFM)	
	Fan #1	Fan #2
0	125	125
0.1	120	115
0.2	115	105
0.3	110	98
0.4	105	85
0.5	100	—

horsepower required to draw air through a filter at a constant or near-constant rate increases with increased resistance to air flow. Fans with higher horsepower ratings are more expensive to purchase and operate. However, an adequately sized fan is essential for proper system performance.

If resistance becomes excessive, performance will decline significantly even when a properly sized fan is used. This occurs if filters are not replaced when they reach their design resistance values. Maximum acceptable resistance values for products are provided by manufacturers, who also provide resistance values for clean filters at their rated air flow.

In many HVAC system applications, a differential pressure gauge monitors pressure differences (ΔP) upstream and downstream of the filter. The pressure drop increases as the filter becomes soiled. In some systems, an alarm sounds when the pressure drop exceeds a predetermined value. The alarm signals maintenance personnel that filters need to be changed to maintain desired system air flows. In other cases, maintenance personnel periodically check pressure gauges to determine filter change requirements.

Sensors or pressure drop indicators are not present in most residential and many HVAC applications. Because of the variability of dust loading on filters, particularly in HVAC systems, it is difficult to know exactly when filters need to be replaced. In such cases, filter replacement is a matter of judgment by service personnel and homeowners. Filters are often replaced on a routine schedule irrespective of their condition. Since excessively soiled filters cause decreased air flows and increased operating costs in HVAC and home heating and cooling systems, it is important that operators implement an appropriate service plan.

B. Electrostatic air cleaners

Electrostatic air cleaners remove airborne particles by electrostatic forces. Three basic designs are used: ionizing plates, and the charged-media ionizing and nonionizing types.

1. Ionizing-plate cleaners

Ionizing-plate electrostatic air cleaners are widely used to collect airborne particles in HVAC systems and residential applications. Their operation is based on the principle that airborne particles can be given a positive or negative charge and then collected on metal plates with the opposite charge. In industrial applications, particles are negatively charged; in indoor applications, they are positively charged.

Both single- and two-stage electronic cleaners are available for use in indoor applications. A two-stage cleaner is illustrated in Figure 12.9. In the first stage, a high electric potential (12,000 volts) is applied to thin, vertical, tungsten wires. Electrons are accelerated toward the positively charged ionizing wires. The accelerated electrons strike air molecules, stripping them of

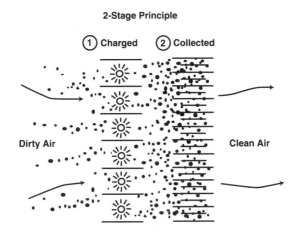

Figure 12.9 Two-stage electronic air cleaner particle collection.

electrons and creating positive ions and additional electrons. The process produces a corona discharge near ionizing wires.

Positive ions become attached to airborne particles by deposition processes. As a consequence, these particles become positively charged. The magnitude of the charge on individual particles depends on the number of charges deposited. Particles with high surface areas gain more charge and therefore have a higher probability of being collected.

Charged particles flow with the airstream into a collector section (second stage) consisting of a series of vertically placed, parallel, thin, metal plates. Alternate plates may be positively and negatively charged by a high DC voltage (6000 V). Positively charged particles are attracted to a negatively charged plate. The magnitude of the electrostatic force acting on a particle depends on its charge, the distance between plates, and the voltage applied.

Particles deposited on collection plates lose their original charge and take on the charge of the collecting surface, where they remain attached to the plate and other collected particles by molecular adhesion and cohesion. As particle buildup occurs, electrostatic forces diminish in magnitude. Collection efficiency decreases as a consequence. Collection plates must be cleaned of their accumulated dust load periodically to restore their initial collection efficiency. Hot water is typically used to clean collection plates; in some systems washing is done automatically.

Electronic air cleaner collection efficiencies depend on particle migration velocity, collection surface area, travel path length (distance through a collection field), and air flow rate. Migration velocity is directly proportional to a particle's charge and the strength of the electric field. In electronic air cleaners used for indoor applications, the travel path length is relatively short, typically 6 to 12" (15.2 to 30.5 cm). In industrial applications, travel path length may be 20 to 25' (6.1 to 7.6 m) or longer, with collection efficien-

Table 12.2   Relationship Between Flow Rate
and Electronic Air Cleaner Dust
Spot Efficiency

Flow rate (CFM)	Dust spot efficiency (%)
400	93
550	90
750	85

cies of approximately 99%. In indoor applications, collection efficiencies are rarely >95%, and are usually less.

As seen in Table 12.2, collection efficiencies are significantly affected by volumetric air flow rates. As air velocity increases, particles have less time to be drawn to and deposited on collection plates. As a consequence, collection efficiency is reduced. Because of the sensitivity of electronic air cleaners to changes in velocity, it is common to use prefilters, perforated plates, etc., to induce some resistance and provide uniform velocity air flow through the collection system.

In single-stage electronic air cleaners, ionizing wires are placed between collection plates. Because of reduced travel path lengths, single-stage cleaners are less efficient than two-stage units. They have the advantage of requiring less space in AHUs.

Electronic air cleaners are available as modular, free-standing units; modular units that can be suspended from a ceiling or mounted on a wall; in-duct units installed in residential heating and cooling systems; or units of various sizes placed in AHUs of mechanically ventilated buildings. A residential in-duct and a portable free-standing electronic air cleaner are pictured in Figure 12.10.

Resistance to air flow in electronic air cleaners is very low; as a consequence, volumetric air flows are constant. Because of low air flow resistance, energy requirements for fan systems are also low. Maintenance is usually limited to cleaning collector plates. It is both desirable and common to use dust stop filters as prefilters to maintain high collection efficiencies for small particles and reduce the need for cleaning.

Electronic air cleaners have one major disadvantage. Because of the high voltages used, they produce ozone (O_3). Ozone production varies among products and product vendors. It is common to smell the sweet odor of O_3 in residences where electronic cleaners are in continuous operation (portable and in-duct units). The presence of O_3 in buildings is of concern because of its known health effects, its ability to oxidize fabric dyes and crack rubber and soft plastics, and its ability to initiate indoor chemical reactions.

2. *Charged-media ionizing cleaners*

In this type of air cleaner, dust in an airstream is charged by passing it through a corona discharge ionizer and then collected on a charged-media filter.

a   b

Figure 12.10 Residential in-duct and free-standing electronic air cleaners. (Courtesy of Honeywell, Inc., Minneapolis.)

3. Charged-media nonionizing cleaners

These cleaners include characteristics of electrostatic and dry-filter particle collection. They consist of a dielectric filter mat made of glass fiber, cellulose, or other fibrous materials supported on or in a gridwork of alternately charged (12,000 V) or grounded members. Consequently, a strong electrostatic field develops in the filter medium. Particles approaching the filter medium are polarized and drawn to it, where deposition takes place. Because of increased resistance to air flow associated with soiling, such filters need to be replaced periodically.

C. Performance measurement

Air cleaners vary in their collection efficiency (as well as overall particle removal) for particles of different aerodynamic diameters, resistance to air flow, service life, and particle-holding capacity. As a consequence, there is a need for uniform evaluation and rating of filter performance. Test procedures prescribed under ASHRAE (American Society of Heating, Refrigerating and Air-Conditioning Engineers) Standards 52.1-76 provide performance characteristics of most commercially available filters and filtration systems. Two different ASHRAE test methodologies were used to evaluate filter and filtration system performance. In the Spring of 2000, ASHRAE published a new standard, 52.2-1999, which is likely to serve as the primary filter test methodology in the future. These methodologies, plus an additional methodology for HEPA filters, are described below.

Table 12.3 Performance of Dry Media Filters

Media type	Arrestance (%)	Dust spot efficiency (%)
Fine open foams/textile nonwovens	70–80	15–30
Mats of glass fiber/multi–ply cellulose/wool felt	85–90	25–40
Mats of 5–10 μm fibers, 6–12 mm thick	90–95	40–60
Mats of 3–10 μm fibers, 6–12 mm thick	>95	60–80
Mats of 1–4 μm fibers, mixtures of fiber types	>95	80–90

Source: From ASHRAE, Air cleaners, in *Equipment Handbook*, ASHRAE, Atlanta, 1983, chap. 10.

1. Arrestance

Arrestance is a measure of the ability of a filter to collect relatively large particles. It is a measure of the performance of panel filters that are used to protect mechanical equipment or serve as prefilters where low-efficiency cleaning is acceptable.

Arrestance values are determined by aerosolizing a standard dust mixture upstream of the filtration system. The ASHRAE dust mixture includes, by weight, 72% standard air-cleaner fines, 23% molacco black, and 5% No. 7 cotton linters ground in a mill. This dust mixture is designed to take into account the large variability in composition and size of particles that enter HVAC system AHUs. Arrestance values are based on the weight of standard dust collected on or in the filter compared to the weight of the standard dust mixture aerosolized upstream of the filter. Arrestance values for a number of filter media/filters are found in Table 12.3. Note that, in general, arrestance values are relatively high, and increase with increasing filter density and smaller fiber size. Because arrestance focuses on weight, it is a measure of large particle (>10 μm) cleaning effectiveness.

2. Dust spot efficiency

The dust spot efficiency test is a better indicator of filter performance over a broad range of particle sizes. It is employed to measure the performance of medium to high efficiency extended-media filters and electronic air cleaners. It measures discoloration differences (as determined by optical density) observed on glass fiber filter tape samples collected both upstream and downstream of filter/filtration units. Dust spot efficiency of the filtration unit is expressed as percent reduction of the optical density of downstream compared to upstream samples.

Dust spot efficiencies for a range of filter media/filters are also summarized in Table 12.3. Note the significant differences between arrestance and dust spot efficiency values. High dust spot efficiencies are associated with filter media with small fiber diameters and high packing densities. Though dust spot efficiency is a relatively good measure of overall dust collection efficiency, it is subject to some error due to differences in optical properties of collected particles. Darker particles absorb more light and therefore result in higher optical density readings than similar concentrations of lighter colored particles.

The dust spot test was designed to measure an objectionable characteristic (ability to soil interior building surfaces) caused by relatively small diameter (<10 μm) particles. It was developed in response to limitations of the arrestance testing method. Because arrestance and dust spot testing measure different performance characteristics, values derived from these tests cannot be used interchangeably. However, filters with medium to high dust spot efficiencies also have high arrestance values.

3 *Minimum efficiency reporting values*

ASHRAE, under its recently published standard 52.2-1999, established a test protocol to test filter performance based on measuring fractional collection efficiencies. These efficiencies are determined by conducting particle counts both upstream and downstream of filters challenged by exposure to laboratory-generated potassium chloride aerosols in 12 different size ranges (smallest 0.3–0.4 μm; largest 7–10 μm).

In addition to measuring the performance of clean filters, particle size fractional efficiency curves are developed at incremental dust loadings. These curves are used to develop composite curves that identify minimum efficiency in each particle size range. Minimum efficiency composite values are averaged in three size ranges (0.3–1.0 [E1], 1.0–3.0 [E2], and 3.0–10.0 [E3] μm) to determine minimum efficiency reporting values (MERVs). Filters with MERV values of 1 to 4 are very low efficiency; they would be used as furnace filters. Pleated filters would have MERV values of 5 to 8; box/bag filters, 9 to 12; box/bag filters, 13 to 16; and HEPA filters, 17 to 20.

4. *DOP smoke penetration tests*

Measurements of particle collection efficiencies >98% are made using the DOP (dioctyl phthalate) smoke penetration method. It is typically used to test the performance of HEPA filters.

In this method, an aerosol of DOP is produced with a uniform particle diameter of 0.3 μm, the size range of minimal collection efficiency for most HEPA filters. DOP particles that pass (penetrate) through the filter or leak around filter-sealing gaskets are determined downstream by means of a photometer that measures light scattering. Their concentration is compared to that measured upstream of the filter. Filter penetration is calculated by using the following equation:

$$P = 100\frac{C_2}{C_1} \qquad (12.2)$$

where P = % penetration
C_1 = upstream concentration
C_2 = downstream concentration

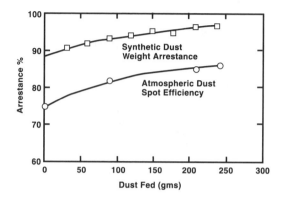

Figure 12.11 Relationship between filter loading and collection efficiency. (From *ASHRAE Equipment Handbook*, ASHRAE, Atlanta, 1983, chapter 10. With permission.)

Because HEPA filters have efficiencies near 100%, penetration rather than efficiency is typically reported. Penetration can be converted to % efficiency by using the following equation:

$$E = 100 - P \tag{12.3}$$

where E = efficiency (%)

5. Filter soiling and efficiency

As a filter becomes soiled, the mat of collected particles serves as a collection medium so that collection efficiency increases with increased service life. This increase in both arrestance values and dust spot efficiency as a function of dust exposure can be seen in Figure 12.11. As a consequence, "dirty" filters have an associated increase in cleaning efficiency. Notice, however, the significant effect on air flow resistance with increasing filter dust loads in Figure 12.12.

D. Use considerations

Air cleaners designed to control airborne particles are used in a variety of applications. In mechanically ventilated buildings, low-efficiency dust stop filters and renewable-media filters are used to protect mechanical equipment. Increasingly, a higher level of particle cleaning and performance is being designed and specified for mechanically ventilated buildings to achieve and maintain cleaner indoor spaces. Medium efficiency extended-surface filters, or bag filters, are commonly specified by mechanical engineers and installed by building owners/operators.

In most residential environments, air cleaning is limited to dust stop filters used in heating and cooling systems to protect mechanical equipment. In the 1980s, a significant air cleaner market targeted to average consumers

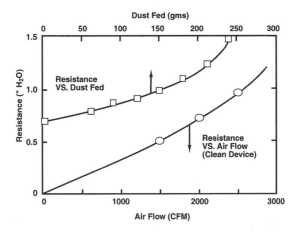

Figure 12.12 Relationships between filter loading and resistance and resistance and volumetric air flow. (From *ASHRAE Equipment Handbook*, ASHRAE, Atlanta, 1983, chapter 10. With permission.)

developed. This market included a variety of products which can be categorized on the basis of cost, potential air cleaning capacity, and efficiency.

1. Residential and consumer air cleaners

a. Fan and filter desktop cleaners. At the low end of the market are devices that range in cost from $20 to $100. They are small desktop devices that use dry, loosely packed, low-density filters located upstream of a high velocity, low air resistance axial fan. Some units utilize electrostatically charged electret filters (thin plastic materials imprinted with high voltage charges). In most cases, such devices have insufficient capacity to clean air in even a single closed room. Performance tests conducted in a 1200 ft³ (33.6 m³) room have indicated that tobacco smoke removal effectiveness of such devices under static chamber conditions (i.e., no air flow in or out of the space) is little better than using no device at all.

b. Negative ion generators. The simplest and least expensive devices generate ions that diffuse out into building air and attach to particles which plate out on building surfaces. More advanced models are designed to eliminate the "dirty wall effect" associated with simple ion generators. In such devices, an attempt is made to draw charged particles into the air cleaner (by using a suction fan), where they can be deposited on an electrostatically charged panel filter. In other ionizers, a stream of negatively charged ions is generated in pulses, and charged particles theoretically are drawn passively to the ionizer, which contains a positively charged cover.

Negative ion generators have long been used as health-promoting devices, particularly by a small population of medical practitioners who treat allergy patients. There is limited evidence that negative ions may affect an individual's sense of well-being and alleviate allergy symptoms. Because such health claims cannot be definitively established, manufacturers cannot promote negative ion generators for such purposes. After being forced from the market in the early 1960s by the Food and Drug Administration (FDA), these devices were reincarnated as air cleaners. They have been shown to be relatively effective in removing smoke from a 1200 ft³ test room under static conditions.

c. Portable extended-media and electronic cleaners. Air cleaners of this type are portable devices which are larger, and more expensive and effective than desktop units. They utilize extended-media high-efficiency filters, HEPA filters, or two-stage electronic systems. A portable electronic air cleaner can be seen in Figure 12.10b. Air flow rates are dependent on the filter type, blower capacity, and fan horsepower. These devices utilize good quality high-efficiency filters/filtration systems and should be relatively effective in cleaning particles from air in a single room with an air volume up to 3200 ft³ (*circa* 90 m³).

As they are widely recommended by allergists, the use of such portable high-efficiency air cleaners is common. They have been demonstrated to be relatively effective in reducing smoke particle concentrations in a single room under static conditions. An approximate 70% reduction in particle levels was observed when an air cleaner with a HEPA filter was placed in the bedrooms of 32 allergy patients for a 4-week period. These studies indicated that portable high-efficiency devices can be used to reduce particle levels in a single room when operated continuously. However, portable electronic air cleaners operated continuously in closed rooms may produce considerable odor (from O_3 and other products) and even irritation effects.

d. In-duct cleaners. In-duct air cleaning devices are widely used in North American residences. In-duct devices employ either extended-media filters or electronic systems (Figure 12.10a). In-duct systems are installed in cold air returns immediately upstream of furnace blower fans. They are designed to clean the air of an entire residence and, except in very large residences, have the capacity to provide whole-house air cleaning. These systems, however, only remove particles when the furnace or cooling system blower fan is activated. In most cases, operation is intermittent since it responds to heating and cooling needs. In-duct systems need to be run continuously to achieve the air cleaning performance they are designed for. Most in-duct filtration media and electronic air cleaners have dust spot efficiencies of 90% or better.

The installation of an in-duct cleaning system can significantly reduce airborne particle levels by the initial action of moving contaminated air through the filter/filtration system and by its continuous recirculation. For maximum cleaning effectiveness, it is desirable to use a high-efficiency filter/filtration system and a high recirculation rate. The effect of two different recirculation rates on cleaning efficiency can be seen in Figure 12.13.

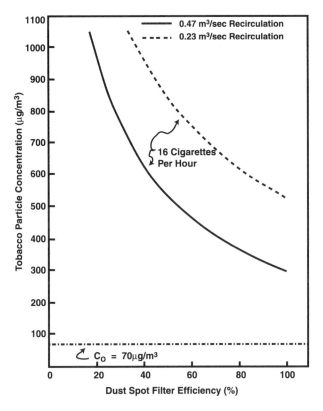

Figure 12.13 Effect of dust spot filter efficiency and recirculation rate on tobacco particle concentration. (From McNall, P.E., *Arch. Environ. Health*, 30, 552, 1975. With permission.)

Performance of an in-duct electronic air cleaner in a residence during a period of smoking can be seen in Figure 12.14. Smoke particle concentration decreased by about 70% in 2 hours to reach a steady-state concentration as occupant smoking continued. It decreased by an order of magnitude after smoking ceased. Other studies with in-duct electronic air cleaners in residences have demonstrated an order of magnitude or more reduction of airborne particles in unoccupied spaces. However, particle concentrations decreased by only 40 to 60% when studied over extended periods of time with unrestricted occupant activity. Occupant activities significantly reduced the apparent performance of filtration units by affecting particle generation and resuspension rates.

2. Air cleaning to control particulate-phase biological contaminants
Air cleaners are used in hospitals to limit infections among surgical and immunocompromised patients. Bacterial contamination of air in operating rooms is a major concern because of the significant potential for postoperative infections. The major cause of such infections is *Staphylococcus aureus* shed by

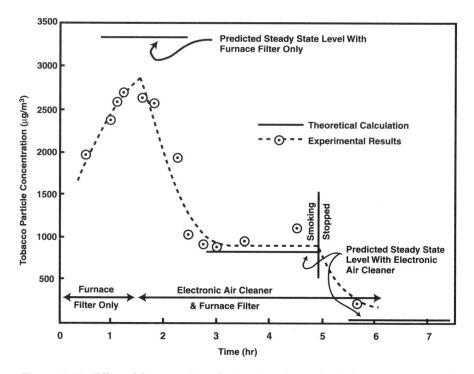

Figure 12.14   Effect of the operation of an in-duct electronic air cleaner on measured tobacco particle concentration in a residence. (From McNall, P.E., *Arch. Environ. Health*, 30, 552, 1975. With permission.)

surgical staff. Typically, surgical rooms use a combination of HEPA filters and laminar air flow. In one study of a surgery room operated at an air exchange rate of 20 ACH, culturable/viable bacterial levels were observed to decrease by approximately 89% when the system was operated under empty-room conditions and approximately 88% while surgery was in progress.

Air cleaners have been evaluated relative to their potential for reducing airborne mold levels in residences. In a study of culturable/viable airborne mold levels in residences, investigators observed significant reductions in total culturable/viable mold counts in residences using in-duct electronic air cleaners. Average reductions in 21 residences were approximately 77% (with a range of 66 to 87%) when the system was operated continuously and 50% (with a range of 0 to 83%) when the air cleaner was operated intermittently. The author has observed reductions in total airborne mold levels (viable and nonviable mold spores and particles) of 90% in a single-family residence using an extended-media filter.

In another study, cat allergen levels were reduced by 38% on the main level and 4% in the basement of a house with an in-duct HEPA filter when the blower system was operated continuously. Air cleaning appeared to be relatively ineffective when a significant active allergen-producing source (two cats) was present.

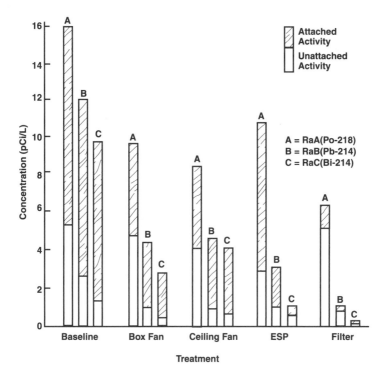

Figure 12.15 Effects of air treatments on radon decay product concentrations. (From Hinds, W.C. et al., *JAPCA*, 33, 134, 1983. With permission.)

The main reason for using an air cleaner in residential environments is to reduce the severity and prevalence of symptoms associated with allergy or asthma. Several studies have shown limited but less-than-definitive diminution of symptoms of allergy and asthma in patients in controlled studies using portable and in-duct high-efficiency air cleaners.

3. Radon control

Because of their particulate nature, radon decay product (RDP) exposures can, in theory, be significantly reduced by using an air cleaner. Radon decay products attached to particles were observed to have been reduced by 70% and 89%, respectively, in a 78 m³ chamber in studies which evaluated the performance of portable electronic and HEPA air cleaners. However, use of HEPA filters resulted in a threefold increase in unattached (to particles) RDP activity. The effects of air cleaning on RDP concentrations can be seen in Figure 12.15.

Though air cleaners are apparently relatively effective in reducing airborne RDP levels, increases in the unattached fraction relative to the attached fraction by HEPA filtration is of unknown, but worrisome, public health consequence since unattached RDPs can penetrate more deeply into the lungs. Consequently, USEPA does not recommend the use of air cleaning to reduce exposure to RDPs in residences.

4. Air cleaner use problems

Air cleaners are often purchased by consumers in response to physician recommendations or as a result of a personal desire to alleviate symptoms of allergy, asthma, or other building-related health concerns. Consumers and physicians are unaware of the limitations of air cleaners and their use in general, and limitations of specific air cleaner types and models. With the possible exception of negative ion generators, desktop devices are apparently ineffective. Portable and in-duct extended-media filters, on the other hand, can be used effectively if selected for the appropriate reason and operated properly.

Individuals often purchase air cleaners for generic use in alleviating allergy symptoms, ostensibly by reducing airborne allergen levels. Unfortunately, only a few airborne allergens are amenable to significant control by air cleaning. These include mold spores and hyphal fragments, and pet allergens when pets are no longer present to actively produce allergen. Dust mite fecal pellets and fragments are very large (>10 μm) and do not remain airborne for more than 10 minutes or so. Exposure occurs on disturbance. The same is likely true for cockroach allergens as well. For such allergens, air cleaning is unlikely to have any measurable benefit.

Though in-duct cleaning devices have the potential to significantly reduce airborne levels of mold and other small particles, air cleaning can only occur when the blower fan is operating. In residences using in-duct systems, the blower fan is usually not wired to allow blower fan operation independent of heating/cooling system activation. In residences where independent operation of blower and heating/cooling systems is possible, residents are rarely aware that good system performance requires continuous or near-continuous operation.

Consumers often purchase a portable air cleaner with the assumption that it can clean the air of a whole house if it is operated in a central location. Such portable devices have relatively low design and actual air flows (compared to in-duct systems). Performance decreases with increasing air volume when flow capacity is insufficient. Portable air cleaners achieve their best performance in single closed rooms (e.g., bedrooms). Good performance depends on air cleaner placement; in most cases, placement near the center of the room or near an allergy patient's nighttime breathing zone is recommended.

5. Ozone production

Electronic air cleaners and negative ion generators employ high voltages which cause ionization. Consequently, they have the capability to produce significant quantities of O_3. Because of its toxicity at low concentrations, its potential effects on rubber products, soft plastics, and fabric dyes, and its role in initiating indoor chemical reactions that produce irritants; O_3 production by such devices is of concern. Products sold as medical devices are regulated by the Food and Drug Administration (FDA). The FDA limits O_3 emission from these products so that an indoor level may not exceed 0.05

ppmv. With intermittent operation, in-duct electronic air cleaners are unlikely to cause any significant elevation of O_3. However, elevated levels may be expected when electronic air cleaning devices are operated continuously, particularly in a closed room.

6. Clean rooms

The need for nearly particle-free environments has led to the development and utilization of clean rooms. They are widely used in the semiconductor industry, pharmaceutical manufacture, aerospace and military applications, and for some medical purposes. Their early history reflects the parallel needs for very high efficiency filters in the nuclear and aerospace industries.

In first-generation clean rooms of the 1950s, very high levels of airborne particle control were achieved by the installation of HEPA filters in AHUs of clean room units. Such filters, even at that time, had DOP efficiency ratings of 99.97% at 0.3 μm. Clean rooms utilizing laminar flow principles were developed in 1962. In laminar flow, air moves uniformly in what is, in essence, parallel streamlines across a space. Laminar flow is used to minimize cross-streamline turbulence. Particles present or generated in the space tend to stay in a streamline until they are removed. Laminar flow patterns are induced by the uniform introduction of low-velocity air through a wall or ceiling area. Laminar flow patterns in a simple clean room design can be seen in Figure 12.16. In an ideal situation, the entire ceiling would serve as a perforated supply air plenum and the floor as a perforated exhaust plenum.

In conventional rooms, particles deposited on room surfaces can be resuspended by foot traffic and other movements that produce turbulent eddies. Under laminar flow conditions, air moves in predictable paths, and turbulent eddy formation is minimized. Laminar flow patterns are disturbed by the presence of people and objects. When a streamline is broken by an object, it may reform some distance downstream. If broken streamlines are not reformed, other streamlines transport airborne particles across the room.

The purpose of laminar flow is to remove (purge) particles generated in a clean room by workers and their activities and to prevent the resuspension of particles from horizontal surfaces, particularly the floor. Laminar flow systems facilitate the use of high purging air flows without the particle-disturbing turbulence typically associated with such air flows. However, as volumetric air flows increase to a critical level, they begin to resuspend more particles than they purge. An optimal flow rate widely used in clean rooms is 20 air changes per hour (ACH).

Clean rooms are limited in size and volume to optimize environmental control. As laminar air flow travel distances increase, particles are purged less effectively.

High recirculation rates are used to move air through HEPA filters. Sufficient makeup air is also provided to adequately dilute human bioeffluents. Recirculation rates on the order of 75% are typically employed.

HEPA filters are located in AHUs on the discharge side of blower fans (rather than the suction side) to prevent inward leakage of particles through

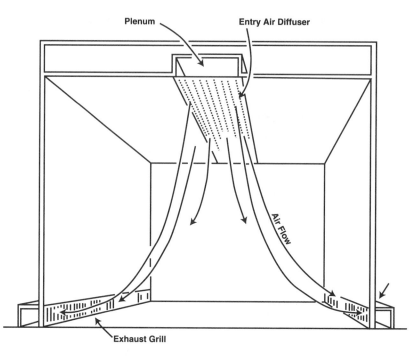

Figure 12.16   Laminar air flow in a clean room.

inadequately sealed ductwork or filter gaskets. Leakage is outward when the filter is on the discharge side. Placing HEPA filters in the last position before the ductwork ensures that unfiltered air will not enter the room. It is critically important to install HEPA filters so that leakage does not occur around them. Achieving an absolutely airtight seal is, however, very difficult.

Pressurization is used to assure that air entering the room is filtered. It is facilitated by air locks which are used by employees for room entry and egress.

Many modern clean rooms have very high cleanliness requirements. As a consequence, HEPA filters with DOP efficiencies on the order of 99.995% or higher are commonly used.

In addition to HEPA filters, laminar flows, and high recirculation rates, clean room users employ a variety of practices to limit particle generation, resuspension, and deposition. These include the use of clothing that covers particle-generating parts of the body and produces few particles itself. It also includes the use of low particle-generating equipment and materials.

II. Gas and vapor-phase contaminants

Application of air cleaning for removal or control of gas/vapor-phase contaminants is, in most cases, a much more difficult undertaking than it is for airborne particles. This is due in part to the fact that the many substances found in indoor air have different chemical properties and, as a consequence, cannot be effectively removed by generic removal processes. The most commonly used technique for removing gas/vapor-phase substances from indoor air and ventilation airstreams is adsorption. Other control approaches that have been attempted include absorption, catalytic oxidation/reduction, botanical air cleaning, and ozonation.

A. Adsorption

Adsorption is a process by which gas, vapor, or liquid-phase substances are physically removed from fluids (including air) by adherence to, and retention on, solid sorbents. This adherence is due to Van der Waal's forces acting on the surface of solids to hold molecules to their surface. The sorbing material is the adsorbent (or sorbent); the adsorbed molecules, the adsorbate (or sorbate). Sorbate condenses (capillary condensation) in the submicroscopic pores of the sorbent.

Though adsorption is a chemical/physical phenomenon, chemical reactions generally do not take place. However, when molecules are adsorbed on a surface, heat is released (heat of adsorption) which is approximately equal to the amount of heat produced when a gas/vapor condenses. As a consequence, sorbate is present on the sorbent as a liquid. Desorption processes require sufficient energy to convert the sorbate to a gas or vapor.

A variety of materials have good sorbent properties. They include activated carbons, activated alumina, silica gel, zeolites, porous clay minerals, and molecular sieves. These materials are widely used in industrial and commercial applications where certain sorbent properties are required. They are used in air cleaning, water softening, and as "kitty litter." Sorbents have high surface-to-volume ratios. Their structure consists of large numbers of submicroscopic pores and channels. Most sorption occurs in pores that have cross-sectional diameters of 10 to 30 angstrom units (Å).

Sorbents may be polar or nonpolar. Metal oxide-, silicaceous-, and active earth-type sorbents are polar. Since polar compounds attract each other, and water is strongly polar, sorbents such as silica gel sorb and retain water preferentially. As such, they cannot effectively remove gases (other than water vapor) in the humid atmospheres common to most air cleaning applications. Activated carbons are nonpolar and have limited affinity for water vapor. They preferentially sorb and retain organic vapors.

B. Activated carbons

Activated carbons are commonly used for solvent recovery and air cleaning in industrial applications, and air cleaning in indoor applications. They are produced in a two-step process in which carbonaceous materials such as wood (primarily hardwoods), coal, coconut and other shells, fruit pits, etc., are heated in a neutral atmosphere and then oxidized at high temperature. Substances that cannot be easily carbonized are volatilized, and numerous submicroscopic pores are produced. Activated carbons originating from different materials vary in their structural properties (e.g., pore size, hardness, density) and, as a consequence, differ in their ability to sorb and retain vapor- (and liquid-phase) substances.

1. Hardness and size

Activated carbons vary in hardness depending on materials and processes used in their production. Hardness is an important use parameter of activated carbons since they must withstand the impact, compression, and sheer forces associated with their use. When air moves through a bed of activated carbon at high flow rates, it causes individual granules of activated carbon to vibrate. Such vibration may cause fragmentation, decrease in granule size, and loss of carbon mass from the bed. This may produce voids in thin-bed adsorption panels and result in reduced sorption efficiency (since air preferentially flows through voids).

Activated carbons are produced in size ranges described by U.S. Sieve Series standard mesh sizes. An 8-14 mesh size, for example, describes activated carbon particles with dimensions of 2.36 × 1.4 mm. Mesh numbers increase with decreasing granule size. A mesh size of 6-14 is typically specified for general purpose air cleaning.

Granule size is a major determinant of air cleaning effectiveness. Effectiveness increases with decreasing granule size. As the sorption bed becomes more tightly packed (as a consequence of smaller granular size), the distance that a sorbate molecule must travel to come into contact with a sorbent surface decreases. As a consequence, the transfer rate of vapor to carbon increases. Though cleaning performance is enhanced when activated carbons with small granule sizes are employed, such use has the same pressure drop problems associated with high-efficiency particle filters.

2. Adsorbability

The degree of physical attraction between a sorbate and a molecule is described as adsorbability. It is a direct function of a sorbate's critical temperature and boiling point. Gases such as oxygen (O_2), nitrogen (N_2), hydrogen (H_2), carbon monoxide (CO), and methane (CH_4) have critical temperatures below $-50°C$ and boiling points below $-150°C$. As a consequence, they cannot be sorbed at ambient temperatures.

Low boiling point gases/vapors such as ammonia (NH_3), hydrogen chloride (HCl), hydrogen sulfide (H_2S), ethylene (C_2H_2), and formaldehyde (HCHO) have critical temperatures between 0 and 150°C and boiling points between -100 and $0°C$. As a consequence, they are moderately adsorbable. However, because of poor retention, activated carbons without special impregnants are not suitable for removal of such gases from air.

Organic vapors that have boiling points >0°C have an increased tendency to be sorbed and retained on activated carbons. These include the higher aldehydes, ketones, alcohols, organic acids, ethers, esters, alkylbenzenes, halocarbons, and nitrogen and sulfur compounds.

In general, the adsorbability of gases and vapors increases with increased molecular size and weight. Small, highly volatile molecules (very volatile organic compounds, VVOCs) have lower adsorbability on activated carbons than larger, lower-volatility compounds (VOCs). In an organic compound series such as paraffins, olefins, and aromatic compounds, adsorbability increases with increased carbon numbers.

3. Adsorption capacity

Activated carbons differ in their ability to sorb and retain sorbate molecules determined on a sorbent weight basis. Adsorption capacity, described as the weight of the sorbate collected per weight of sorbent, depends on (1) sorbent surface area; (2) active sorbent pore volume; (3) gas/vapor sorbate properties; and (4) environmental factors such as temperature, relative humidity, and pressure.

Surface areas for commonly used activated carbons range from 500 to 1400 m^2/g. An inverse exponential relationship exists between pore size and surface area; surface area available for sorption decreases dramatically as pore size increases.

Adsorption capacity of activated carbons is rated relative to their ability to sorb carbon tetrachloride (CCl_4) vapors. A standard weight of activated carbon is exposed to a saturated dry stream of CCl_4 at 68°F (20°C) until the sorbent no longer increases in weight. The ratio of the weight of sorbed CCl_4 to the weight of activated carbon is the maximum possible sorption of CCl_4. This adsorptive capacity, expressed as CTC% (or g CTC/g carbon), ranges from a low of 20 to a high of 90% for different activated carbons.

Though adsorption capacity is determined by using CCl_4 as a standard, other chemical substances will differ in their adsorption capacity on a given activated carbon. As an example, reported adsorption capacities for carcinogenic vapors such as 1,2-dibromomethane, CCl_4, and 1,1-dimethyl hydrazine are 1.020 g/g, 0.741 g/g, and 0.359 g/g, respectively.

As indicated, adsorption capacity can be affected by environmental conditions such as temperature, pressure, and relative humidity. Under elevated temperatures and low pressures, significant loss in adsorption capacity occurs. Fortunately, such temperature and pressure extremes do not occur in nonindustrial air cleaning applications.

Under ordinary ambient (outdoor) conditions, relative humidity is the most likely factor to affect adsorption capacity. Nonpolar activated carbons can nevertheless sorb water vapor from the atmosphere. At high relative humidities (>50%) significant reductions in adsorption capacities have been reported for activated carbons.

4. Retentivity

Retentivity is a property of both the sorbate and sorbent, described as the maximum concentration of vapor retained by a sorbent when the vapor content of an airstream is reduced to zero. It is measured by passing clean,

Figure 12.17   Thin-bed activated carbon panel.

dry air at constant temperature and pressure over a bed of sorbent previously saturated with a specific vapor. Retentivity is expressed as the ratio of the weight of the retained vapor to the weight of the sorbent. The retentivity ratio is always less than the adsorption capacity. This indicates that sorbents have a higher capacity to sorb vapors than to retain them. Though activated carbons have a relatively high adsorption capacity for water vapor, retentivity is low. Because of this low water vapor retentivity, sorbed gases and vapors flowing through an activated carbon bed will cause sorbed water to leave the sorbent and progressively reduce its sorptive capacity for water vapor. Activated carbons are poor sorbents for low-molecular-weight substances such as HCHO and ethylene (C_2H_2) because they are poorly retained.

5.   Carbon beds and filters

Gas/vapor air cleaning in industrial and indoor applications is often accomplished by passing contaminated air through a bed of activated carbon. Carbon beds used in solvent recovery and industrial air cleaning have depths of circa 1 to 2 meters (3 to 6'). In indoor applications, such bed depths are impractical. As a consequence, thin-bed panels with bed depths of 25 to 30 mm (1 to 1.25") are used (Figure 12.17). Because of severe pressure drop problems with such thin-bed filters, they are usually inserted in a module with multiple V configurations, much like that in an extended-surface pleated filter.

Increasingly, activated carbons are being introduced into pleated extended-surface fiber filters. Dry-processed carbon, composite-based adsorption filters utilize very fine activated particles evenly dispersed throughout the filter mat. Like other extended-surface media filters, they have the advantage of low pressure drops and can be used in many applications without expensive filter modules and thin-bed panel filters.

6.   Gas/vapor removal in carbon beds

Gas/vapor-phase contaminants are removed in a distinct pattern of adsorption waves (Figure 12.18). As air moves through an activated carbon sorbent

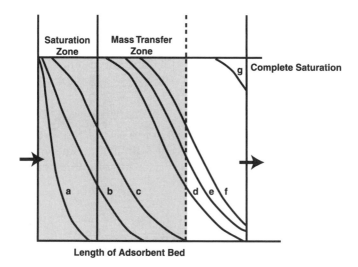

Figure 12.18 Adsorption patterns in an activated carbon bed. (From Turk, A.C., Adsorption, in *Air Pollution*, 3rd ed., Vol. IV, Stern, A.C., Ed., Academic Press, New York, 329, 1977. With permission.)

bed, concentrations of contaminants in the airstream fall rapidly to zero at some finite distance downstream. As sorption continues, carbon granules near the face gradually become completely saturated. In the saturation zone, a dynamic equilibrium becomes established between saturated sorbent and in-coming contaminants. Downstream of this saturation or equilibrium zone, carbon granules are actively sorbing influent gases/vapors. This is described as the mass transfer zone. It is an area of the bed between zero concentration and complete saturation. It moves progressively from one face of the bed to the other with time. It can be described over time by curves a–g. When the bed reaches the condition described by curve e, breakthrough occurs, and the bed has reached the end of its useful life. Carbon in the bed must now be replaced or reactivated.

7. Reactivation and replacement
In industrial and commercial applications, saturated sorbents are reactivated once they become saturated. They are reactivated (desorbed) by passing low-temperature steam or hot air through the bed. In most indoor applications, thin-bed filters are replaced with new filters when they become saturated. They are often discarded but may be returned to the manufacturer for reactivation.

8. Residence time
Because of the limited bed depth associated with thin-bed filters used in indoor applications, residence time of air moving through such filters is a critical factor in their performance. It would take approximately 0.08 seconds to traverse a 25 mm (1″) thin-bed filter under standard air velocity condi-

tions. In closely packed thin-bed activated carbon filters, the half-life of a contaminant (time required to reduce the concentration by 50%) flowing though it is 0.01 seconds. In theory, four half-lives would be required to achieve 90% contaminant removal and require a minimum bed depth of 10 to 13 mm (0.4 to 0.5″).

9. Service life

When contaminant breakthrough (breakpoint) occurs, the sorbent is becoming saturated and needs to be replaced. It is difficult to know with certainty when the breakpoint is reached. The service life of a thin-bed filter can be calculated using the following equation:

$$T = \frac{6.43(10^6)SW}{EQMC} \tag{12.4}$$

where T = service life, hr
   S = fractional saturation of sorbent (retentivity)
   W = sorbent weight, lb (kg)
   E = fractional sorption efficiency
   Q = air flow rate, CFM (L/sec)
   M = average molecular weight of sorbate
   C = average vapor concentration, ppmv

Applying this equation to the removal of toluene (the most common and abundant indoor contaminant) by a small room air cleaner, let us make the following assumptions:

$$E = 0.95$$

$$C = 0.20 \text{ ppmv}$$

$$Q = 100 \text{ CFM } (47.5 \text{ L/sec})$$

$$S = 0.30$$

$$W = 6 \text{ lb } (2.27 \text{ kg})$$

then

$$T = \frac{6.43 \times 10^6 \times 0.30 \times 6}{0.95 \times 100 \times 92 \times 0.20}$$

$$= 6621 \text{ hr}$$

$$= 276 \text{ days}$$

If one knew the average concentration of individual contaminants and their sorption and retention potentials, service life could be calculated from Equation 12.4. For toluene alone at an average concentration of 0.20 ppmv, the service life is predicted to be approximately 9 months. An equivalent TVOC (total VOC) concentration would be approximately 0.77 mg/m³, a moderate level of VOCs in indoor spaces. Under such use conditions, the sorbent would require replacement after approximately 9 months of continuous operation. At concentrations of 1.5 mg/m³ TVOC (a high TVOC concentration), sorbent life would be about 4.5 months. If in the initial case the TVOC level were 0.77 mg/m³ and flow rate were increased to 200 CFM (to clean a larger air volume), service life would be approximately 4.5 months.

The service life of thin-bed carbon filters is difficult to determine under real-world conditions. A variety of approaches have been suggested to determine when carbon filters applied for odor control should be replaced. These include (1) detection of odor when the filter is saturated, (2) removal of sample carbon granules to determine their degree of saturation in the laboratory, and (3) challenge with odoriferous compounds such as isoamyl acetate (banana oil) or wintergreen.

10. Catalytic properties

Activated carbons are often used to remove objectionable gases from emission sources, indoor air, and other media, based on their catalytic abilities. Activated carbons have been used in southern California to remove O_3 from ambient air serving greenhouses and, in some cases, building HVAC systems. In addition to catalytically destroying O_3, activated carbons can destroy other oxidants such as ozonides, peroxides, and hydroperoxides. Activated carbon filters are used to reduce O_3 emissions from O_3-generating equipment such as electrostatic photocopiers and other high-voltage devices.

In the presence of O_2, activated carbons can catalytically oxidize H_2S to elemental sulfur. H_2S is a common malodorant associated with sewage treatment, oil/gas extraction and refining, and a variety of decomposition processes. It is a common contaminant in groundwater supplies used for domestic and industrial purposes, and can be removed from water by activated carbon filters.

In addition to their natural catalysis of compounds such as O_3 and H_2S, activated carbons can be impregnated with catalysts for specific applications.

C. Chemisorption

The large surface area associated with sorbents provides an optimum environment for a variety of chemical reactions. To facilitate such reactions, sorbents are coated or impregnated with selected chemicals that will react with target substances which come into contact with the chemical impregnants. The process is called chemisorption. It is the process by which contaminant levels are determined from gas sampling tube measurements (see Chapter 9).

Chemisorption media are produced by impregnating activated carbon, activated aluminum, silica gel, etc., with catalysts that include bromine, metal oxides, elemental sulfur, iodine, potassium iodide, and sodium sulfide. Bromine-impregnated activated carbons are used to chemically sorb ethylene, which reacts catalytically to produce ethylene bromide, a sorbable gas. Carbons impregnated with metallic oxides are used to oxidize H_2S under low O_2 conditions. Elemental sulfur- and iodine-impregnated activated carbons can be used to sorb mercury by producing stable mercuric sulfide and mercuric iodide, respectively. Such impregnated carbons can be used to reduce exposure to mercury associated with accidental spills of mercury and elemental mercury-releasing compounds. Sodium sulfide and other impregnants have been used to remove HCHO and have been used in residential applications.

Activated alumina impregnated with potassium permanganate ($KMnO_4$) has been widely used in industrial and commercial air cleaning systems and has seen limited residential use. Though activated alumina does not have the sorptive capacity of activated carbons, it can be used effectively for the control of low-molecular-weight gases, such as HCHO and C_2H_2, and odors in buildings. Organic vapors are sorbed on the surface of the activated alumina where they are oxidized by $KMnO_4$ in a thin film of water. If oxidation is complete, CO_2 and H_2O vapor will be produced as by-products. If incomplete, it has the potential to produce a variety of oxidized compounds such as aldehydes and ketones, and in the case of halogens, substances such as hydrogen chloride (HCl). $KMnO_4$-impregnated activated alumina is used commercially to remove C_2H_2 from fruit storage facilities. On use, the pink-colored $KMnO_4$ is reduced to the brown-colored $KMnO_2$, manganese oxide.

D. Performance studies

Performance evaluations of air cleaning systems in nonindustrial, nonresidential buildings, and residential environments have been limited. Several studies have been conducted to evaluate the performance of thin-bed activated alumina filters on HCHO in residences. In a study using a portable cleaner with an air flow of 130 CFM (61 L/s), HCHO levels were reduced by 25 to 30% in a mobile home and 35 to 45% in a urea–formaldehyde-foam-insulated (UFFI) house. In another study using a cleaner with an 1800 CFM (849 L/s) flow rate, HCHO levels were reduced on the order of 75%. A study of thin-bed activated carbon filters showed that they were effective (on the order of 80%) in removing VOCs in the low ppbv range found in buildings, with a projected service life of 60 to 466 days. The effectiveness of dry-processed carbon composite-based filters has been evaluated under laboratory conditions. Such filters have been reported to have 10 times the adsorption capacity of large mesh activated carbon, but they apparently have much shorter service lives.

E. Absorption

Absorption is widely used to remove gas-phase contaminants from industrial waste streams. In such processes, contaminants are brought into contact with water or chemically reactive liquid, slurry, or solid media. The process is called scrubbing. Depending on the application, cleaning efficiencies for target contaminants are on the order of 70 to 99%. The process has been evaluated for use in controlling HCHO under laboratory conditions.

F. Room temperature catalysts

Low-temperature catalysts were developed by a U.S. company for use in residential air cleaners. The room temperature catalyst, which included a mixture of copper and palladium chlorides, was mixed with an equal quantity of activated carbon in a thin-bed filter. Performance studies conducted in a 1152 ft^3 (32.3 m^3) room under static conditions indicated that this system was capable of significantly reducing levels of O_3, H_2S, sulfur dioxide (SO_2), CO, and ammonia (NH_3), and relatively ineffective in reducing concentrations of nitric oxide (NO), nitrogen dioxide (NO_2), and benzene. Its performance under real-world conditions has not been reported.

G. Botanical air cleaning

Proposed by a NASA scientist, botanical air cleaning is a relatively novel approach to removing vapor-phase contaminants from indoor air. Based on this concept, air would be cleaned by the active uptake and metabolism of contaminants by plant leaves. In initial static chamber studies, significant reductions in HCHO and other VOCs were reported. Other studies indicated this uptake was mostly associated with potting soil. Though uptake of various VOCs and other gases by plant materials can be shown in chamber investigations, no studies have been conducted to demonstrate the efficacy of plant use in indoor environments under real-world conditions. Because absorption is passive and contaminant generation is dynamic, the use of plants to control indoor air contaminants effectively is not probable. Nevertheless, it has caught the fancy of many Americans who have attempted to use plants for air cleaning purposes.

H. Ozonation

Ozone is a powerful oxidizing substance produced by high-voltage systems incidentally or for special purposes (e.g., water treatment). Ozone generators are commonly used by contractors as an odor control measure in post-fire rehabilitation of buildings. In the past two decades, a number of companies have sold O_3 generators as air cleaning devices to homeowners, schools, and a variety of building owners. The assumption has been made by manufacturers that O_3 generated in a space will react with aldehydes and other VOCs to produce harmless CO_2 and H_2O vapor, and that the levels of O_3 used are not harmful to the health of human occupants. Studies conducted on these devices indicate that, when operated as recommended, O_3 concentrations generated are not adequate to significantly reduce VOC concentrations; they are, however, above ambient air quality and occupational exposure limits (particularly for individuals in the near vicinity of operating devices). As a consequence of potential consumer exposure hazards, a number of states have sued manufacturers of O_3 generators, alleging they pose a substantial risk to public health. Ozone generators continue to be widely sold in North America.

The deliberate introduction of O_3 into indoor spaces poses several additional concerns. As a powerful oxidant, O_3 bleaches fabric dyes and cracks rubber and plastic products. As indicated in Chapter 4, it initiates indoor chemical reactions that produce a variety of new irritants.

III. Air cleaners as contaminant sources

Though air cleaners are designed to clean air, they have been reported to be sources of indoor contaminants such as VOCs and microbial organisms and products. There have been reports of significant emissions of toluene and xylene from oil-treated particle filters, and a variety of odorous aldehydes (hexanal, heptanal, octanal, nonenal) and small organic acids from filters infested with microorganisms. Filters can also be a source of fibrous particles shed through HVAC systems and, as indicated previously, electronic air cleaners can be a significant source of O_3.

Filters in HVAC systems can capture and retain fungal spores and mycelial fragments which subsequently grow on collected dusts. Such growth can penetrate filter media, contaminating other filters, air, and surfaces downstream. These fungi can produce allergens that can be detected in building air independent of mold spore/particle concentrations.

Biocides are commonly used in filters. Though there is some potential for them to become airborne, biocide levels in indoor air associated with filter use have not been evaluated and reported.

Oxidizing media such as $KMnO_4$ have the potential to produce a variety of small aldehydes and ketones in addition to CO_2 and H_2O vapor. They also have the potential to produce HCl from the oxidation of chlorinated solvents such as methyl chloroform, commonly found in indoor environments. Studies on such potential contaminants have not been conducted. As indicated previously, ozonation is the deliberate introduction of O_3 for what is described as air cleaning purposes. It poses both direct and indirect indoor air quality/indoor environment concerns.

Readings

American Lung Association, *Residential Air Cleaning Devices: Types, Effectiveness and Health Impact*, Washington, D.C., 1997.

American Society of Heating, Refrigerating and Air-Conditioning Engineers, *Equipment Handbook*, ASHRAE, Atlanta, 1983, chap. 10.

Ensor, D.S. et al., Air cleaner technologies for indoor air pollution, *Engineering Solutions to Indoor Air Problems, IAQ '88*, ASHRAE, Atlanta, 1989, 111.

Godish, T., *Indoor Air Pollution Control*, Lewis Publishers, Chelsea, MI, 1989, 247.

Institute of Environmental Sciences, *Recommended Practice for HEPA Filters*, IES RP-CC-00-1-86, Institute of Environmental Sciences, Mt. Prospect, IL, 1986.

Maroni, M., Siefert, B., and Lindvall, O., *Indoor Air Quality — A Comprehensive Reference Book*, Elsevier, Amsterdam, 1995, chap. 31.

McNall, P.E., Practical methods of reducing airborne contaminants in interior spaces, *Arch. Environ. Health*, 30, 552, 1975.

Spengler, J.D., Samet, J.M., and McCarthy, J.F., Eds., *Indoor Air Quality Handbook*, McGraw-Hill Publishers, New York, 2000, chaps. 9–11.

USEPA, *Residential Air Cleaning Devices: A Summary of Available Information*, EPA 400/1-90-002, USEPA, Washington, D.C., 1990.

USEPA, *Residential Air Cleaners — Indoor Air Facts #7*, EPA 20A-4001, USEPA, Washington, D.C., 1990.

USEPA, *The Inside Story: A Guide to Indoor Air Quality*, EPA 402-K-93-007, USEPA, Washington, D.C., 1995.

USEPA, *Should You Have the Air Ducts in Your Home Cleaned?*, EPA 402-K-97-002, USEPA, Washington, D.C., 1997.

USEPA, *Ozone Generators That Are Sold as Air Cleaners. An Assessment of Effectiveness and Health Consequences*, www.epa.gov/iaq/pubs/ozongen.html, 1999.

Viner, A.S. et al., Air cleaners for indoor air pollution control, in *Indoor Air Pollution — Radon, Bioaerosols, and VOCs*, Kay, J.G., Miller, G.E., and Miller, J.F., Eds., Lewis Publishers, Boca Raton, FL, 1991, 115.

Questions

1. What is the primary use of air cleaning in indoor spaces? Why?
2. How do fibrous media filters collect dust particles?
3. What is the relationship between particle size and the ability of media filters to collect particles efficiently?
4. What factors affect the performance of fiber media filters?
5. Describe the properties and performance of HEPA filters.
6. Describe the relationship between filter collection efficiency and pressure drop.
7. Why are extended media filters used in many building applications?
8. Describe uses for viscous and renewable media filters.
9. Describe the relationship between filter soiling, air flow resistance, and cleaning efficiency.
10. Describe the principle of operation and cleaning effectiveness of electronic air cleaners.
11. What are the advantages and disadvantages of using extended media and electronic air cleaners as in-duct devices in residential air cleaning?
12. How is the performance of dust air cleaning devices determined?
13. What is the difference between arrestance and dust spot efficiency?
14. How effective are free-standing air cleaning devices commonly available in the market and purchased by consumers?
15. How do ion generators work? How effective are they?

16. If you were to use an air cleaner at home to reduce your exposure to allergens, what factors should you consider before you purchase it?
17. What are the advantages and disadvantages of using in-duct dust air cleaners?
18. Describe air cleaning principles used in clean rooms.
19. What chemical/physical processes can be used to remove gas/vapor contaminants in indoor air?
20. Describe sorbent use in air cleaning.
21. What properties of activated carbons affect their performance in air cleaning applications?
22. Describe the movement of contaminant vapors through an activated sorbent bed.
23. How can one determine when an activated carbon sorbent bed filter needs to be replaced?
24. Describe how the catalytical properties of activated carbons can be used for specific air cleaning applications.
25. What is chemisorption? Describe a specific application.
26. What factors limit the performance of free-standing air cleaners using sorption/chemisorption principles in residential air cleaning?
27. Describe botanical air cleaning and its limitations.
28. Why is ozonation an undesirable air cleaning practice?
29. How may air cleaners be a source of indoor contaminants?

chapter thirteen

Regulatory and nonregulatory initiatives

I. Introduction

Indoor environments are subject to a wide variety of contamination problems associated with natural or anthropogenic sources that may adversely affect the health and well-being of building occupants. Consequently, some form of individual or collective efforts is needed to identify, prevent, and in many cases mitigate, indoor air quality (IAQ) and other indoor environment (IE) problems.

In the U.S. and other developed countries in western Europe and Asia, identification of individual environmental problems such as those involving air, water, and waste has been followed by a pattern of initial slowly evolving government involvement, with subsequent significant regulatory requirements. Government action to solve or attempt to solve environmental problems through regulations or some type of public policy initiatives has been very common in the past three decades.

With the exception of some specific and limited cases, the traditional model of governmental regulatory involvement in controlling/mitigating environmental problems cannot easily be applied to indoor environments. Ambient (outdoor) air pollution control focuses on the free-flowing air of the atmosphere that becomes contaminated from a variety of stationary and mobile sources. Ambient air is not confined to an individual's property. Its contamination by anthropogenic sources imposes potential risks to humans and the environment that are involuntary. As such, government regulatory action is essential.

The history of environmental regulation in North America and other developed countries has been to use regulation as a tool to reduce exposures that result in involuntary risks to the public and adversely affect the environment. For indoor environments, regulatory initiatives/requirements have been promulgated at the federal level for asbestos, lead, and formaldehyde (HCHO); at state and local levels for environmental tobacco smoke (ETS).

Federal and state authorities, not unsurprisingly, have been reluctant to impose significant regulatory requirements on building owners and those who have control over other indoor spaces. In many cases, the nature of risks to homeowners are not clear-cut and are almost entirely limited to occupants. Since private homeowners have significant control over their own environments, exposure risks to contaminants such as combustion-generated pollutants, radon, asbestos, lead, those of biological origin, and even HCHO are, to various degrees, subject to homeowner control. As such, risks may be both voluntary and involuntary. A case for regulatory action cannot be easily made except when it involves the sale of dangerous or potentially dangerous products or property.

Regulatory requirements imposed on residences would, in most cases, be impractical. In the U.S., there are over 70 million single-family, as well as millions of multifamily dwellings leased to private individuals. Therefore, an enormous number of structures and individuals would be subject to regulation. Respecting individual property rights is a significant regulatory concern. Private property and its use and individual privacy are among the most cherished privileges in the U.S.

Regulatory actions by governmental agencies are more likely to apply to public-access buildings and interior spaces. Public-access buildings can be publicly or privately owned. They are public-access in that they are open to employees and members of the public in the normal course of providing services and doing business. These would include schools, colleges and universities, hospitals, municipal buildings, private office buildings, motels/hotels, restaurants, retail establishments, planes, trains, etc. In such spaces, exposure to contaminants that could affect an individual's health and well-being would be, in most cases, involuntary. Numerous precedents, particularly at state and local levels, have been set in regulating various aspects of building/indoor environments for the purpose of ensuring public safety (e.g., fire and other safety codes). Indeed, ventilation requirements designed for comfort purposes are a part of most state and local building codes. The imposition of regulatory requirements to protect or enhance the quality of air and other environmental aspects of public-access indoor spaces would not be precedent-setting.

II. Regulatory concepts

A. Air quality standards

The setting, promulgation, and enforcement of air quality standards (AQSs) is the primary regulatory mechanism used to reduce exposures to targeted contaminants in the ambient air environment in the U.S. An AQS is the maximum permissible air concentration of a regulated pollutant. This numerical limit is selected to provide health protection (with an adequate margin of safety) to both the general population and those who are at special risk. These health-based standards are based on the assumption that there is a threshold dose (concentration as a function of time) below which no adverse effects occur. Standard setting is a difficult activity since the scientific literature is often insufficiently definitive in supporting both threshold values and adequate margins of safety. Due to the economic burdens involved, the regulated community, through due-process procedures involved in rule-making, usually challenges the validity of studies used in decision making as well as proposed and promulgated rules and standards. This is often done through extraregulatory political efforts as well.

The standard-setting process attempts to set acceptable numerical limits on airborne contaminant concentrations in order to protect public health. It is based on a review of the scientific literature by regulatory staff and outside review panels within the context of uncertainties as to what those limits should be, economic considerations, and the general and detailed criticisms of the regulated community. In theory, the only consideration in setting numerical limits should be the protection of public health. In reality, scientific judgment as well as economic and political considerations play a role. As a consequence, AQSs may not be sufficiently protective.

In theory, AQSs could be used to regulate air quality in public-access buildings/environments, and possibly residences. In the early 1980s, Wisconsin and Minnesota attempted to control HCHO levels in new mobile homes using indoor air quality (IAQ) standards of 0.40 and 0.50 ppmv, respectively. These were later rescinded to conform with a federal preemption in regulating HCHO emissions from wood products used in mobile home manufacture.

Development and promulgation of AQSs and other regulatory activities associated with toxic contaminants in the ambient environment is a long administrative process. Initially, health risks are assessed by regulatory agency staff.

The risk assessment process includes (1) hazard identification, (2) exposure assessment, (3) assessment of potential dose–response relationships, and (4) risk characterization. Hazard identification and dose–response assessment involve determining potential causal relationships between observed health effects and specific contaminant exposures. Human exposures under real-world conditions are characterized in exposure assessment. The magnitude and uncertainty of risks associated with an individual contaminant are evaluated in risk characterization. Risk assessment for a single chemical is a long process, easily involving a half a decade or more of evaluating health risk.

Use of IAQ standards to control human exposures in indoor air would be subject to the slow timetable common for ambient air pollutants. It would also be subject to the political and economic considerations which compromise health protection when setting AQSs. A notable example was the attempt by the Minnesota Department of Health to require a 0.1 ppmv

HCHO IAQ standard in new mobile homes. As a result of industry lobbying efforts, the state legislature significantly weakened (to 0.5 ppmv) the standard. In another instance, the Department of Housing and Urban Development (HUD) used a target level of 0.4 ppmv HCHO as a *de facto* standard, claiming (without benefit of a risk assessment) that it provided reasonable health protection to occupants of new mobile homes. Scientific studies, however, have shown that HCHO exposures well below 0.4 ppmv may cause serious health effects in those exposed in residential environments.

Standards, as interpreted by many professionals and the lay public, convey a perception of implicit safety when measured values are below the standard and implicit danger if above the standard. These perceptions result in a false sense of security in the former case, and excessive fear in the latter. The true nature of a standard, incorporating the uncertainties and political compromises involved, is generally not understood.

Compliance with ambient AQSs is determined by monitoring community air in fixed sampling locations and/or modeling specific sources. Though manageable, monitoring ambient air quality to assess compliance with AQSs requires significant personnel and resources; this is true when evaluating compliance with most environmental standards.

Assessing compliance with IAQ standards would pose significant difficulties in its implementation due to the enormous resource requirements as well as a variety of practical problems. If applied to residences, effective monitoring using dynamic integrated sampling would be intrusive and, in many cases, homeowners would not be receptive to it. Passive monitoring would be less intrusive but less reliable. Results would depend on the integrity of those using the passive monitoring equipment. In addition to intrusiveness, privacy issues, property rights, and maintaining the integrity of passive samplers, it would be physically impossible to monitor compliance for even one contaminant in a targeted subset of 80+ million residences in the U.S.

Monitoring public-access buildings would be a less formidable undertaking. It would pose fewer privacy and access issues and there would be fewer buildings to monitor. Nevertheless, the task would still be enormous and could not be achieved without requiring building owners to take on the task themselves. An AQS approach to control air quality in buildings would require that standards be self-enforced, as has been the case for smoking restrictions. Though the latter has been effective, it would likely be less effective in the case of IAQ standards.

B. Emission standards

Emission standards are used in ambient air pollution control programs to control emissions from all new or significantly modified existing sources (New Source Performance Standards, NSPS) and have been used for pollutants regulated under National Emissions Standards for Hazardous Air Pollutants (NESHAP). In both cases, emission limits are uniform for all sources in a source category, regardless of air quality in a region. Emission standards are also used to achieve ambient AQSs. Depending on existing air quality, emission standards on individual sources may vary from place to place.

1. Product emissions

An emission standard is a numerical limit on the quantity of a contaminant that can be emitted from a source per unit time (e.g., lbs/hr, gm/sec, etc.). A variant of the emission standard concept has been used to control HCHO emissions from urea–formaldehyde (UF)-bonded wood products such as particle board and decorative wood paneling produced for use in the construction of mobile/manufactured homes in the U.S. These limits are better described as product standards. They are not specified as an emission rate (e.g., $mg/m^2/hr$) but as the maximum acceptable air concentration in a large, environmentally controlled chamber at a loading rate (m^2/m^3) typical of a mobile home environment. Product standards are used in western and north European countries, e.g., Germany, Denmark, and Finland, to conform with indoor air guideline values for HCHO (see Section IV.A).

Product standards have considerable potential for improving air quality in buildings and other environments. They could conceivably be used to limit emissions of volatile and semivolatile organic compounds (VOCs and SVOCs) from products such as carpeting, vinyl floor and wall coverings, paints, varnishes, lacquers designed for indoor use, adhesives and caulking compounds used in building construction, and coating materials used in arts and crafts.

Product standards in the regulatory context have a very important attribute; they are relatively simple to implement, administer, and assess compliance. The burden of compliance is placed on manufacturers, who must verify that their product meets emission limits before the product is sold. A special application of the product standard concept has been employed by the state of Washington in its office building construction program. Vendors who contract with the state must provide products that do not exceed an air concentration of 0.05 ppmv HCHO, 0.5 mg/m^3 TVOCs, 1 ppbv 4-PC and 50 $\mu g/m^3$ particles at the anticipated loading conditions (m^2/m^3) within 30 days of installation. In addition, any substance regulated as an ambient air pollutant must meet emission limits that will not exceed the USEPA's primary or secondary AQSs, and one tenth the Threshold Limit Value (ACGIH occupational guideline value for an 8-hour time-weighted exposure) of other substances of concern.

2. NSPSs for wood-burning appliances

USEPA's emission standard program for new ambient sources (NSPS) has had an unintended but positive impact on IAQ. USEPA, in an attempt to reduce the impact of wood-burning appliance emissions on ambient air quality, promulgated an NSPS for wood-burning stoves to reduce emissions of PM_{10} (particles) and CO. These performance standards, applied nationwide, have had the effect of improving the emission performance of all new wood-burning stoves to both the ambient and indoor environments.

3.   *VOC emission limits*

Many sources of both total and specific VOCs are required to limit emissions to the atmosphere under programs designed to achieve compliance with AQSs or hazardous/toxic pollutant standards (e.g., for benzene, styrene, HCHO, etc.). One of the primary means to achieve compliance with such limits is to use one or more "clean manufacturing" or pollution prevention techniques. These include changing manufacturing processes and product formulations to limit the use of regulated substances. Such practices limit, and in some cases eliminate, emissions to both the ambient and indoor environments (if the product is used indoors). A number of USEPA research programs on IAQ are based on pollution prevention principles.

C. Application standards

Significant IE contamination problems occur when products are misapplied. Standards of performance and certification may be required of corporations and individuals who apply or install products that have the potential to cause significant indoor contamination as a result of poor application procedures. Pest control service providers are the most notable example of this. In New Jersey, for example, onsite supervision of certified pesticide applicators, and conditions under which organochlorine compounds can be used, are specified. Application standards for termiticides and other pesticides vary from state to state, with some states having none.

In the United Kingdom, urea–formaldehyde foam insulation (UFFI) has been used to retrofit insulate millions of residences. Unlike the U.S. and Canada where UFFI has been viewed as inherently dangerous, U.K. authorities approach UFFI, and HCHO emissions from it, as a manageable health concern. A British standard specifies the formulation of UFFI and mandates a code of practice for its installation to minimize HCHO exposure levels associated with its misapplication. Companies installing UFFI are required to have the necessary expertise, suitably trained personnel, and a properly formulated foam product.

Application standards can be required by regulatory authorities who enforce compliance. They can also be established by a trade association or by collective industry agreements. Such voluntary application standards are self-enforced and depend on the integrity of individual installers and corporate management. Application standards were proposed by the Formaldehyde Institute and UFFI companies. Their petition was denied by the Consumer Product Safety Commission (CPSC) before CPSC promulgated its UFFI ban (see below).

D. Prohibitive bans and use restrictions

Prohibitive bans are commonly used to help achieve ambient AQSs. Examples include prohibitions on open burning of trash and leaves, use of apartment house incinerators, and use of high-sulfur coal and fuel oil in steam boilers.

Bans or use restrictions may be applied to products that have the potential for causing indoor contamination and contributing to health risks. Most notable of these are bans on the use of (1) paints containing >0.06% lead and (2) hand-friable and, more recently, mechanically friable asbestos-containing materials (ACM), in building construction. These bans have effectively reduced the potential for both ambient and indoor contamination by lead and asbestos in buildings constructed after 1978 and 1980, respectively.

Initial NESHAP bans on hand-friable or potentially hand-friable asbestos-containing materials in building construction were promulgated to reduce emissions of asbestos fibers to ambient air during building renovation or demolition. It had the unintended consequence of raising concerns about potential exposures of building occupants to airborne asbestos associated with ACM used in construction.

Urea–formaldehyde foam insulation was banned for use as an insulating material in walls and ceilings of residences in Canada in 1980. A similar ban promulgated in the U.S. by the CPSC was voided by a federal appellate court in response to an industry appeal. A ban on the use of UFFI for residential applications remains in effect in Massachusetts and Connecticut.

Bans or use restrictions have been placed on methylene chloride in paint strippers, chlordane for termite control, pentachlorophenol as a wood preservative, chlorpyrifos for broadcast flea control, and mercury biocides in latex paint by regulatory actions or voluntary industry agreements. California has placed use restriction on kerosene heaters.

Partial or complete bans can be applied to products whose use is discretionary (such as tobacco smoking). Since the 1986 Surgeon General's report on involuntary smoking, total or partial bans on smoking in public-access buildings and public transportation have been imposed by regulatory action or management in most public-access environments in North America.

Prohibitive bans, like product standards, are an attractive tool to improve existing air quality in some cases and prevent future indoor exposures in others. They are simple to implement and require no assessment of compliance with numerical limits.

Application of a ban, or a proposed ban, on "bad products" can have significant actual or perceived economic repercussions on affected industries. As a consequence, an industry can be expected to use all legal and political means to overturn the ban. Federal regulatory agencies in the U.S. must conform to the Administrative Procedures Act, which is designed to ensure that parties with an interest in proposed regulatory actions are accorded full due process. They also have a right to appeal regulatory actions. As a consequence, final disposition after appeals to state or federal courts following the regulatory imposition of a ban or restriction on use of a product often takes years. In two notable cases, federal courts in the U.S. voided the ban on UFFI and greatly limited USEPA's phase-out rule on a number of asbestos-containing products. To reduce such time delays, USEPA often negotiates voluntary use restrictions with an industry or industry group.

E. Warnings

If a product is hazardous or potentially hazardous, the manufacturer has a common law duty to warn potential users. In the case of pesticides and other toxic/hazardous substances, manufacturers are required by law to place warning labels on products. Such warnings describe conditions under which the product can be safely used and hazards and health risks if it is not. Paint strippers, oil-based paints and varnishes, and cleaning solvents have warning labels advising consumers to use them only in ventilated areas. Kerosene heater labels warn consumers of potential fire hazards and advise consumers to use only in ventilated areas. Warning labels are required on all chemicals and chemical formulations subject to regulation under the Occupational Safety and Health Administration's (OSHA) hazard communication standard (HCS). The HCS is designed to protect workers. Wood product manufacturers producing particle board or hardwood plywood apply warning labels (for HCHO) to their product in addition to the standard mill stamp. Under HUD regulations, a specific warning label which describes potential health risks associated with HCHO exposures must be displayed in a prominent place inside new mobile homes and be included in the owner's manual. The required warning is illustrated in Figure 13.1.

The basic premise of a warning is that by being informed of the hazards or potential hazards, users can make informed decisions in order to protect themselves and their families. In practice, few consumers read warning labels and even fewer respond to them in a way that reduces exposure risks. Warning labels on cigarette packages are a classic example. Despite warnings of serious health effects associated with tobacco smoking, tens of millions of Americans smoke, and several million children begin smoking each year. HUD warnings required on new mobile homes since 1986 had no apparent effect on sales. Despite warning labels on pesticides and pesticide formulations, misapplication and illness symptoms associated with home pesticide use are common.

Warnings required by law or voluntarily placed on products by manufacturers have limited effectiveness. They have one unintended consequence: they have apparently reduced manufacturers' legal liability in many claims involving personal injury (as interpreted by judges or juries).

F. Compulsory HVAC system performance evaluations

A regulatory mandate for the regular inspection of ventilation system performance has been legislated by the Swedish Parliament for all nonindustrial buildings (except single-family residences with mechanical exhaust and natural ventilation). The inspection intervals vary from 2 to 9 years depending

---

Important
Health Notice

Some of the building materials used in this home emit formaldehyde. Eye, nose, and throat irritation, headache, nausea, and a variety of asthma-like symptoms, including shortness of breath, have been reported as a result of formaldehyde exposure. Elderly persons and young children, as well as anyone with a history of asthma, allergies, or lung problems, may be at greater risk. Research is continuing on the possible long-term effects of exposure to formaldehyde.

Reduced ventilation resulting from energy efficiency standards may allow formaldehyde and other contaminants to accumulate in the indoor air. Additional ventilation to dilute the indoor air may be obtained from a passive or mechanical ventilation system offered by the manufacturer. Consult your dealer for information about the ventilation options offered with this home.

High indoor temperatures and humidity raise formaldehyde levels. When a home is to be located in areas subject to extreme summer temperatures, an air-conditioning system can be used to control indoor temperature levels. Check the comfort cooling certificate to determine if this home has been equipped or designed for the installation of an air-conditioning system.

If you have any questions regarding the health effects of formaldehyde, consult your doctor or local health department.

Figure 13.1 Warning label required by HUD to be posted in new mobile homes and included in owner's manuals.

on occupants and system principles. Inspected systems that meet performance criteria are approved and issued a compliance certificate. Inspections that identify minor faults require that they be remedied before the next inspection; serious faults must be corrected and followed by a new inspection before the system is approved and certified. The performance evaluation requirements appear to work well, with high approval/certification rates for schools and day nurseries (>85%) but lower rates for offices (40%), hospitals (40%), and apartments (65 to 70%).

Performance requirements for HVAC systems in Canadian federal office buildings, along the lines of those currently being developed by the American Society of Heating, Air-Conditioning and Refrigeration Engineers (ASHRAE), have been incorporated into the Canadian Labor Code. The amended Code requires that records of a building's HVAC system operation, inspection, testing, cleaning, and maintenance, written by a qualified person, be maintained. The Code also requires the conduct of IAQ investigations using recognized investigative protocols.

Though the principle of compulsory inspections of ventilation systems has enormous potential to improve IAQ in buildings, it is doubtful that such a regulatory requirement could be imposed in the U.S. Its use is more likely in countries with a strong social welfare tradition.

III. Regulatory actions and initiatives

Indoor contaminants subject to significant federal, and in some cases, state regulatory initiatives to protect the health and safety of building occupants include asbestos and lead and, to a lesser degree, HCHO and radon.

A. Asbestos

In 1973, USEPA designated asbestos a hazardous air pollutant and promulgated regulations to reduce community exposures. An area of major concern was the release of asbestos fibers into ambient air as a result of building-related renovation and demolition activities which disturb hand-friable asbestos-containing (ACM) building materials. As a consequence, USEPA required use of wet techniques to remove friable ACM from buildings prior to renovation or demolition activities. To prevent future potential releases of asbestos fibers from hand-friable ACM, use of asbestos-containing fire-proofing, acoustical plaster, and molded insulation products was banned by USEPA in the period 1973–1978. The regulatory history of asbestos in buildings is summarized in Table 13.1.

In 1978, significant public health concern arose as a consequence of the emerging awareness of the extensive use of friable ACM in school buildings. Millions of children in the U.S. were believed to be at risk of asbestos fiber exposure from damaged or deteriorating ACM, and asbestos-related disease

Table 13.1   Public Policy and Regulatory History of Asbestos in Buildings

Year	Actions
1973	USEPA designates asbestos as a hazardous air pollutant under NESHAP; USEPA bans use of friable ACM in U.S. buildings and requires removal of friable ACM before demolition or renovation.
1978	USEPA bans use of asbestos in acoustical plaster and molded thermal system insulation; USEPA develops technical guidance documents for ACM in schools.
1980	Congress enacts Asbestos School Hazard and Detection Act.
1982	USEPA promulgates "asbestos in schools" rule; school inspections required.
1986	Congress enacts Asbestos Hazard Emergency Response Act (AHERA); requires school inspections, etc.
1987	USEPA promulgates regulations to implement AHERA.
1988	OSHA promulgates asbestos construction industry standard, requires use of engineering controls and respiratory protection for abatement workers, and requires application of work practices to protect building occupants from asbestos exposure.
1990	USEPA issues advisory on use of O&M to manage ACM in place.
1992	USEPA revises asbestos NESHAP, extends accreditation requirements for all indoor asbestos work, expands ACM materials regulated.
1994	OSHA revises construction industry standard; requires building owners to presume certain materials contain ACM and develop programs to ensure service workers are not unduly exposed; reduces PEL.

such as lung cancer and mesothelioma. In response to these concerns, USEPA developed and implemented a program of guidance and technical assistance to school districts and state and local public health and environmental authorities in identifying and mitigating potential asbestos hazards. This program was conducted in cooperation with the Public Health Service and the Occupational Safety and Health Administration (OSHA). A guidance document which provided detailed information on how to identify and control friable ACM in schools was developed, published, and distributed to school officials and other interested parties.

After the technical assistance program was implemented, USEPA in 1979 initiated a process of rule-making in response to citizen petitions and a lawsuit. The rule-making process was completed in 1982.

In 1980, Congress enacted the Asbestos School Hazard and Detection Act. It authorized the Secretary of Education to establish procedures to make federal grant money available to (1) assist state and local education agencies (LEAs) in identifying ACM in school buildings and (2) provide low-interest loans to abate asbestos hazards.

In 1982 USEPA promulgated the "asbestos in schools" rule. It required that all public and private elementary and secondary schools implement programs to identify friable ACM, maintain records, notify employees of the location of friable ACM, provide instructions to employees on how to reduce asbestos exposures, and notify the school's parent–teacher association of inspection results. The response of LEAs to USEPA's 1982 asbestos in schools requirements was one of considerable uncertainty. Questions arose concerning the adequacy of inspection procedures, the need to manage asbestos problems, and the cost to individual LEAs. Because of these uncertainties and failure of Congress to appropriate sufficient money for the program, the "asbestos in schools" rule failed to achieve its objectives.

Because of the failure of the asbestos in schools rule to adequately address asbestos exposure concerns in school buildings, Congress amended the Toxic Substances Control Act (TSCA) in 1986. The new amendments, described as the Asbestos Hazard Emergency Response Act (AHERA), mandated that USEPA promulgate rules regarding (1) inspection of public and private schools in the U.S. for ACM; (2) a description of response actions, circumstances in which they would be required, and their implementation; (3) establishment of operation and maintenance (O&M) programs for friable ACM; (4) establishment of periodic surveillance and reinspection programs for ACM; (5) notification of state governors of asbestos management plans; and (6) transportation and disposal of waste ACM.

Final rules to implement AHERA were promulgated by USEPA on October 17, 1987. Regulatory requirements not specifically addressed in the AHERA statute included: (1) development of a model accreditation plan specifying minimum training requirements for building asbestos inspectors, management planners, abatement workers, project designers, and supervisors/contractors; (2) bulk sampling using specified procedures to identify/confirm the presence of asbestos fibers in suspect building materials;

(3) visual inspection and assessment of the physical condition of friable ACM; (4) development and implementation of asbestos management plans; (5) identification of a designated person in an LEA responsible for the implementation of asbestos management plans; and (6) minimum training requirements for custodial staff and maintenance workers who might disturb asbestos.

AHERA required all schools K–12 to be inspected by an accredited inspector. It also required the preparation and submission of an asbestos management plan for each building to an authorized state agency by October 2, 1988 (postponed to May 1989).

It was widely assumed in the late 1980s that USEPA would subsequently develop and promulgate rules requiring the inspection of other nonresidential, nonindustrial buildings for asbestos. USEPA evaluated the much larger problem (in terms of the number of buildings that would be involved) of asbestos in public-access buildings. The review indicated that ACM was present in such buildings, but was less prevalent than in schools. USEPA officials, for a variety of reasons, deferred action on requiring AHERA-type inspections and management plans in nonresidential, non-school buildings indefinitely.

Under authority granted under AHERA, USEPA extended OSHA asbestos worker protection rules (which are limited to construction and general industry) to public employees. As a consequence, school employees were provided OSHA worker protection for the first time.

Under AHERA, USEPA required that all major abatement projects that disturb ACM must be visually inspected and pass a clearance standard of 0.01 f/cc (fibers per cubic centimeter) based on aggressive sampling prior to the completion of asbestos abatement projects. Though only required for schools, these clearance standards have become the accepted practice in asbestos abatement activities in buildings subject to subsequent use.

In 1988, more than a decade after the promulgation of the USEPA NESHAP, which required removal of friable ACM before renovation or demolition, OSHA promulgated a construction industry standard for asbestos. It was designed to protect asbestos abatement workers as well as workers in other asbestos-related construction trades. Covered activities included removal, encapsulation, enclosure, repair/maintenance, transportation, disposal, and storage of ACM. This standard required use of administrative and engineering controls and respiratory protection to protect workers from excessive asbestos exposures. Administrative controls included the demarcation of regulated areas where abatement activities were to occur and access restriction for nonabatement personnel. Abatement activities required a "competent person" who was capable of identifying asbestos hazards and selecting appropriate control strategies for reducing asbestos exposure, and who had the authority to take prompt corrective measures to eliminate asbestos hazards to workers and building occupants. Engineering controls included use of negatively pressurized enclosures/containments which isolated the abatement area from other building spaces. Abatement workers were required to wear approved respirators designed to protect them from exposures above the then-permissible exposure limit (PEL) for asbestos of 0.2 f/cc.

In the early 1990s, it was evident there was a need for accredited personnel with a minimum level of standardized training in all asbestos abatement work. As a consequence, AHERA was amended to require that all asbestos professionals working in public and commercial buildings be trained and accredited according to a revised model accreditation plan (MAP). The USEPA NESHAP for asbestos was also amended to require that buildings be inspected for regulated ACM prior to renovation or demolition activities. Regulated ACM includes friable ACM which, when dry, can be crumbled and pulverized by hand pressure, and nonfriable ACM, which can be reduced to powder by mechanical means. Under the revised NESHAP, nonfriable ACM is regulated, and in many cases must be removed prior to renovation or demolition. USEPA identified and designated two categories of nonfriable ACM. Category I and category II nonfriable ACM can be distinguished from each other by their potential to release fibers when damaged. Category II ACM is more likely to become friable when damaged. It includes asbestos cement shingles and fibrocement boards or panels. Category I ACM includes asbestos-containing gaskets, packings, resilient floor covering, mastics, and roofing products. Unlike AHERA, under which ACM inspections are limited to indoor materials, NESHAP requires inspectors to locate and identify ACM in both interior and exterior environments.

During the early 1990s, it became evident that school occupants such as students and nonmaintenance staff were at relatively low risk of asbestos exposure and disease in buildings in which ACM was present. As a consequence, USEPA concluded that expanding inspection and management plan requirements to public and commercial buildings was not warranted. However, there was increasing scientific evidence that service workers were at special risk of exposure and developing asbestos-related disease. Therefore, in 1994, OSHA revised its construction industry standard to require building owners to designate all thermal system insulation and surfacing materials installed prior to 1980 as presumed ACM (PACM). Building owners have a duty under the revised OSHA construction industry standard to inform employees and workers who work or will work in areas with PACM or known ACM. They must be informed of its presence and location and employ appropriate work practices to ensure PACM/ACM will not be disturbed. The Occupational Safety and Health Administration requires that building owners post signs at the entrance to mechanical rooms and rooms where service workers can reasonably be expected to enter.

Designation of PACM, or its rebuttal by conducting a full AHERA-type inspection, is in good measure a *de facto* OSHA asbestos inspection/management requirement in public and commercial buildings which is designed to protect service workers. It is, for the most part, a self-enforcing rule.

The revised OSHA construction industry standard includes a 0.1 f/cc PEL and a 1 f/cc excursion limit. It also defines four levels of work activities in buildings that require different degrees of building and worker protection.

When asbestos exposure and health risk concerns for individuals in schools were initially raised, regulators faced an unknown but potentially significant health risk to children and other building occupants; they assumed the worst. Significant resources were expended conducting inspections, preparing management plans, and abating potential asbestos hazards. Based on the current scientific understanding of asbestos risks in buildings, the regulatory response was much greater than it needed to be. Contemporary asbestos exposure concerns in buildings focus appropriately on maintenance workers, the individuals who are at greatest risk of exposure.

B. Lead

It became increasingly evident to public health officials in the 1950s that lead poisoning observed in many children was associated with deteriorated lead-based paint (LBP) in old housing. As a consequence, a number of U.S. cities including Chicago, Baltimore, Cincinnati, New York, Philadelphia, St. Louis, Washington, Jersey City, New Haven, and Wilmington banned LBP intended for use in building interiors. The paint industry voluntarily limited the lead content in interior paints to 1% by dry weight in 1955. These early public efforts and subsequent regulatory and policy actions related to LBP hazards are summarized in Table 13.2.

In the 1950s and 1960s, several cities initiated childhood lead screening programs and developed programs to educate parents whose children were at risk of significant lead exposure on ways to minimize that risk.

The first federal LBP legislation was enacted by Congress in 1971. The Lead-based Paint Poisoning Prevention Act (LBPPPA) authorized the Secretary of Health, Education and Welfare (DHEW) to prohibit use of LBP (defined as containing more than 1% lead by weight) in residential dwellings constructed or rehabilitated under federal programs. The LBPPPA also authorized development of a national program to encourage and assist states, counties, and cities to conduct mass screening programs to identify children with elevated blood lead levels (EBLs), refer them for treatment, investigate homes for lead sources, and require LBP abatement where deemed necessary.

At that time, the public health understanding of childhood lead poisoning was that EBLs resulted when unsupervised children ate paint chips; lead poisoning in children was seen as a problem of deteriorating indoor paint that contained high lead levels.

In 1972, HUD promulgated regulations prohibiting the use of LBP in public housing or HUD-financed housing. The LBPPPA was amended in 1973 to lower the permissible paint lead content to 0.5% until December 31, 1974, and to 0.06% thereafter unless the Consumer Product Safety Commis-

Table 13.2 Regulatory and Public Policy History of Lead-Based Paint and Lead
Contamination of Building Environments

Year	Actions
1955	Paint manufacturers voluntarily reduce lead content in interior paints.
1956–1970	Cities begin to develop childhood lead screening programs.
1971	Congress enacts Lead-based Paint Poisoning Prevention Act (LBPPPA): authorized (1) prohibition of LBP in federally financed housing, (2) mass screening programs, and (3) investigations of EBLs.
1972	HUD promulgates regulations prohibiting use of LBP in public housing.
1973	LBPPPA amended: lowers permissible lead to 0.5%; requires HUD to eliminate lead hazards in pre-1950 public housing.
1976	LBPPPA amended: HUD required to eliminate LBP hazards; lead content in paint limited to 0.06%.
1978	CPSC banned sale of LBP with content >0.06%.
1987	LBPPPA amended: intact paint described as immediate hazard; inspection of random sample of pre-1978 public housing by 1994; abatement of LBP hazards in public housing.
1992	HUD publishes interim guidelines for identification and control of LBP hazards; Residential Lead-Based Paint Hazard Reduction Act (Title X) enacted by Congress: focuses on lead-based paint hazards, training requirements for professionals, grants special authorities to USEPA.
1995	HUD publishes guidelines document on identification and control of lead hazards
1996	USEPA promulgates rules describing accreditation requirements for lead professionals; disclosure of known LBP hazards in real-estate transactions.

sion (CPSC) determined that a higher level was safe. CPSC concluded at that time that 0.5% lead in paint was safe.

Under the 1973 amendments, HUD was required to eliminate, to the extent that was practical, LBP hazards in pre-1950 public housing, subsidized housing, and houses covered by Federal Housing Administration (FHA) mortgage insurance. Regulations to achieve the congressionally mandated requirements were promulgated in 1976.

Amendments to the LBPPPA of 1976 again limited paint lead content to 0.06% unless CPSC determined that a higher level not exceeding 0.5% was safe. This time, CPSC declined to make such a finding. As a consequence, after June 1977, any paint that had a lead content above 0.06% was considered to be LBP. The CPSC in 1978 banned the sale of all LBPs (>0.06% Pb) to consumers, and the use of LBP in residences and other areas where consumers and their families may have direct access to it. This ban did not apply to paint products used on bridges and industrial building materials.

Under the 1973 LBPPPA amendments and its 1976 rules, HUD was required to eliminate LBP hazards in pre-1950 public and public-financed housing. It focused its abatement activities on deteriorated paint, which it considered an immediate hazard. It was challenged in federal court to define an "immediate lead hazard" under its rules. Plaintiffs alleged that HUD rules were deficient in failing to define intact LBP surfaces as an immediate hazard requiring treatment/abatement. As a result of the federal court ruling on behalf of plaintiffs, HUD promulgated new rules that defined intact LBP as an immediate hazard and changed the construction cutoff date for housing subject to its rules from 1950 to 1973.

Congress, in 1987, amended the LBPPPA to require (1) inclusion of intact paint in the definition of an immediate hazard; (2) a construction cutoff date of 1978; (3) inspection of a random sample of units in pre-1978 public housing to be completed by December 6, 1994, and abatement of LBP with >1.0 mg lead/cm^2; (4) an extensive research and demonstration program; and (5) reports on the feasibility and cost of LBP abatement in privately owned housing. Amendments in 1988 required HUD to develop a comprehensive workable plan for lead abatement in public housing.

Congress, in 1992, enacted the Residential Lead-Based Paint Hazard Reduction Act, better known as Title X. Title X shifted the focus of LBP poisoning control programs from the presence of high lead levels in paint to LBP hazards that would more likely cause significant exposure to young children and result in EBLs. These included deteriorated paint; lead paint on surfaces subject to friction, impact, or chewing; and high lead levels in house dust and soils near dwelling units. Title X required USEPA to issue specific training and accreditation requirements for lead professionals including inspectors, risk assessors, abatement workers, supervisors, and contractors. It was intended to provide a mechanism by which the public could be educated about potential lead hazards in housing and to abate LBP hazards from federal housing.

Responsibility for issuing regulations under Title X falls to both USEPA and HUD. Rules that described federal accreditation requirements for training programs and lead professionals were promulgated in 1996 and went into effect in 1999. Rules were also promulgated in 1996 that require disclosure of LBP hazards in real-estate transactions and leasing contracts involving pre-1978 housing. Building owners are to provide prospective buyers with a USEPA-written hazard brochure, disclose any "known" LBP hazards, and allow buyers 10 days to have an inspection conducted prior to finalizing a purchase agreement. Landlords are also required to disclose any known lead hazards. A copy of a model disclosure form for home buyers is illustrated in Figure 13.2.

Employers involved in LBP abatement activities are subject to provisions of OSHA's interim final Lead in Construction Standard of 1993. It prescribes a PEL of 50 µg/m^3 over 8 hours and an action level of 30 µg/m^3, respiratory and personal protection equipment, hygienic work practices, training requirements, record-keeping, medical surveillance, medical removal requirements, and medical treatment in the case of excessive exposure.

HUD, in 1990, issued comprehensive technical guidelines on testing, abatement, and disposal of LBP in public and Indian housing. These interim HUD guidelines were updated, expanded, and issued as *Guidelines for Eval-*

Disclosure of Information on Lead-Based Paint and/or Lead-Based Paint Hazards

Lead Warning Statement

Every purchaser of any interest in residential real property on which a residential dwelling was built prior to 1978 is notified that such property may present exposure to lead from lead-based paint that may place young children at risk of developing lead poisoning. Lead poisoning in young children may produce permanent neurological damage, including learning disabilities, reduced intelligence quotient, behavioral problems, and impaired memory. Lead poisoning also poses a particular risk to pregnant women. The seller of any interest in residential real property is required to provide the buyer with any information on lead-based paint hazards from risk assessments or inspections in the seller's possession and notify the buyer of any known lead-based paint hazards. A risk assessment or inspection for possible lead-based paint hazards is recommended prior to purchase.

Seller's Disclosure

(a) Presence of lead-based paint and/or lead-based paint hazards (check (i) or (ii) below):
    (i)___Known lead-based paint and/or lead-based paint hazards are present in housing (explain).

_____

    (ii)__Seller has no knowledge of lead-based paint and/or lead-based paint hazards in the housing.

(b) Records and reports available to the seller (check (i) or (ii) below):
    (i)___Seller has provided the purchaser with all available records and reports pertaining to lead-based paint and/or lead-based paint hazards in the housing (list documents below).

_____

    (ii)__Seller has no reports or records pertaining to lead-based paint and/or lead-based paint hazards in the housing.

Purchaser's Acknowledgment (initial)

(c)___Purchaser has received copies of all information listed above.
(d)___Purchaser has received the pamphlet *Protect Your Family from Lead in Your Home.*
(e)___Purchaser has (check (i) or (ii) below):
    (i)___received a 10-day opportunity (or mutually agreed upon period) to conduct a risk assessment or inspection for the presence of lead-based paint and/or lead-based paint hazards; or
    (ii)__waived the opportunity to conduct a risk assessment or inspection for the presence of lead-based paint and/or lead-based paint hazards.

Agent's Acknowledgment (initial)

(f)___Agent has informed the seller of the seller's obligations under 42 U.S.C. 4852d and is aware of his/her responsibility to ensure compliance.

Certification of Accuracy

The following parties have reviewed the information above and certify, to the best of their knowledge, that the information they have provided is true and accurate.

Seller	Date	Seller	Date
Seller	Date	Seller	Date
Seller	Date	Seller	Date

Figure 13.2 Model residential lead disclosure form.

uation and Control of LBP Hazards in Housing in 1995. These guidelines are good practice documents that describe inspection, risk assessment, and abatement practices. Though not rules, they are, for all practical purposes, standards of care by which professional activities associated with LBP hazards are to be conducted.

C. Formaldehyde

Regulatory involvement in the problem of HCHO-related odor and health complaints began in the late 1970s. These complaints were reported for UFFI houses, mobile homes, and stick-built homes using UF-bonded wood products. Complaints were directed to state and local health, consumer, and environmental agencies, and, at the federal level, to CPSC. Health authorities in many states conducted investigations of complaints and provided HCHO air testing services. As a result of such investigations, public health departments in several states instituted efforts to control one or more aspects of the "formaldehyde problem" through regulatory initiatives.

In 1980, Massachusetts banned the sale and installation of UFFI in residential structures and required installers, distributors, or manufacturers to repurchase the product at homeowner request. Though initially overturned on appeal, the ban was sustained by the Massachusetts Supreme Court and remains in effect. A ban on UFFI for residential applications also went into effect in Connecticut.

The Minnesota state legislature enacted an HCHO statute in 1980 authorizing the Health Commissioner to promulgate rules regulating the sale of building materials and housing units constructed with UF-containing materials. The Minnesota law required written disclosure prior to the sale of new homes and construction materials containing UF resins. After an attempt to establish a 0.1 ppmv IAQ standard for new homes was unsuccessful, the Minnesota Health Commissioner adopted an IAQ standard of 0.5 ppmv for new housing units and UFFI installations. The standard was appealed and upheld by the Minnesota Supreme Court, which remanded the 0.5 ppmv level back to the state health department for reconsideration. Subsequently, an IAQ standard of 0.4 ppmv was adopted and went into effect in 1985. Because of preemption by HUD rules (see below), the Minnesota HCHO statute was amended to establish product standards; the IAQ standard was then repealed. In the early 1980s, Wisconsin promulgated an IAQ standard for new mobile homes. Because of legal and political problems, the standard was never enforced.

In response to numerous complaints associated with UFFI installations and its own investigations of the problem, CPSC proposed rules in 1980 that would have required UFFI installers, distributors, and manufacturers to notify prospective purchasers of the potential adverse effects associated with the product. It subsequently concluded that such disclosure would not adequately protect the public and, as a result, imposed a ban on UFFI in 1982. In promulgating the ban, CPSC denied a petition by the Formaldehyde

Institute (an umbrella group of HCHO manufacturers, UF-wood product manufacturers and users, etc.) to establish a mandatory standard for UFFI resin formulation and installation. The CPSC reasoned that the product was inherently dangerous and that application standards would be insufficient to protect public health. The ban was appealed to the First U.S. Circuit Court of Appeals, where it was voided on procedural and technical grounds. The CPSC elected not to appeal to the Supreme Court. The ban, adverse publicity, and litigation caused the UFFI industry to collapse.

In response to requests from wood-product and manufactured housing industries and public interest groups, HUD initiated rule-making that required product emission standards for particle board and hardwood plywood used in construction of mobile homes. The HUD product standards prescribe that under standardized large chamber conditions of product load (load factors similar to those in mobile homes), temperature (78°F, 25°C), relative humidity (50%), and air exchange rate (0.5 ACH), emissions from particle board and hardwood plywood paneling shall not cause chamber concentrations to exceed 0.3 ppmv and 0.2 ppmv, respectively. Under combined loading conditions, it was projected that HCHO concentrations in new mobile homes would not exceed a target level of 0.4 ppmv, a level that HUD administrators concluded would provide a reasonable degree of health protection. These product standards reflected what the wood products industry was capable of achieving at that time. Minnesota product standards were similar to HUD standards except that Minnesota standards applied to medium-density fiber board as well.

The HUD rule, which went into effect in 1985, requires that mobile home manufacturers prominently display a specifically worded health warning in the kitchen and owner's manual (Figure 13.1).

D. Smoking in public places

Significant regulatory efforts have evolved in North America over the last two decades to limit smoking in public places. These regulatory initiatives reflected changing attitudes toward smoking and the acceptability of smoking in public places since the issuance of the 1964 Surgeon General's report. Antismoking efforts were given significant momentum with publication of the Surgeon General's 1986 report on involuntary smoking and USEPA's 1992 document on respiratory effects of exposure to environmental tobacco smoke (ETS).

With few exceptions, restrictions on smoking in public places has resulted from state, local, and private initiatives. Restrictions on smoking in commercial aircraft was, of course, a federal action.

Regulation of smoking in public places by state and local governments evolved from permitting a no-smoking section to requiring that nonsmoking was the assumed case. Legislative language made it clear that its intent was to safeguard health and contribute to the general comfort of building occupants.

One of the most notable pieces of early smoking-restrictive legislation was Minnesota's Clean Indoor Air Act of 1975. It prohibited smoking in public places and meetings except in designated smoking areas. It covered restaurants, private worksites, and a large number of public places. The Minnesota law served as a model for smoking restriction legislation and ordinances throughout the U.S.

At this writing, smoking restrictions in public places due to state and local regulatory actions and actions by building owners are nearly universal in the U.S. The extent of smoking prohibition varies from completely smoke-free buildings and modes of public transport to smoking in very limited designated areas. The major area where smoking restrictions are resisted is in the food and beverage trade, where there is an apparent link between smoking and food and beverage consumption. Not surprisingly, fewer restrictions on smoking in public buildings have been enacted in tobacco-growing states such as Kentucky and North Carolina.

Enforcement of smoking-restriction laws and ordinances is the responsibility (in most cases) of state and local health departments. Many laws and ordinances include penalties that may be imposed on those who violate smoking restrictions and on building owners/managers who fail to designate smoking/nonsmoking areas. Such penalty-backed enforcement has not been necessary, as most smokers and building owners have willingly complied. Compliance has resulted from a combination of individual smokers' sense of duty to obey the law, peer disapproval, and an assured right to smoke in designated smoking areas.

Smoking restriction legislation/ordinances were opposed by the tobacco industry, the food and beverage industry, and libertarian smokers who believed they had the right to smoke anywhere. The tobacco industry lobbied against legislation/ordinances restricting smoking in public places because it anticipated that such legislation would reduce tobacco consumption.

Laws and ordinances restricting smoking in public places have been implemented with few, if any, problems. They have been, for the most part, well accepted by both smokers and nonsmokers. Nonsmoking is now the norm in most public spaces in North America. Smoking is still the norm in many countries of Europe and Southeast Asia where a large percentage of the adult population smokes. Regulation of smoking in public spaces in many European countries and Japan is still in its early stages.

E. OSHA actions and proposals

In the U.S., worker health and safety is the primary responsibility of OSHA. Acting within its authority, OSHA designated ETS a class 1A carcinogen (known human carcinogen) and began proposed rule-making that would protect workers from exposure to ETS and provide for a more acceptable work environment in nonindustrial buildings. As a result, in 1994 OSHA published notice of its proposal to promulgate an IAQ standard.

The proposed Air Quality Standard would have required employers in both industrial and nonindustrial environments to take steps to protect employees from ETS by either eliminating smoking or restricting it to separately ventilated spaces. It would also have required employers in nonindustrial environments to (1) establish a written IAQ compliance program, (2) designate an individual for implementing IAQ programs, (3) maintain and operate HVAC systems to conform with original design specifications and consensus standards on outdoor air flow rates, (4) establish an employee complaint record, (5) use general or local exhaust ventilation where maintenance and housekeeping activities could cause other areas to be exposed to potentially hazardous substances or particulate matter, (6) conduct periodic IAQ-related building inspections, (7) establish a program of IAQ-related record-keeping, and (8) form an employee information and training program.

Because of the storm of criticism from the tobacco industry and other interest groups subject to regulation (e.g., Building Owners and Managers Association), OSHA deferred action on the proposed IAQ standard indefinitely. The proposed OSHA IAQ standard is notable in that a federal agency concluded that IAQ was a health issue of sufficient magnitude to warrant regulatory action. It is notable also in the degree of opposition that the proposed rule-making engendered and the difficulties that federal efforts to regulate IAQ would face in any future regulatory attempts.

F.   Other actions and authorities

The USEPA has primary rule-making and enforcement authority for the use of pesticides under the Federal Insecticide, Fungicide and Rodenticide Act (FIFRA). As such, it has authority to approve the use of pesticides for various applications as well as to restrict their use should it determine that such use poses a risk to public health. Under this authority, USEPA reconsidered the use of mercury biocides in paints and worked out a voluntary agreement with paint manufacturers to allow the use of mercury biocides only in exterior latex paints. It has also used this authority to effect a voluntary phaseout and elimination of the use of chlorpyrifos for termiticidal treatments and residential indoor and lawn use.

Pesticides designed to be used indoors must be approved by USEPA. For example, benzoic acid esters used as acaricides in European countries to control dust mite populations and reduce allergy and asthma risk cannot be legally used in the U.S. since they have not been approved for such use by USEPA.

The USEPA has authority to regulate the use of a large number of substances under TSCA. Under this authority, it has placed restrictions on the use of methylene chloride, a suspected human carcinogen present in paint strippers commonly used indoors. USEPA's authority to regulate various aspects of asbestos and LBP in the U.S. is also derived from TSCA.

Under the 1988 Radon Reduction Act, USEPA has limited authority to regulate various aspects of radon testing and mitigation efforts. The USEPA requires laboratories providing radon analytical services to meet specific proficiency requirements and be accredited. Radon professionals who provide either testing or mitigation services must also meet minimum training requirements and be accredited to provide services. In most states, they must be licensed as well.

IV. Nonregulatory approaches

A variety of nonregulatory principles and concepts have been used over the past two decades in response to various IAQ/IE problems and concerns. These include development and publication of consensus health guidelines by state and federal government agencies, world bodies, and professional organizations; development and publication of ventilation guidelines by professional organizations and, in some cases, governments; development and publication of performance guidelines for the operation of building systems and for other building environmental quality concerns; voluntary initiatives directed to selected industries and potential IE problems; development of public information and education programs; and civil litigation.

A. Health guidelines

An alternative to using AQSs to achieve and maintain acceptable IE quality is to use health guidelines developed by government agencies, world bodies such as the World Health Organization (WHO), or professional groups such as ASHRAE. Health guidelines do not have regulatory standing. As such, compliance is voluntary. Their development and publication is, in most instances, less cumbersome than regulatory standard-setting processes. They are, in the main, less subject to the political and economic compromises common to standard setting. As such, they have the potential to better reflect true health risks and public health protection needs.

Though not enforceable, guideline values for contaminant levels have considerable value. They have the power of scientific consensus and, in the case of WHO- and government-published guidelines, convey a sense that contaminant levels above the guideline are unsafe (and that levels below the guideline are safe). This is particularly the case for the USEPA-recommended guideline level for indoor radon. That guideline value of 4 pCi/L annual average concentration is used (and misused) by home and other building owners as a reference value in interpreting results of radon testing and determining the need for remedial measures. It was adopted in the late 1970s based on research work associated with uranium mill tailings and open-pit phosphate mining spoils. It reflected (1) a need to provide reasonable health protection and (2) the practical limits of mitigation measures applied to houses with high radon levels (>200 pCi/L). Guideline values for radon in dwellings recommended by other countries and organizations are summarized in Table 13.3.

Guidelines may also affect manufacturing decisions. Though corporations may not agree that a guideline value is necessary, or believe that it is

Table 13.3   Guideline Values for Radon in Dwellings
(pCi/L annual average exposure)

Country/organization	Existing buildings	Future buildings
Belgium	6.8	6.8
Canada	21.6	21.6
CEC[a]	16.8	5.4
USEPA (USA)	4.0	4.0
Finland	10.8	5.4
Germany	6.8	6.8
ICRP (1993)[b]	5.4–16.2	—
Norway	5.4	5.4
Sweden	3.8	—
United Kingdom	5.4	5.4
WHO[c]	5.4	5.4

[a] CEC — Commission of European Communities.

[b] ICRP — International Council on Radiation Protection.

[c] WHO — World Health Organization.

too stringent, they are nevertheless under both moral and marketing constraints to be seen as producing safe products.

The HUD guideline values of 1 mg Pb/cm^2 in paint, 100 $\mu g/ft^2$ floor dust, 400 $\mu g/ft^2$ windowsill dust, and 800 $\mu g/ft^3$ soil have regulatory standing only in the context that if they are known by the homeowner they must be disclosed to potential purchasers. Homeowners who ignore the guidelines in real-estate transactions, or abatement contractors who disregard them, are at considerable civil liability risk because the guideline values express what is considered to be good practice, the standard of proof in personal injury or property damage claims. They are also subject to regulatory liability.

The Canadian Department of Health and Welfare (Health Canada) developed numerical exposure guideline values for HCHO and radon. For HCHO, guideline values were formulated in the context of an action level of 0.10 ppmv and target level of 0.05 ppmv. If measured levels are above the action level, homeowners and other building owners are advised to take steps to reduce HCHO levels to the target level or below. Similar guideline values and recommendations were developed and published by the California Department of Health and California Air Resources Board. Guideline values for HCHO have also been developed and published by a variety of countries (Table 8.7).

Guideline values for a number of indoor contaminants were published by ASHRAE in their ventilation guidelines document, Standard 62-1981. They were developed to support use of the Indoor Air Quality Procedure by building designers to provide adequate ventilation air. Guideline values included CO_2 (2500 ppmv), HCHO (0.10 ppmv), ozone (0.05 ppmv), chlordane (5 $\mu g/m^3$), and radon (0.01 WL). Guidelines were also recommended for contaminants that might be drawn into a building in ventilation air. These were six contaminants for which national ambient AQSs had been promulgated and an additional 27 substances regulated under occupational safety and health rules. ASHRAE guideline values for HCHO proved to be very controversial. Due to threatened lawsuits by HCHO producers and users, ASHRAE, in its revision of 62-1981, deleted HCHO from its IAQ guidelines and relegated other guideline values to the document's appendix.

In response to residential exposure concerns, a committee of the National Research Council developed and published interim exposure guidelines for pesticides used to control termites. These included guideline values of 5 $\mu g/m^3$ chlordane, 2 $\mu g/m^3$ heptaclor, and 10 $\mu g/m^3$ chlorpyrifos; values that were one tenth of the permissible limits used to protect occupationally exposed persons.

B. Ventilation guidelines

Ventilation guidelines for mechanically ventilated buildings have a long history of use in the U.S. and other developed countries. Historically, these have been developed through a consensus process by ASHRAE. Ventilation guidelines (or standards, as they are called) are recommended values. They have the force of law only when incorporated into building codes (as they often are). ASHRAE guidelines are considered to be good practice design values; therefore they are used by architectural firms whether or not they are included in state and local building codes. They usually specify some minimum ventilation rate that will provide occupants with a reasonably comfortable environment at design capacity, with minimum sensory perception of human odor and discomfort.

Guideline values for ventilation in mechanically ventilated buildings have changed over the past four decades. In 1973, ASHRAE, in response to energy conservation concerns, reduced its ventilation guideline for office environments from 10 CFM (4.76 L/sec) per person to 5 CFM (2.37 L/sec) per person. This was contained in ASHRAE Standards 62-73 and 62-81, and used by building design engineers for approximately 15 years. In Standard 62-89, ASHRAE, recognizing that a 5 CFM per person ventilation standard was not adequate, increased the ventilation guideline to 20 CFM (9.52 L/s) per person for office buildings and 15 CFM (7.14 L/sec) per person for schools. A building designed for a 5 CFM per person ventilation rate at design capacity would be expected to have maximum CO_2 levels of 2500 ppmv; one for 20 CFM would have 800 ppmv.

Guideline values are used by design engineers to determine air flow capacity required in new buildings at design occupancy. It is expected that facility managers can then operate ventilation systems to achieve and maintain adequate ventilation during occupancy. Unfortunately, in many cases ventilation systems are not operated to conform to guideline values, as evidenced in complaint investigations. This has been due to management concerns associated with ventilation air energy costs, malfunctioning equipment, and poor technical and operational understanding of HVAC system operation by facility staff.

C. Public health advisories

In 1987, USEPA, in conjunction with the Surgeon General, issued a public health advisory recommending that all homeowners voluntarily conduct radon testing in their homes to determine radon levels and take appropriate remedial action if radon levels were excessive. Subsequent to this, USEPA issued a public health advisory to school districts recommending they do the same in their schools.

Response to USEPA's public health advisories was strong, with millions of homes and tens of thousands of school buildings tested for radon. Though not all homes or school buildings were tested, as the advisories recommended, they nevertheless had a significant impact on public awareness of potential radon exposure and hazards in buildings.

D. Performance guidelines and requirements

1. ASHRAE proposals

In its attempt to revise ventilation guidelines, ASHRAE's IAQ committee changed the focus of its ventilation standard-setting process from designating numerical values to describing performance guidelines for the operation and maintenance of HVAC systems. These performance recommendations include provision of adequate ventilation and thermal control, and maintenance of system components to ensure adequate air filtration and prevent problems associated with microbial growth. These and other proposed revisions provoked a storm of controversy and, as a consequence, ASHRAE refused to ratify the recommendations of its IAQ committee on Standard 62-1989R. The proposed revisions would have introduced a standard of care for the design, operation, and maintenance of HVAC systems. As such, system designers and building facility managers would have had a duty, under penalty of civil litigation, to design and operate systems according to the consensus principles adopted by ASHRAE.

Because of the lack of support for ASHRAE Standard 62-1989R, ASHRAE changed its approach to its standards revision process. It put its existing standard under "continuous maintenance," by which Standard 62-1989 is to be modified by proposing and approving incremental changes through the use of addenda. This "continuous maintenance" process is intended to serve as a vehicle for revision of Standard 62-1989. Revisions will include a code-intended standard, a user's manual, and guidelines containing additional information that is not appropriate for a minimum standard.

The code-intended standard is being written primarily for building code organizations, design professionals, construction and property managers, and other building professionals. It is intended for adoption by the American National Standards Institute (ANSI). While the standard's goal is to achieve acceptable IAQ, it will state that compliance may not necessarily provide health, comfort, or occupant acceptability. Standards requirements are to be written in mandatory enforceable language that describes what must be done to comply and facilitates the determination by enforcing bodies of whether compliance has been achieved.

Under its "continuous maintenance" process, ASHRAE has adopted and proposed a number of addenda. In the former category, addenda to Standard 62-1989: (1) clarify the intent of the standard to "minimize the potential for adverse health effects," (2) change the standard's designation to indicate its status as an ANSI Standard, (3) indicate that the standard may not achieve acceptable IAQ because of the presence of some contaminants and varied susceptibilities of occupants, (4) eliminate language that the standard could accommodate a moderate amount of smoking, and (5) clarify language on CO_2. Proposed addenda, which at the time of this writing are under public review, include (1) requirements for classification, signage, and separation of areas where smoking is permitted; (2) an updated version of the IAQ Procedure, a performance-based approach in which a building and its ventilation system are designed to maintain contaminants below acceptable levels and indicate where the IAQ Procedure can be used (also see Chapter 11); (3) revisions on how natural ventilation can be used; (4) description of how the standard applies to new and existing buildings; (5) recommendations and requirements on construction and ventilation system startup, operation, and maintenance; (6) revisions in the procedure for determining design ventilation rates; (7) recommendations for providing ventilation for smoking rooms and other rooms where smoking is permitted; and (8) recommendations on how to provide combustion air to dilute water vapor, CO_2, and other contaminants produced by indoor combustion appliances and how to provide sufficient ventilation air.

In its operation and maintenance addendum, ASHRAE is proposing the use of Standards of Performance (SOPs) for HVAC systems, and maintenance requirements pertinent to acceptable IAQ in buildings. Standards of performance are defined by ASHRAE as standards that ensure a building system or component is operating properly — determined through a series of measures. The measure may be a property, process, dimension, material, relationship, concept, nomenclature, or test method. ASHRAE has developed a consensus SOP structure to ensure consistency for three levels of evaluation: building, system, and component. Standards of performance documents include seven sections, the first of which describes basic information on the building, system, or component; the others describe how to evaluate system performance and provide information on corrective action. An example of an IAQ SOP for a central AHU is indicated in Figure 13.3. Such SOPs can be evaluated by comparing them to measurements and calculations.

More extensive maintenance requirements are being developed to reflect research results and recognition that HVAC system maintenance can affect IAQ (see the relationship between maintenance level and perception of IAQ degradation in Figure 13.4). The ASHRAE 62 Projects Committee has determined that maintenance requirements for the code version of the document should address fundamental maintenance activities applicable to a wide range of HVAC systems. These would include filters/air cleaning

---

Indoor Air Quality Standard of Performance
Central Air-Handling Unit (AHU-1)

GENERAL INFORMATION

System type: Variable air volume (VAV) built-up air handling unit with chilled water cooling, hot water heating, and an economizer cycle. Conditioned and ventilation air is supplied to the space through VAV terminal diffusers. Supplemental heat is provided by a perimeter baseboard hot water radiation system.

Area served: This two-story building is served by AHU-1, 2, 3 (all identical). Each room has individual control, and each floor is divided into two primary suites for the purpose of minimum air control.

Description: AHU-1, 2, and 3 are equipped with a chilled water coil (440 Mbh, 88 gpm, 45°F entering, 55°F leaving), a hot water coil (500 Mbh, 33 gpm, 120°F entering, 90°F leaving), an economizer section, filters and a supply fan (20,000 cfm), and return (18,000 cfm) fan. Each air handler provides a variable volume of air to maintain a constant static pressure (0.4"w.g.) in the ductwork, based on a static pressure sensor located in the supply ductwork above room 125, 225, and 280, respectively. Individual zone VAV terminal units vary the air to maintain the space setpoint based on integral thermostats.

The supply air temperature is reset based on the return air temperature. At a return air temperature of 76°F, 50°F supply air is provided, and at a return air temperature of 72°F, 58°F supply air is provided. A central direct digital control (DDC) system maintains the discharge air temperature by varying the position of the chilled and hot water coil valves. The controls also operate the economizer when the outdoor air temperature is below 65°F. When the outdoor air is above 63°F, the outdoor air dampers go to their minimum position. The minimum position is reset based on the total supply airflow to maintain 4000 cfm of outdoor air. At full system flow the outside air dampers are set to 20% flow and at minimum system flow (5000 cfm), the dampers are set to 80% of flow (4000 cfm).

Cooling is disabled below an outdoor air temperature of 55°F, and Heating is disabled above an outdoor air temperature of 53°F.

Figure 13.3 Model standard of performance for a central air handling unit. (From Dorgan, C.E. and Dorgan, C.B., *Proc. 8th Internatl. Conf. Indoor Air Quality & Climate*, Edinburgh, 4, 218, 1999.)

devices, outdoor air dampers, humidifiers, dehumidification coils, drain pans, outdoor air intakes, sensors, cooling towers, and floor drains. Maintenance activities beyond these core requirements would be included in a guidance document.

2. Lead

The 1995 HUD *Guidelines for the Evaluation and Control of Lead-Based Paint Hazards in Housing* describe good practice procedures to use when conducting: building inspections for lead, risk assessments, and lead abatements employ-

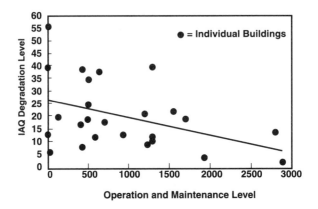

Figure 13.4 Relationship between perceived IAQ degradation and operational and maintenance levels in buildings. (From Dorgan, C.E. and Dorgan, C.B., *Proc. 8th Internatl. Conf. Indoor Air Quality & Climate*, Edinburgh, 4, 218, 1999.)

ing a variety of interim and permanent control measures. They describe LBP hazards using numerical values; testing procedures to be used in conducting paint, dust, and soil sampling, and analysis for lead; and clearance levels that should be achieved before abatement actions can be completed. Because they are performance guidelines, they describe good professional practice and, as a consequence, have a high probability of being followed by lead professionals in the absence of specific regulatory requirements.

3. VOCs

Voluntary guidelines for reduction of VOC exposures in newly constructed or renovated office buildings were developed by the California Department of Health's Indoor Air Quality Program. Guidelines included recommendations for design, construction, commissioning, and operation during initial building occupancy. They included source control recommendations such as (1) selecting building materials to avoid strong VOC emission sources, (2) isolating construction zones in partially or fully occupied buildings, (3) scheduling construction or furnishings installation to minimize buildup of elevated contaminant levels before occupancy, (4) installing adsorptive materials before applying wet coating and other similar materials, and (5) using low-emitting maintenance and housekeeping materials. Guidelines suggested the use of building bakeouts (see Chapter 10) prior to occupancy.

E. Governmental voluntary initiatives

Federal and state government agencies can request that an industry, group of industries, or building owners/managers voluntarily participate in programs designed to improve the quality of air in buildings. Most notable of these has been USEPA's carpet initiative and dialogue.

1. Carpet initiative

As a consequence of a reported outbreak of illness among some USEPA staff that occurred following installation of new carpeting, USEPA published a notice in the *Federal Register* requesting that carpet and rug manufacturers voluntarily conduct emission testing of their products to identify individual contaminants and determine emission rates. Carpet and rug manufacturers responded to this request by conducting extensive testing of their products as well as adhesives. The industry then developed a low emissions labeling program. Carpets and rugs that met TVOC emission limits set by the industry could indicate their compliance with the industry standard by having a green label affixed to them. The green label itself does not guarantee that carpet/rug products pose no risk of harmful exposures to sensitive individuals. Nevertheless, USEPA's voluntary carpet initiative and the industry's green label program have resulted in a substantial reduction in TVOC emissions from both carpeting and carpet adhesives.

2. Labeling

The concept of product labeling has been given serious attention by the Danish Ministry of Housing. The Danish program includes evaluating building materials and furnishings using qualitative headspace analysis of emitted VOCs followed by quantitative measurement of single compounds under controlled laboratory conditions. Emitted compounds are selected on the basis of odor detection and mucous membrane irritation thresholds and occupational guidelines. A model is used to determine whether the emission profile of products is above one or more acceptable values. In such a case, labeling could be voluntary or required by regulatory authorities. The advantage of a labeling program is that it provides information to manufacturers about emissions from their products and may indirectly induce them to modify manufacturing processes to lower VOC emissions.

F. Citizen initiatives

Citizens or groups of citizens who have concerns about their health and that of others often initiate a variety of public policies that may result in local ordinances, state/provincial laws and regulations, or voluntary operating practices. Bans or restrictions on smoking in public spaces began with citizen initiatives at the local level and evolved into policies employed nationwide.

Citizen activists in North America are involved in moving public policy on a number of fronts which affect indoor environments and, potentially, public health. Among these is the so-called "no scents" movement. Individuals involved believe that their health is adversely affected by scents used in perfumes, colognes, soaps, fabric softeners, etc. As a result of their lobbying activities a "no scents" policy has been adopted in the province of Nova Scotia; fragrance-free environments such as public transit systems, library, school, and university buildings, and medical and dental offices are becoming increasingly common. The policy focuses on educating building occupants and making visitors aware of its purpose and necessity. Enforcing the "no scents" policy is the responsibility of building staff. Citizen initiatives directed to limiting the use of lawn pesticides are becoming more common in the U.S., with somewhat spotty success. Antipesticide initiatives are increasingly being directed at school corporations. The citizen initiative that is achieving some level of success in school corporations is the use of integrated pest management (IPM) to control cockroaches and other insects in buildings. These IPM initiatives have the full support of USEPA and, in many cases, state environmental and public health agencies.

In response to citizen initiatives on pesticide use in schools, the U.S. Senate passed an amendment to the Affordable Education Act of 1999 which, if passed by the House of Representatives as well, would require that schools that receive federal funding notify parents at least 48 hours before using pesticides. School districts would also have to supply USEPA with a list of pesticides they use which are known carcinogens, developmental or reproductive toxins, or neurotoxins. Whether or not this amendment will become law is not known at the time of this writing. However, it illustrates the power of citizen initiatives.

Citizens who characterize themselves or their children as having environmental/ecological illness are increasingly demanding that employers and school corporations provide a variety of accommodations. Most notably these include limiting such activities as painting or floor waxing (during work/school hours), use of scented soaps in lavatories, etc.

G. Public information and education programs

Public information/education programs are low-cost tools used by USEPA and other agencies to provide individual Americans, building owners/managers, school officials, etc., with basic information on specific indoor environment concerns. In the public information arena, USEPA has developed and distributed small, simple-to-read brochures on asbestos, radon, and lead; established dedicated hotlines on each of these issues; and established clearinghouses for both lead and IAQ/IE.

The USEPA has developed and distributed major guidance documents for in-house facilities personnel to assist them in preventing, diagnosing, and mitigating IAQ/IE problems in nonresidential buildings. USEPA's *Building Air Quality Manual* has been widely distributed and used. In an initiative to improve IAQ in schools, USEPA developed its *Tools for Schools* program, which provided a kit for school facilities managers to better manage air quality in schools and assist them in solving IAQ problems. (See www.epa.gov for information on obtaining these documents.) These documents are of excellent quality and are used by building facilities personnel as well as professionals in the field. The Illinois Department of Public Health has also developed and distributed documents on IPM.

USEPA maintains public information and education programs in 10 federal regions. Each region dedicates one or more staff who are responsible for IAQ/IE programs and responding to citizen information requests. All regions have several staff members who serve as regional asbestos coordinators and others who serve as lead coordinators. These coordinators assist state agencies and others who have either asbestos- or lead-related concerns.

A model national education program to improve the quality of indoor air in homes and protect public health from home-related indoor air contamination has been developed as a collaborative effort between the Cooperative Extension Service (CES) of the U.S. Department of Agriculture (USDA) and USEPA. The *Healthy Indoor Air for America's Homes* program uses university faculty as state program managers and is delivered through the CES's network of trained professional and volunteer community leaders. The program has developed extensive teaching materials and attempts to provide accurate up-to-date information on IAQ/IE problems such as radon, carbon monoxide, home remodeling hazards, etc., to homeowners/leasers.

H. Civil litigation

In our civil legal system in the U.S., individuals who have suffered some form of personal injury or property damage as a result of the wrongful action of others have a right to seek legal redress. One can, under a number of legal theories, file suit seeking monetary, and in some cases punitive, damages for injuries claimed. These include expressed warranties, implied warranties, negligence, and strict liability.

Expressed warranties are made by sellers of products and appear in sales contracts, labels, advertising, or samples. They are positive representations of products. Liability depends only on the falsity of the representation and not on any particular knowledge of fault by sellers. Because of their relative simplicity, liability claims based on breach of an expressed warranty are often a desirable means of pursuing redress of personal injury.

Implied warranties are interpreted by courts to exist even when no expressed warranties are made by product manufacturers/sellers. Products are assumed to be fit for the normal purposes for which they are intended. In the case of mobile homes and other residential structures, the courts assume that they are fit for human habitation. One may not be fit for human habitation if it is structurally unsound or heavily contaminated with such substances as HCHO, pesticides, other toxic substances, or mold such as *Stachybotrys* that pose a substantial risk of serious health problems to those who would dwell in it. Habitability has been a major warranty claim in law suits involving HCHO in mobile homes, and misapplication of pesticides or contamination by toxic substances such as mercury in residences.

Negligence is one of the most widely applied principles in product liability claims, as well as claims involving problem/sick buildings. Negligence is legally defined as failure by a defendant to exercise due care, the level of care that would be exercised by a "reasonable person." The level of care in providing a service depends on what is accepted in the professional community as good practice. If a defendant in the conduct of his/her profession or job deviates from what is deemed good practice, he/she may be liable for negligence in claims involving personal injury or property damage. If a manufacturer produces a product without exercising due care that the product is safe for its intended purpose, the manufacturer is liable. Plaintiffs, through their legal counsel, must prove that personal injuries and/or property damage claimed were due to the unreasonable or negligent acts of defendants.

Strict liability applies to claims related to defective products. Such products may be defective as a result of their design or problems in manufacturing. Under strict liability, there is no determination of "fault" as is the case in negligence claims. Strict liability claims challenge the safety of products and not the conduct of manufacturers. Strict liability has historically been used in cases involving the safety of products. However, in some judicial districts, it has been applied to buildings as well. Buildings could, in many cases, be described as products. Because the burden of proof is relatively limited, strict liability claims may have a higher probability of success (than those based on other legal theories) in recovering damages.

Over the past two decades, plaintiffs have filed numerous suits alleging personal injury, property damage, or both, associated with IE-associated problems. These have included hundreds of claims filed against installers, distributors, and manufacturers of UFFI; sellers and manufacturers of mobile homes; and manufacturers of urea–formaldehyde (UF)- bonded wood products. Most of these claims were filed in the early to mid-1980s and, for the most part, have now been resolved. Most claims alleged personal injury from HCHO exposures. In UFFI cases, claims were often limited to property damage.

Personal injury, and in some cases property damage, claims have been filed in cases alleging misapplication of pesticides in and around residential structures. Hundreds of claims alleging personal injury associated with the misapplication of chlordane were filed in the 1980s. Claims involving misapplication of a number of pesticide products in indoor environments continue to be common.

Occasionally, claims are filed against architects who are deemed to be negligent because they: failed to include exhaust ventilation in school buildings where solvents could reasonably be expected to be used; designed buildings without adequate general ventilation; or specified unsafe products. Architects are sued for malpractice (failure to perform one's job to the standards of the profession). Claims are occasionally filed against owners and operators of buildings who are alleged to have operated ventilation systems improperly, used pesticides unsafely, etc.

Claims against installers and vendors of combustion appliances and associated systems are often brought in cases involving flue gas spillage and alleged CO poisoning, particularly when appliances are new or changes are made in flue-gas venting systems. Claims have also been made against manufacturers of high-efficiency furnaces who failed to anticipate corrosion problems that resulted from the use of combustion air that was contaminated with chlorine from the use of chlorinated water and other chlorine compounds. Claims were, on occasion, filed against manufacturers of unvented space heaters.

Claims have been filed, and continue to be filed, against landlords in housing units where children have EBL levels and symptoms of lead poisoning. These allege that such exposures were due to the landlord's negligence. Sellers and leasers of pre-1978 housing are liable if they fail to disclose any known lead hazards to buyers/tenants. Sellers of residences who have conducted radon testing are liable if they fail to disclose to buyers that radon levels were elevated. Similarly, sellers of residences are liable if they fail to disclose the known presence of any contamination problem that would pose a health risk to the new owner.

The use of toxic torts to seek legal redress associated with indoor environment problems poses significant challenges to would-be plaintiffs. Litigation is expensive, and if the case goes to trial, the outcome is unpredictable. In most cases, individuals who allege injuries cannot afford the costs involved. Their access to the courts is often facilitated by contingency fee arrangements with trial attorneys who agree to accept the case with fees contingent on the outcome of the claim. Typically a trial attorney would receive 25 to 40% of settlement or jury awards after expenses are deducted. Expert witness fees, environmental testing costs, and costs of discovery may be borne "up front," in whole or in part, by plaintiff's counsel or by the plaintiff.

Most (90+%) personal injury/property damage claims are settled without going to trial. Such settlements reduce the burden on the courts and costs to defendants, trial attorneys, and plaintiffs. In many cases, out-of-court settlement provides an equitable resolution of claims. It also reduces the often lengthy time period (up to 6 years) required to conclude a case, and relieves the plaintiff of the emotional anxiety and vagaries of the trial process.

A successful outcome of a claim, particularly in court, requires more than the truth of a claim that a wrong has been done to the plaintiff. The legal process involves two parallel, but disparate, tasks on the part of those involved. These are: determining the truth and dispensing justice and, not surprisingly, winning and losing. Assuming that the case has merit, the outcome will depend on the relative skill and preparation of plaintiff's and defendant's counsel; the plaintiff's ability to testify, respond to cross-examination, and evoke jurors' empathy/sympathy; the abilities of the plaintiff's and defendant's experts; the judge's ruling on the admissibility of witnesses, evidence, and testimony; and the general attitudes of the jury pool in a given judicial district toward toxic torts.

Plaintiffs have a variety of disadvantages when a case goes to trial. Foremost of these is economics. In many cases, defendants have significant resources to engage very capable defense counsel and expert witnesses. Plaintiffs have two very different burdens of proof. Through the preponderance of evidence, plaintiff's counsel must prove in personal injury cases that the exposure caused the alleged injuries/health problems and that the defendant was legally responsible. The defendant, on the other hand, only has to raise sufficient uncertainty in the minds of jurors or the trial judge. Defense counsel tends to focus on limiting the admissibility of evidence and the scope of an expert's testimony. Defendant's ability to limit the scope of expert testimony has been significantly advanced by a U.S. Supreme Court decision known as the Daubert rule. Under Daubert, trial judges are given considerable discretion in admitting the testimony of experts. Some trial judges will only admit evidence that has been subject to publication in the peer-reviewed literature, a very high standard, particularly for a newly emerging problem.

Civil litigation is often characterized disparagingly by critics of personal injury claims and by product manufacturers. Though flawed, like any other system, civil litigation has had, and will continue to have, positive impacts on American society. In the absence of regulation, it is a critical tool in both redressing wrongs and promoting changes that may have significant public health benefits. Litigation against installers, distributors, and manufacturers of UFFI was one of the main factors in the demise of this industry. Litigation against manufacturers of mobile homes and UF-bonded wood products was responsible for the significant reductions in HCHO emissions from wood products and indoor HCHO concentrations that have occurred in the past two decades.

The threat, as well as use, of civil litigation is an important tool in encouraging manufacturers to produce safe products and improve their products when safety problems arise. Unfortunately, the filing of numerous meritorious claims indicates safety considerations did not have the high priority they should have had in the design, manufacture, or use of a product (including buildings).

Readings

Asbestos Hazard Emergency Response Act, PL 98-469, USC Title 15: Subchapter 11, Sections 2641–2654, 1986.

Department of Labor, Occupational Safety and Health Administration, Indoor air quality — Notice of proposed rulemaking, *Fed. Reg.*, 59, 15968, 1994.

Godish, T., *Indoor Air Pollution Control*, Lewis Publishers, Chelsea, MI, 1989, chap. 7.

Hirsch, L.S., Behind closed doors: indoor air pollution and governmental policy, *Harvard Environ. Law Rev.*, 6, 339, 1982.

Maroni, M., Siefert, B., and Lindvall, T., *Indoor Air Quality. A Comprehensive Reference Book*, Elsevier, Amsterdam, 1995, chaps. 32, 34, 37.

Sexton, K., Indoor air quality: an overview of policy and regulatory issues, *Sci. Tech. Human Values*, 11, 53, 1986.

Spengler, J.D., Samet, J.M., and McCarthy, J.F., Eds., *Indoor Air Quality Handbook*, McGraw-Hill Publishers, New York, 2000, chap. 71.

USEPA, Asbestos-containing materials in schools: final rule, *Fed. Reg.*, 52, 41826, 1987.

Questions

1. What are the advantages and disadvantages of using air quality standards in regulating air quality in buildings?

2. What is a public-access building? What is the relevance of public-access buildings in the context of regulating air quality in buildings?
3. What advantages are there in using product standards in achieving an indoor air quality objective?
4. What are the advantages and disadvantages of product bans for the purpose of protecting indoor air quality and public health?
5. How do emission standards applied to industrial sources differ from the products standards used to reduce formaldehyde emissions from particle board and hardwood plywood?
6. Describe how emissions standards designed to protect ambient air have had or have the potential to have positive effects on indoor air quality.
7. What are application standards? How may they be used to minimize indoor air quality contamination?
8. What responsibilities do schools have under Asbestos Hazard Emergency Response Act-mandated regulations?
9. What federal agency or agencies have regulatory authority associated with asbestos and lead in buildings?
10. What obligations do owners of pre-1978 housing have under Title X, the Residential Lead-based Paint Hazard Reduction Act?
11. Why do federal agencies regulate training requirements for lead, asbestos, and radon-related professional activities?
12. What is the value of indoor air quality guidelines established by international agencies, federal and state governments?
13. What are ventilation guidelines? What is their regulatory status?
14. What are performance guidelines? Why would their establishment by a professional association prove to be controversial?
15. What are the limitations on using warning labels on products that may cause indoor air quality problems?
16. What are the advantages of voluntary initiatives?
17. Describe the role of citizen initiatives in determining public policy relative to indoor air quality concerns.
18. What are the standards of proof required on the part of plaintiff's counsel in successfully pursuing litigation alleging adverse health effects associated with a building environment?
19. Under what theories can lawsuits be filed claiming personal injury associated with illness symptoms in a building?
20. What advantages and disadvantages are there to allowing the legal system to improve air quality in buildings?

Index

A

Absorption, 269, 393, 401
Acaricides, 126, 324
ACBM (asbestos-containing building
　　materials)
　airborne fiber concentrations,34–35
　miscellaneous materials, 31–32, 34
　surfacing materials, 29–30, 32
　thermal system insulation, 29–31,
　　315–316, *see also* TSI
Accuracy, 266–269, 271–272, 281
Acremonium, 178, 181
Acetaldehyde
　biological effects, 97–98, 107
　chemical characteristics, 96–97
　emission sources, 67, 78, 105, 108, 221
Acetic acid, 199, 221
Acetone, 110–111, 221
ACGIH , *see* American Conference of
　　Governmental Industrial
　　Hygienists
ACM (asbestos-containing materials),
　　29, 315–318, 414–415, 417
Acoustical plaster, 315, 317, 414
Acquired immune deficiency syndrome,
　　see AIDS
Acrolein
　biological effects, 97–98, 108
　chemical characteristics, 96–97, 105,
　　107
　emission sources, 78, 88
Actinomycetes, 147–150, 153, 200
Activated
　alumina, 393, 400
　carbon(s), 393–397, 399–400

Adhesion, 379
Adsorbability, 394–395
Adsorption, 269, 392–399
Aerosols, 121
Aflatoxins, 175–176
AHERA, *see* Asbestos Hazard
　　Emergency Response Act
AHUs (air handling units), 19, 21,
　　328–329, 343, 354, 370, 380, 382
AIDS (acquired immune deficiency
　　syndrome), 180, 265
AIHA, *see* American Industrial Hygiene
　　Association
Air
　cleaners
　　as contaminant sources, 402
　　electronic/electrostatic, 378–381,
　　　387–391
　　general, 304, 369
　　in-duct, 386–387
　　residential, 385–387, 389
　　use problems, 390
　cleaning, 369–370, 393
　emission standards, 408–410
　exchange rates, 338–342
　flow rate, 371
　fresheners, 111, 121
　handling units, *see* AHUs
　infiltration, *see* Infiltration
　movement, 208–211
　Quality Procedure, Indoor, 349–351, 427,
　　430
　quality standards, 406–408, 422,
　　424–425
　testing, 231–233, 236
Airplanes, commercial, 22–23